Maritime Surveillance with Synthetic Aperture Radar

Related titles on radar

Advances in Bistatic Radar
Willis and Griffiths
Airborne Early Warning System Concepts, 3rd Edition Long
Bistatic Radar, 2nd Edition Willis
Design of Multi-Frequency CW Radars
Jankiraman
Digital Techniques for Wideband Receivers, 2nd Edition Tsui
Electronic Warfare Pocket Guide Adamy
Foliage Penetration Radar: Detection and characterisation of objects under trees Davis
Fundamentals of Ground Radar for ATC Engineers and Technicians
Bouwman
Fundamentals of Systems Engineering and Defense Systems Applications
Jeffrey
Introduction to Electronic Warfare Modeling and Simulation Adamy
Introduction to Electronic Defense Systems Neri
Introduction to Sensors for Ranging and Imaging
Brooker
Microwave Passive Direction Finding Lipsky
Microwave Receivers with Electronic Warfare Applications Tsui
Phased-Array Radar Design: Application of radar fundamentals
Jeffrey
Pocket Radar Guide: Key facts, equations, and data Curry
Principles of Modern Radar, Volume 1: Basic principles
Richards, Scheer and Holm
Principles of Modern Radar, Volume 2: Advanced techniques
Melvin and Scheer
Principles of Modern Radar, Volume 3: Applications
Scheer and Melvin
Principles of Waveform Diversity and Design
Wicks *et al.*
Pulse Doppler Radar
Alabaster
Radar Cross Section Measurements Knott
Radar Cross Section, 2nd Edition
Knott *et al.*
Radar Design Principles: Signal processing and the environment, 2nd Edition
Nathanson *et al.*
Radar Detection
DiFranco and Rubin
Radar Essentials: A concise handbook for radar design and performance Curry
Radar Foundations for Imaging and Advanced Concepts
Sullivan
Radar Principles for the Non-Specialist, 3rd Edition
Toomay and Hannan
Test and Evaluation of Aircraft Avionics and Weapons Systems McShea
Understanding Radar Systems
Kingsley and Quegan
Understanding Synthetic Aperture Radar Images
Oliver and Quegan
Radar and Electronic Warfare Principles for the Non-specialist, 4th Edition Hannen
Inverse Synthetic Aperture Radar Imaging: Principles, algorithms and applications
Chen and Marotella
Stimson's Introduction to Airborne Radar, 3rd Edition
Baker, Griffiths and Adamy
Test and Evaluation of Avionics and Weapon Sytems, 2nd Edition McShea
Angle-of-Arrival Estimation Using Radar Interferometry: Methods and applications Holder

Biologically-Inspired Radar and Sonar: Lessons from nature
Balleri, Griffiths and Baker
The Impact of Cognition on Radar Technology
Farina, De Maio and Haykin
Novel Radar Techniques and Applications, Volume 1: Real aperture array radar, imaging radar, and passive and multistatic radar
Klemm, Nickel, Gierull, Lombardo, Griffiths and Koch
Novel Radar Techniques and Applications, Volume 2: Waveform diversity and cognitive radar, and target tracking and data fusion
Klemm, Nickel, Gierull, Lombardo, Griffiths and Koch
Radar and Communication Spectrum Sharing
Blunt and Perrins
Systems Engineering for Ethical Autonomous Systems
Gillespie
Shadowing Function from Randomly Rough Surfaces: Derivation and applications
Bourlier and Li
Photo for Radar Networks and Electronic Warfare Systems
Bogoni, Laghezza and Ghelfi
Multidimensional Radar Imaging
Martorella
Radar Waveform Design Based on Optimization Theory
Cui, De Maio, Farina and Li
Micro-Doppler Radar and its Applications
Fioranelli, Griffiths, Ritchie and Balleri

Maritime Surveillance with Synthetic Aperture Radar

Edited by
Gerardo Di Martino and Antonio Iodice

The Institution of Engineering and Technology

Published by SciTech Publishing, an imprint of The Institution of Engineering and Technology, London, United Kingdom

The Institution of Engineering and Technology is registered as a Charity in England & Wales (no. 211014) and Scotland (no. SC038698).

First published 2020

The Institution of Engineering and Technology
Michael Faraday House
Six Hills Way, Stevenage
Herts, SG1 2AY, United Kingdom

www.theiet.org

British Library Cataloguing in Publication Data
A catalogue record for this product is available from the British Library

ISBN 978-1-78561-601-3 (hardback)
ISBN 978-1-78561-602-0 (PDF)

Typeset in India by MPS Limited

Contents

About the editors

Gerardo Di Martino is a tenure-track Assistant Professor of Electromagnetic Fields in the Department of Electrical Engineering and Information Technology of the University of Naples Federico II, Italy. He is a senior member of IEEE and an Associate Editor for IEEE Access and a section board member for Remote Sensing (MDPI). His research interests are in the field of microwave remote sensing, wireless propagation, and electromagnetics.

Antonio Iodice is a Full Professor of Electromagnetics at the Department of Electrical Engineering and Information Technology of the University of Naples Federico II, Italy. He has been a principal investigator for research projects on microwave remote sensing and wireless propagation, and he has received the "2009 Sergei A. Schelkunoff Transactions Prize Paper Award" from the IEEE AP-S. He is a senior member of the IEEE and chairs the South Italy Geoscience and Remote Sensing Chapter.

Foreword

Nowadays, maritime surveillance is a topic of great relevance and is one of the key priorities of military and civil institutions. Maritime surveillance systems are developed based on the integration of data coming from different sources: within this framework, remote-sensing systems, and, in particular, synthetic aperture radar (SAR), play a prominent role. In the last decade, many advances occurred. The new generation of SAR sensors has been the driving factor of a huge development in maritime applications, exploiting the potentialities of high-resolution X-band imaging. Moreover, in the last years, new sensors based on wide-swath acquisition modes have become operative: the TOPSAR mode of Sentinel-1, whose data are provided by the European Space Agency at no cost, is the most relevant example. The availability of these systems also stimulated the development of advanced techniques for sea observation. This book is intended to cover all the main issues regarding the use of SAR for maritime surveillance applications, providing a comprehensive source of material on the subject, intended for SAR system engineers, private and public corporations, oceanographers, and remote-sensing researchers and end-users.

The book is divided into two parts. The first one deals with models and techniques, thus focusing on the SAR system viewpoint. In this part, the specific requirements of SAR systems for maritime observation will be analyzed; moreover, this part provides the necessary background for a complete understanding of the second part, which is devoted to maritime surveillance applications. For all the considered topics, the basic principles are illustrated. The critical review of the state of the art on each application represents the core of each chapter. In the single chapters, written by leading experts in the respective subjects, the last advances in each field are detailed.

In the editors' intention, the book should become a reference text for scientists and end-users working on maritime surveillance, providing the necessary information for a self-consistent understanding of each of the considered topics. At the same time, each chapter is closed by a wide reference list, which will guide the demanding reader in finding more specific answers to her/his questions.

Gerardo Di Martino and Antonio Iodice
Università di Napoli Federico II,
Via Claudio 21, Napoli, Italy

Chapter 1

Introduction

Gerardo Di Martino[1] *and Antonio Iodice*[1]

1.1 Maritime surveillance

According to the United Nations (UN) [1], maritime-based trade is expanding at very high rates in recent years (4% in 2017), with forecasts of a compound annual growth rate of 3.8% between 2018 and 2023. Regarding the sole European Union (EU), almost 90% of the external and 40% of intra-EU freight trade are seaborne. Moreover, more than 400 million passengers pass through European ports each year. Regarding fuel exchange, 90% of oil is already transported by sea, and the transport of natural gas by tankers is experiencing relevant increases [2]. Regarding fisheries, the European fishing fleet was made up of almost 83,000 vessels in 2017 [3], even if its number has been decreasing in the last decade. In this scenario, efforts to guarantee the safety of navigation and, more in general, of the maritime environment are crucial: indeed, all these factors contributed to making maritime surveillance an increasingly hot topic.

In this context, one of the major challenges lies in the remote monitoring and tracking of ships. Several means have been developed to pursue this objective, even though no one by itself can guarantee night and day geolocation and tracking of all kinds of ships in every weather and sea-state condition. Nowadays, maritime surveillance systems are mostly based on the use of automatic identification system (AIS). Indeed, in the European Union, AIS is mandatory for all vessels of 300 gross tonnages and above on international voyages, cargo ships of 500 gross tonnages and above, and passenger ships irrespective of size. The AIS was originally developed for safety reasons and mainly for collision avoidance. It is a shipborne very high frequency (VHF)-based system used by the ships to share with other ships and with coastal receivers their identification, position, speed, direction, and so on. However, the range of transmission is quite limited (about 40 km) and, most importantly, the system is aimed at cooperative ships, mounting properly working AIS hardware. To overcome the coverage limitation of traditional AIS, in recent years, the use of satellite-based AIS has been explored, leading to the deployment

[1]Dipartimento di Ingegneria Elettrica e delle Tecnologie dell'Informazione, Università di Napoli Federico II, Naples, Italy

of SatAIS (Satellite AIS) [4]. Thanks to this system, the ship AIS signal can be received in very wide areas: for a satellite orbiting at an altitude of 650 km, the average field of view is above 20 million square kilometers [4]. A SatAIS system has been implemented by the German Space Agency (DLR) in 2011, and several other countries may follow in next years. Even if SatAIS can help in extending the coverage of traditional AIS, it cannot improve the limited reliability (due to the intrinsic human-based nature of the transmitted information) of AIS data, which represents another important drawback of this technology. Indeed, errors are very frequent in AIS data, especially in ship length: it would be enough to say that for the AIS reports analyzed in [5], 6.4% shows a length of 0 m, 36.3% presents an error between 1 m and 5 m, and 4.3% presents an error larger than 5 m (further details are provided in Chapter 10 of this book). Another long-range satellite-based alternative to the AIS is the long-range identification and tracking (LRIT) system [6]. LRIT system is mandatory for all passenger ships, high-speed craft, mobile or shore drilling units, and cargo ships of over 300 gross tonnes. Unlike AIS, the management of LRIT messages is centralized: the ships must send information only to one communication service provider via satellite link; the LRIT data are then transmitted to the LRIT Data Centre and stored. LRIT data access is based on the data distribution plan, managed by International Maritime Organization (IMO) [6]. Finally, the LRIT systems also require cooperative vessels, mounting properly working LRIT hardware.

Regarding fisheries control, in the European Union, the vessel monitoring system (VMS) is mandatory for all fishing vessels of 15 m length and above [7]. VMS is a satellite-based system, which enables fishing ships to transmit their identification, position, speed, direction, and so on to a fishery monitoring center (FMC). Unlike AIS, VMS has no strict limitations in coverage (being a satellite system), but it still requires that properly working hardware is mounted on the ships. Moreover, VMS is a proprietary technology that can be used only by selected public agencies, thus imposing a limit on the rapid exchange of data.

Currently, cooperative systems, like AIS and VMS, are supported in the monitoring by coastal ground-based radars, which provide all-weather all-time capabilities. Despite this, it is evident that current maritime surveillance systems strongly rely on a cooperative behavior of vessels. This is especially true for small-size vessels since ground-based radars have limited resolution and can hardly detect small vessels, especially in crowded areas or under rainy/windy weather conditions (i.e., high clutter conditions). Indeed, this limitation is exploited for illegal activities, such as traffic of drugs, humans, and weapons in small, hardly detectable boats. Moreover, this may result in collisions with large vessels in crowded harbor scenarios, due to the absence of external guidance from the coast. For these reasons, the development of techniques suitable for the monitoring of noncooperative ships represents a very relevant issue.

In this context, remote sensing technologies can provide valuable support, and, indeed, their prominent role is gaining increasing attention. Among remote sensors, synthetic aperture radar (SAR) is particularly well suited for all-weather all-time sea monitoring, and, in fact, satellite SAR data have been used to foster ocean applications since the first satellite SAR instrument was mounted on the SeaSAT platform. In addition, modern SAR sensors offer wide-area imaging capabilities (also through dedicated acquisition modes, see Chapter 2 for more details) and low revisit times, since they mostly operate in constellations. However, along with the mentioned advantages, SAR has also relevant limitations: (1) it cannot be used for precise identification of ships, even though with high-resolution systems good indications of the type of vessels can be obtained; (2) it cannot be used for real-time monitoring since currently the highest revisit time is in the order of some days; and (3) regarding moving ships, accurate estimation of their location is hampered by the presence of Doppler-induced shifts in the final image. In addition, SAR images must be geocoded before use, which requires reprojection and interpolation to change coordinate system. Recent studies are trying to exploit the complementarity of SAR and AIS data, integrating them in order to improve AIS data reliability or analyzing the behavior of noncooperative ships: more details on this topic are provided in Chapter 10.

Maritime surveillance is not restricted to ship identification and tracking. It includes a broad range of requirements that are intended to guarantee national sovereignty, in terms of law enforcement, search and rescue, environmental issues, and resource management. In this context, sensor integration is crucial for obtaining a comprehensive knowledge of the surveyed areas. Many applications fall in the wide domain of maritime surveillance, such as monitoring of potentially dangerous objects on the sea surface (e.g., sea ice and icebergs, see Chapter 8), monitoring of coastal habitats (see Chapter 7), identification of pollutants on the sea surface (e.g., oil slicks, see Chapter 9), and monitoring of currents and surface winds (see Chapter 11). From this viewpoint, SAR represents an unparalleled tool, providing capabilities for the observation of all the mentioned phenomena, as will be discussed in detail in the second part of this book. This is the reason why many projects in the European Union (and not only) are focused on the use of SAR data for maritime monitoring: this trend is also supported by the new European Space Agency (ESA) policy of free data distribution (indeed, Sentinel 1 SAR data are freely available for the public), which is aimed at spreading the use of data. In fact, the European Maritime Safety Agency (EMSA) has already developed SAR-based tools for oil spills detection [8] and ship detection [9]. More specifically, these services are aimed to fight against the phenomenon of accidental and deliberate discharge of pollutants from ships, providing support in locating and identifying the polluters.

For all these reasons, it is of paramount importance to review and discuss the potentialities of SAR in the wide range of maritime surveillance applications, which is the main reason that led to the publication of this book.

1.2 Synthetic aperture radar systems

1.2.1 SAR principles and main SAR missions

A synthetic aperture radar (SAR) is a microwave imaging sensor that consists of a radar mounted on a moving platform. Its name is due to the fact that a high resolution (in the along-track, i.e., azimuth direction) is achieved by properly coherently combining echoes of different pulses transmitted as the platform moves, so to generate a synthetic antenna much longer than the real one [10,11]. More technical details on SAR principles will be provided in Chapter 3. We here want to emphasize the advantages of this sensor over optical and infrared ones. Since it operates at microwave frequencies (usually, in L, i.e., 1–2 GHz; C, i.e., 3.75–7.5 GHz; and X, i.e., 7.5–12 GHz bands), and since microwaves can penetrate clouds, a SAR system can acquire images of the ground or sea surface even in the presence of a cloud cover. In addition, since it is an active sensor (i.e., it has got an independent illumination source, in fact, the radar transmits electromagnetic pulses toward the imaged ground), it can operate even during the night. In addition to these all-weather, day-and-night operating capabilities, SAR systems share the global-monitoring capability with other satellite remote sensing systems. Accordingly, since the first SAR systems were developed, it was soon realized that they are particularly suitable for remote sensing of the sea (monitoring of sea state, winds, currents, coastal areas, and sea ice) and for maritime surveillance (detection of ships, icebergs, and oil slicks). As a matter of fact, the first civilian SAR satellite, the L-band NASA/JPL Seasat, that was launched in 1978 and operated only for 3 months, was specifically devised for remote sensing of the sea [12]. Subsequent SAR missions, and in particular currently operating ones, are usually designed by accounting for the requirements of both terrain and sea remote sensing applications. Therefore, SAR data acquired by all of these missions can be used for maritime surveillance. Tables with complete lists of currently available SAR satellite missions are reported elsewhere in this book (see Chapters 3, 4, and 9). We here recall that continuous global monitoring of the planet surface at C band is available since 1991 thanks to the SAR satellite missions of the European Space Agency (ESA): ERS-1/2, then Envisat, and now Sentinel-1, whose data are free of charge [13]. C-band data covering the years from 1995 are also available thanks to the Radarsat 1 and 2 missions of the Canadian Space Agency (CSA) [14]. L-band coverage since 1992 (although with some gaps) is provided by Japanese SAR satellite JERS-1, then ALOS/PalSAR and now ALOS-2 [15]. Finally, high-resolution global coverage with short revisit times at X band since 2007 is provided by the COSMO/SkyMed satellite constellation of the Italian Space Agency (ASI) [16] and by the TerraSAR-X/TanDEM-X mission of the German Aerospace Agency (DLR) [17]. Future satellite SAR systems at P (250-500 MHz) and S (2-3.75 GHz) bands are currently being planned [18–20].

An SAR image is characterized by a number of parameters: frequency, viewing angle, geometric and radiometric resolutions, noise-equivalent sigma zero (NESZ), and employed polarization.

With regard to the choice of *frequency*, one has to keep in mind that (1) microwave penetration depth in illuminated media increases as frequency decreases, and (2) microwaves mainly interact with objects whose size is of the order of wavelength. Actually, the first point mainly involves land applications, particularly on vegetated areas: high-frequency (X-band, wavelength of about 3 cm) SAR return is mainly from vegetation top (i.e., tree branches and leaves), whereas low-frequency (L-band, wavelength of about 30 cm) SAR return also comes from the underlaying ground. However, with regard to sea water, penetration depth is negligible at all employed SAR frequencies. In this case, the second point is more important: it tells us that microwave SAR return from the sea is mainly due to sea surface spectral components at centimetric wavelengths, that is, to capillary and gravity-capillary waves (see Chapter 2). Generally, in maritime applications, the choice of frequency is less critical, with the only exception of oil-spill monitoring, for which detection ability may be dependent on frequency because a surface oil film damps differently capillary waves of different wavelengths (see Chapter 9).

With regard to *viewing angle*, SARs are side-looking sensors. For spaceborne SAR sensors, viewing angle (i.e., the angle between nadir and viewing directions, also termed off-nadir angle) may vary from about 25° to about 50°. In scattering from the sea, viewing angle, together with frequency, determines the sea surface wavelength that mainly contributes to SAR backscattered signal (see Chapter 2).

Geometric resolution is the minimum distance between two targets in the illuminated scene such that they can be separated in the image. A surface area of linear size equal to the geometric resolution is called a *resolution cell*. Geometric resolution of SAR systems may vary from a fraction of meter to several tens of meters, depending on transmitted pulse bandwidth and acquisition geometry (see Chapter 3).

Radiometric resolution can be roughly defined as the uncertainty on the measurement of the radar cross section (RCS) of a target. In SAR images, it is degraded by the so-called speckle noise, a multiplicative noise strictly related to the coherent nature of the SAR transmitted signal (see below). Speckle filtering techniques are available [21,22] that can improve radiometric resolution of SAR images. However, this improvement is often achieved at the expense of geometric resolution.

Very calm sea areas (because of very low wind, or areas dampened by oil slicks) produce a very low radar return, so that the signal from those areas may be dominated by receiver thermal noise. In SAR literature, the level of receiver thermal noise is expressed in terms of *NESZ*, that is, the surface normalized radar cross section (NRCS), or sigma zero (i.e., the surface RCS per square meter), that would produce a received backscattered power equal to the receiver thermal noise power.

Finally, as better described in Chapter 4 and briefly summarized below, *polarizations* of transmitting and receiving antennas may be either vertical (V) or horizontal (H) so that VV, HH, VH, and HV polarization channels can be obtained. SAR systems may have one (*single-pol*), two (*dual-pol*), or all four (*quad-pol*) polarization channels.

1.2.2 Coherent nature of SAR systems: polarimetry, interferometry, and speckle

SAR systems are fully coherent systems. The transmitted signal is coherent, in the sense that each transmitted pulse has known, deterministic amplitude and initial phase and a known, deterministic, polarization state, either vertical or horizontal (in some particular cases, either right-handed circular or left-handed circular). At the receiver, coherent demodulation is achieved, so that both in-phase (I) and quadrature (Q) channels are received, and hence both amplitude and phase of the received signal are measured; in addition, both the vertically and horizontally polarized components of each backscattered pulse can be separately detected.

The coherent capability of transmitter and receiver is necessary for the functioning itself of SAR systems since it allows coherently combining echoes of different pulses transmitted as the platform moves, so to generate the synthetic antenna. In addition, it has allowed devising polarimetric and interferometric techniques, as shown in the following. The price to be paid is the presence of speckle noise, that is intrinsic and unavoidable when coherent illumination is used.

As mentioned above, each transmitted pulse has a well-defined polarization state, usually either vertical or horizontal. However, the scattering process changes the wave polarization state, so that the backscattered field is always a combination of vertical (V) and horizontal (H) polarizations. If all transmitted pulses have the same polarization, and only one polarimetric component of the scattered field is selected, the system is termed single-polarization (or single-pol) SAR. Usually, the same antenna is employed for transmission and reception, so that a single-pol SAR may have HH or VV polarization. If again all transmitted pulses have the same polarization, but both polarimetric components of the scattered field are measured, the system is termed dual-polarization (or dual-pol) SAR. According to the polarization of the transmitted signal, we may have dual-pol SAR systems with either HH and HV or VV and VH polarization channels. Finally, if H and V polarized pulses are alternately transmitted and both polarimetric components of the scattered field are measured, the system is termed fully polarized (or quad-pol) SAR, in which case all four polarimetric channels HH, VV, VH, and HV are available. As shown in Chapter 4, use of multipolarization SAR systems allows recovering more information about the scattering surface.

The fact that also the phase of the received signal is measured allows using SAR interferometric techniques. In fact, the phase of the signal scattered by a SAR resolution cell has a deterministic component, related to the distance from the sensor to the center of the resolution cell, and a completely random component related to the microscopic details of the scattering surface at the wavelength scale. In a single SAR image, these two phase components cannot be separated, so that no useful information can be extracted from the phase. However, if we consider two SAR antennas flying along two parallel, slightly separated, trajectories, and we take the difference of phases of the two SAR images, the phase random component (almost) cancels out, and the remaining phase difference is related to the difference of antenna-to-target distances. From this distance difference, the elevation of the target can be measured. Accordingly, the topography of the imaged area can be obtained (cross-track interferometry). The accuracy of the retrieved surface height increases as the spatial separation of antennas (the spatial baseline) increases. However, such separation cannot exceed a critical value, above which random phase components do not cancel out (baseline decorrelation [23]). Accordingly, this accuracy turns out to be of the order of a few meters. This is sufficient for terrain topography retrieval, but often it is not for maritime applications, although some applications to sea ice monitoring have been explored, see Chapter 8. However, useful phase information can be obtained also if we consider two SAR antennas moving along the same trajectory and acquiring two SAR images of the same area in two subsequent times. In fact, if in the time between the two acquisitions (temporal baseline), the microscopic details of the scattering surface are unchanged, then the random phase again cancels out and the phase difference is again related to the difference of antenna-to-target distances, that this time represents the line-of-sight component of the surface displacement. In case of terrain surfaces, the temporal baseline can be of the order of days or even months, and the technique, termed differential SAR interferometry (DInSAR) [24] is used to monitor slow terrain movements (subsidence, landslides, down- or up-lift related to volcanic activity, etc.). Conversely, the sea surface microscopic details change very rapidly so that in this case the temporal baseline must be of the order of fractions of a second. This can be obtained by using two antennas mounted on the same platform and spaced along the line-of-flight direction, and the technique, termed along-track interferometry [25] can be used to monitor sea surface currents or ship motion, see Chapter 6.

As already mentioned, the price to be paid for the advantages of the coherent nature of SAR systems is the appearance of speckle noise. The signal backscattered by a resolution cell is the coherent combination of signals backscattered by the several objects belonging to the resolution cell. These signals will interfere constructively or disruptively according to the microscopic (i.e., at the wavelength scale) details of the objects' shapes and positions. Accordingly, in a SAR image a macroscopically homogeneous area appears to be composed of pixels of randomly varying intensity. The overall effect is a "salt-and-pepper" look of SAR images, that reduces their interpretability by a humane eye and affects the performance of automatic information retrieval algorithms. This effect can be modeled as a multiplicative noise over the SAR amplitude image. It can be reduced by spatially averaging neighboring pixels (*spatial multilook*, so that the nonfiltered original image is called *single-look* image), but this of course worsens the geometrical resolution. Several speckle filtering techniques have been proposed in literature, with the aim of reducing speckle noise while preserving image details [22].

1.3 Book organization

This book is divided into two parts. The first one deals with models and techniques, thus focusing on the SAR system viewpoint. This part provides the necessary background for a complete understanding of the second part, which is devoted to maritime surveillance applications. In the following, a brief description of book chapters is provided.

1.3.1 Part I: Models and techniques

Chapter 2. Scattering models

To relate the received SAR signal to sea state parameters and/or to ship features, electromagnetic scattering models have been developed and are available in literature. With regard to scattering from the sea, the air-sea interface is often modeled as a randomly rough surface with proper power spectrum (Pierson-Moskowitz,

JONSWAP, Elfouhaily spectra), and the scattered electromagnetic field may be evaluated by using both approximate analytical solutions (Kirchhoff approximation, small perturbation method, integral equation method, and two-scale model) and numerical methods (method of moments). With regard to scattering from the ship, an important role is played by multiple reflections, due to interactions between sea and ship, and among different parts of the same ship. Asymptotic (geometrical optics, physical optics, and geometrical theory of diffraction) or numerical (method of moments) methods may be used to evaluate the scattered electromagnetic field in this case. This chapter focuses on closed-form solutions that can be more efficiently employed in sea surface parameter retrieval and ship detection techniques.

Chapter 3. Acquisition modes

In addition to the standard, stripmap acquisition mode, SAR systems may operate in different modalities, based on a trade-off between resolution and coverage. The spotlight mode favors high resolution at the expense of coverage, and vice versa the ScanSAR mode, and its more advanced version, TOPSAR, allow for wide coverage at the expense of resolution. Other modalities, specifically developed for ocean monitoring and/or ship detection applications, are available to reduce the necessary data rate (wave mode) and/or increase coverage while preserving resolution (low pulse repetition frequency mode, compressive sensing SAR, staggered SAR, Coprime SAR, all recently proposed and still at an experimental stage). In this chapter, main acquisition modes are described, their advantages and drawbacks are highlighted, and their applicative domains are indicated.

Chapter 4. SAR polarimetry

A fully polarimetric (or "quad-pol") SAR alternately transmits two orthogonally (usually, horizontally and vertically) polarized pulses and receives separately both vector components of the backscattered pulse. In this way, four polarimetric channels are obtained (HH, VV, HV, and VH), from which the complete relationship between scattered and incident polarizations can be derived from the observed scene. Dual-pol SAR sensors, that transmit a single polarization and receive both, are also available: they are simpler and less expensive, but they only partially capture the relationship between scattered and incident polarizations. In any case, polarimetric SAR systems are helpful to improve performances of ocean remote sensing (oil spill detection, ship detection). In this chapter, basic theory of SAR polarimetry is provided, and its main applications to maritime surveillance are described.

Chapter 5. Ambiguity problems and their mitigation

Due to the specific characteristics of the SAR system, peculiar artifacts may appear on SAR images. In particular, finite pulse repetition frequency (PRF) and nonideal antenna pattern give rise to azimuth ambiguity, with the possible presence of "ghosts" on the image. They are due to the replicas of strong targets located outside of the antenna main beam, superposed onto low-intensity areas of the imaged scene, such as the sea surface. It is then very important, to avoid false alarms in ship detection application, to properly design SAR systems to reduce azimuth ambiguity and to use processing methods to mitigate it. In this chapter, the formation mechanisms of this and other artifacts (e.g, range ambiguity) are summarized, and available methods to filter them out are described.

1.3.2 Part II: Applications

Chapter 6. Ship detection

In this chapter, the problem of ship detection from SAR images is introduced. This topic is of great significance in maritime surveillance, since it can help in the identification of noncollaborative ships. The techniques used for ship detection are mainly based on a two-step procedure: a first pre-screening detection phase, frequently based on a constant false alarm rate (CFAR) approach, followed by a discrimination phase, necessary to lower the false alarm rate due to ship-alikes (e.g., ambiguities and icebergs). In the chapter, the main frameworks for ship detection are illustrated with emphasis on its main applications.

Chapter 7. Monitoring of intertidal areas and coastal habitats

Intertidal zones are coastal areas that fall dry once during each tidal cycle. This chapter demonstrates that high-resolution SAR imagery of coastal zones can be used to study morphodynamical changes in these zones since exposed intertidal flats show up as dark or bright patches on SAR imagery. In addition, submerged sandbanks can cause SAR image signatures through variations of the surface current field. Multifrequency SAR data can be used for the derivation of surface roughness parameters, and the polarimetric decomposition of dual-co-polarization SAR data helps in identifying coastal habitats.

Chapter 8. Sea ice and icebergs

In this chapter. the characteristic signatures of sea ice and icebergs on SAR sea images are discussed. The detection of icebergs on the sea surface is of key importance for safety reasons, especially in navigation applications. Moreover, icebergs represent a source of false alarms in ship-detection applications. Space-borne SAR surveillance of ocean regions that are permanently or seasonally covered with ice is also of fundamental importance both for navigation safety and for environmental studies. In this chapter, the main techniques used for retrieving information about ice conditions and for detecting the presence of icebergs using SAR images of the sea surface are illustrated.

Chapter 9. SAR oil spill imaging, interpretation, and information retrieval techniques

In this chapter, the response of SAR backscattered signal to the presence of oil pollutants on the sea surface is discussed. The presence of oil films dictates damping of sea surface spectra in the capillary wave region, involved in Bragg scattering. Therefore, these polluted areas appear as dark spots on the images. The detection of oil spills is hampered by the potential presence of look-alikes, that is, physical phenomena producing similar signatures on the images, such as plankton or low-wind areas, or other natural sea surface slicks. In the chapter, the main techniques for oil spill detection and monitoring are illustrated.

Chapter 10. Joint use of SAR and collaborative signals

There is no sensor or technology that does not present limitations and that can be used as a unique source of information for all surveillance applications in the maritime domain. It is therefore indispensable, in many applications, to jointly use all available sources of information: self-reporting (i.e., collaborative) system data, observation-based data, and static databases. In this chapter, the main focus is on the joint use of observation-based SAR data and collaborative signals (e.g., automatic identification system, AIS). Basic principles are presented, and examples of applications are provided.

Chapter 11. Sea state and wind speed

The knowledge of sea state parameters and wind speed is very important for maritime safety and surveillance. This knowledge is also necessary for appropriate tuning of most of the techniques described in the previous chapters. In this chapter, the retrieving of sea-state parameters and wind speed from SAR data is discussed. The main theoretical aspects are firstly described, with the introduction of appropriate direct models linking specific oceanographic phenomena to their SAR signatures. Then, the inversion techniques used for the estimation of these physical parameters are illustrated.

References

[1] United Nations (UN). *Review of maritime transport*. New York and Geneva: United Nations; 2018.

[2] European Commission (EC). Energy policy and maritime policy: Ensuring a better fit [online]. 2007. Available from https://ec.europa.eu/transparency/regdoc/ [Accessed September 2019]

[3] Eurostat. Fishing fleet, 2008 and 2017 [online]. 2019. Available from https://ec.europa.eu/eurostat/ [Accessed September 2019]

[4] Bošnjak R., Šimunović L., and Kavran Z. "Automatic identification system in maritime traffic and error analysis." Transactions on Maritime Science. 2012; 1(2): 77–84.

[5] Brusch S., Lehner S., Fritz T., Soccorsi M., Soloviev A., and van Schie B. "Ship surveillance with TerraSAR-X." IEEE Transactions on Geoscience and Remote Sensing. 2011; 49(3): 1092–1103.

[6] International Maritime Organization (IMO). Long-range identification and tracking (LRIT). 2014. Available from http://www.imo.org/ourwork/safety/navigation/pages/lrit.aspx [Accessed September 2019]

[7] European Commission (EC). Vessel monitoring system (VMS) [online]. Available from https://ec.europa.eu/fisheries/cfp/control/technologies/vms_en [Accessed September 2019]

[8] European Maritime Safety Agency (EMSA). Cleanseanet [online]. 2013. Available from http://www.emsa.europa.eu/csn-menu.html [Accessed September 2019]

[9] European Maritime Safety Agency (EMSA). Vessel traffic monitoring in EU waters (Safe-SeaNet) [online]. 2013. Available from http://www.emsa.europa.eu/ssn-main.html [Accessed September 2019]

[10] Curlander J. C. and McDonough R. N. *Synthetic aperture radar: Systems and signal processing.* New York, NY: Wiley, 1991.

[11] Franceschetti G. and Lanari R. *Synthetic aperture radar processing.* Boca Raton, FL: CRC Press, 1999.

[12] Jordan R. L. "The Seasat-A synthetic aperture radar system." IEEE Journal of Oceanic Engineering. 1980; 5(2): 154–164.

[13] Torres R., Snoeij P., Geudtner D., *et al.* "GMES Sentinel-1 mission." Remote Sensing of Environment. 2012; 120: 9–24.

[14] Dabboor M., Iris S., and Singhroy V. "The RADARSAT constellation mission in support of environmental applications." *Proceedings of 2nd International Electronic Conference on Remote Sensing.* Vol. 2, paper no. 223, 22 March 2018–5 April 2018.

[15] JAXA – Japan Aerospace Exploration Agency. *ALOS-2/PALSAR-2 calibration and validation results Ver. 2018.08.0* 7 [online]. Available from https://www.eorc.jaxa.jp/ALOS-2/en/calval/PALSAR2_CalVal_Results_JAXA_201808.pdf [Accessed 31 July 2019]

[16] Italian Space Agency. *COSMO - SkyMed Mission and Products Description.* ASI, ASI-CSM-PMG-NT-001, Issue 2, 2016.

[17] Fritz T. and Eineder M. (ed.). *TerraSAR-X ground segment. Basic product specification document.* DLR, TX-GS-DD-3302, Issue 1.5, 2008.

[18] Hélière F., Fois F., Arcioni M., Bensi P., Fehringer M., and Scipal K. "Biomass P-band SAR interferometric mission selected as 7th Earth Explorer Mission." *Proceedings of EUSAR 2014,* Aachen (Germany), 2–4 June 2014.

[19] NASA and ISRO. *2015 NISAR Applications Workshop* [online]. Available from https://nisar.jpl.nasa.gov/files/nisar/2015_NISAR_Application_Workshop_Report_20160926.pdf [Accessed 31 July 2019]

[20] Iervolino P., Guida R., and Whittaker P. "NovaSAR-S and maritime surveillance." *Proceedings of IGARSS* 2013. Melbourne, Australia, 21–26 July 2013. pp. 1282–1285.

[21] Touzi R. "A review of speckle filtering in the context of estimation theory." *IEEE Transactions on Geoscience and Remote Sensing.* 2002. 40(11): 2392–2404.

[22] Argenti F., Lapini A., Bianchi T., and Alparone L, "A tutorial on speckle reduction in synthetic aperture radar images." *IEEE Geoscience and Remote Sensing Magazine.* 2013; vol. 1(3): 6–35.

[23] Zebker H. A. and Villasenor J. "Decorrelation in interferometric radar echoes." *IEEE Transactions on Geoscience and Remote Sensing.* 1992; 30: 950–959.

[24] Gabriel, A. K., Goldstein R. M., and Zebker H. A. "Mapping small elevation changes over large areas: Differential interferometry". *Journal of Geophysical Research.* 1989; 94: 9183–9191.

[25] Goldstein R. M. and Zebker H. A. "Interferometric radar map of ocean currents." *Nature.* 1987; 328: 707–709.

Part I

Models and techniques

Chapter 2

Scattering models

Daniele Riccio[1], Giuseppe Ruello[1], Pasquale Iervolino[2] and Raffaella Guida[2]

2.1 Introduction

The analytical description of the scattering of electromagnetic fields from the sea surface is an old but still open problem [1–6]. Initially devoted to removing the sea clutter from radar acquisitions [7], sea scattering models gained growing interest as instruments to investigate the sea physical characteristics and today new challenges call scientists to enrich their models with the presence of ships on the sea surface [8,9]. The opportunity of remotely retrieving information about geometrical and electrical properties of the sea surfaces, the possibility of using radar satellites to monitor ships in oceans and coastal waters as well as the efficient analysis of microwave links in marine environment depend on the capability of modeling the interaction of the electromagnetic field with the sea surfaces [10,11]. Such an analysis has to account for the involved dependence of the scattered field on incident wavelength, surface roughness and dielectric properties, polarization, look angle, and so on. In addition, time variance and hydrodynamic wave–wave interactions affect the power distribution in sea scattering phenomena [1–3,12,13].

Both numerical and analytical solutions have been proposed in literature with the goal to fulfill competing application requirements of accuracy, efficiency, simplicity, and generality. In this chapter, attention is mainly focused on analytical methods, whose potential of expressing the scattering in terms of the physical parameters is very attractive for the comprehension of the physical phenomena at hand.

The scattering complexity is mainly inherited by the complexity of the sea surface. Therefore, as in all the scattering problems, it is crucial to face the scattering study in two steps: first, the sea surface must be properly described; then, electromagnetic methods must be developed to model the field-waves interaction. Therefore, the chapter is organized as follows. First, we present a review of the techniques employed to obtain a reliable and amenable description of the ocean surface for scattering model purposes. Then, the electromagnetic models relevant to clean and clear sea surfaces are recalled.

[1]Department of Electrical Engineering and Information Technology, University of Napoli, Italy
[2]Electronic Engineering Department, University of Surrey, Guilford, UK

The problem of describing the scattering from a region where a ship floats on the sea surface is finally faced, proposing solutions of particular interest in maritime surveillance applications.

2.2 Sea surface models

The sea surface is a nonlinear dynamic system, whose physics is mainly governed by the Navier–Stokes nonlinear differential equations. The main forces that govern the wave physics are due to gravitational, rotational, and atmospheric agents. Despite the fact that the human interest in the sea surface dynamics dates back to the beginning of the mankind era, there is still a lack of adequate knowledge on several phenomena occurring in the sea [12].

Given the complexity of the forces acting to shape the rough sea surface, statistical time-variant models are required to describe it. The most used approach is the spectral description [1–5,14]. Several spectra have been proposed in literature, as brilliantly summarized in [15], where a unified approach was presented and is still used today in several radar-related applications.

The sea spectrum wavelengths are usually classified according to their peak-to-peak length as long (several hundred meters), intermediate (tenths of meters), and short waves (fractions of meters). In literature, the ocean waves are also classified according to the physical phenomenon that dominates their formation: long and intermediate waves are usually referred to as gravity waves; the range of short waves includes the shortest gravity waves, and gravity-capillary (few centimeters to decimeters) and capillary (less than one centimeter) waves [12].

Though the different scales are reciprocally linked via nonlinear relationships, in some applications adequate results can be obtained by modeling only the waves that directly interact with the sensors, whose length is selected by the electromagnetic wavelength that, at the microwave, is of the order of centimeters. This modeling can be done by selecting only appropriate portions of the sea spectrum or using a description of the sea surface as a predictable stochastic process. The advantage of the last description is that the sea surface height can be expressed as a series of functions that contain statistical parameters and allows for the evaluation of the field statistics from the first order onwards, therefore it will be hereinafter referred to as the "first-order description". In the following, we recall the characteristics of both the spectral and the first-order description of the sea surface.

2.2.1 Spectral representation

Sea waves are mainly formed as a result of the action of the wind on the sea surface. When the wind blows over the ocean surface, capillary waves grow up first as a result of a resonance phenomenon between the wind and the sea surface. Then, the capillary waves transfer energy to waves with a longer wavelength, via wave–wave nonlinear mechanisms. If the winds keep blowing for a significant time on an adequate fetch, the equilibrium is reached forming the fully developed sea [5,12,14]. Such an equilibrium depends on wind strength and duration. The long waves propagate far from their source forming a swell [12]. On a swell, local winds can produce new short waves with different directions of propagation. Though not directly measured by radars, intermediate and long waves also modulate the amplitude of the short waves [13]. This complexity was faced with different spectral models, usually able to well match only limited parts of the sea spectrum. We recall the unified directional spectrum proposed by Elfouhaily, built as a sum of two spectral regimes for high and low wavenumbers [15].

The low wavenumbers spectrum is built as a modification of the Pierson-Moskowitz [14] and the JONSWAP [16] spectral models, as reported in [15]; the short wavenumber spectrum is reported here with the formulation recently used in [17]. The bidimensional spectrum $W_{2D}(\kappa)$ is expressed as the product of an omnidirectional spectrum $W(\kappa)$ and a directional spreading function $\phi(\kappa, \varphi)$ symmetric around the wind direction: $W_{2D}(\kappa) = W(\kappa)\phi(\kappa, \varphi)$, where κ is the sea wavenumber and φ is the wave direction.

The short wave omnidirectional spectrum has been expressed in [17] as

$$W(\kappa) = \frac{\pi \alpha_m c_m}{\kappa^4 c} e^{-0.25*(\frac{\kappa}{\kappa_m}-1)^2} \tag{2.1}$$

where α_m is a parameter depending on the friction velocity and c_m is the minimum phase speed in correspondence with the expected peak of the gravity-capillary spectrum (values reported in [15] are κ_m= 370 m^{-1} and c_m= 0.23 m/s); the wave phase velocity c can be inferred by the dispersion relation:

$$\omega^2 = g\kappa + \frac{\tau_s}{\rho}\kappa^3 \tag{2.2}$$

where ω is the angular frequency of the ocean wave, g (m/s^2) is the gravity acceleration, τ_s (N/m) is the surface tension, and ρ (Kg/m^3) is the water density. The dispersion (2.2) clearly shows that long waves are governed by the gravity force, at variance of short waves, related to the surface tension.

The directional spreading function $\phi(\kappa, \varphi)$exhibits a dependence on ϕ of a cosine type, possibly raised to an appropriate exponent.

2.2.2 *First order representation*

In several applications, only the cm scale roughness is included in the scattering model, considering the effect of the long wave affecting only the local incidence angle and a description of the sea surface as a predictable stochastic model is preferred. Several electromagnetic methods have been developed using the classical Gaussian model for the sea surface representation. Gaussian models are widely used thanks to their simplicity that allows them to obtain closed-form expressions for the electromagnetic scattering problem in several applications. Gaussian models, for example, allow elegant descriptions for the scattering of ships on the sea surface, as presented in Section 2.4.

Anyway, (2.1) exhibits a power law decaying, that is, a typical feature of fractals and leads us to recall the most suitable fractal model for sea surface description. Another key characteristic that makes fractals attractive for sea surface description is that fractal parameters describe intrinsic properties of the observed surfaces. Classical parameters (standard deviation, correlation length, and autocorrelation function) measured at one position are generally poorly representative of their surroundings. In remote sensing, this means that fractal parameters are expected to be independent of the sensor. A relationship between classical and fractal parameters for physical band-limited stationary fBm fractal surfaces can be found in [18].

As already argued in [17], the spectrum in (2.1) can be obtained as the spectrum of a fractional Brownian motion, whose Hurst coefficient $H = 0.75$ and whose spectral parameter can be evaluated with the formula reported in (10) of [17]. Such a relationship provides a direct link between fractals and sea surfaces. The main limit of the spectral description is the fact that only second-order statistics can be inferred on the surface and, as a consequence, on the scattered field.

The Weierstrass–Mandelbrot (WM) model is a model able to reproduce the power law spectrum exhibited by an fBm and, as a consequence, by sea surfaces at cm scales, and to provide a representation of the height profile in terms of a superposition of infinite sinusoidal tones whose periods are spaced in the wave-number domain according to a geometric progression regulated by an irrational factor [18–20]. The WM function $z(x,y)$ can be expressed as

$$z(x,y) = b\sum_{p=0}^{P-1} C_p \nu^{-Hp} \sin\left[k_0 \nu^p \left(x \cos\ \psi_p + y \sin\ \psi_p\right) + \phi_p\right] \tag{2.3}$$

In (2.3), the WM is described in terms of five independent parameters: b is the overall amplitude and H is the Hurst coefficient are intrinsic surface parameters [19]; k_0 is the fundamental tone wavenumber, P is the number of tones, dictated by the highest tone wavenumber, and ν is the irrational wavenumber scaling factor are extrinsic parameters that depend on how an external observer looks at the sea surface. In addition, C_p, ϕ_p, and ψ_p are random variables, whose values determine the amplitude, direction, and phase behavior of each tone.

In remote sensing applications, the extrinsic parameters are determined by the sensor. In radar applications, the larger and smaller spatial scales that interact with the electromagnetic field are provided by the illuminated area and the electromagnetic wavelength, respectively. The number of tones P is dictated by the highest tone wavenumber and by the irrational parameter ν

In Figure 2.1, we show a two-dimensional (2D) representation of a WM process with parameters set to synthesize a power law sea spectrum at a centimetric scale. Details on the synthesis procedure are provided in [19].

2.3 Electromagnetic scattering from the sea surface

In this section, we consider that the incident field can be expressed as a plane wave with amplitude E_0 and propagation vector $\mathbf{k}_i$:

$$\mathbf{E}_i = \widehat{p} E_0 e^{-j\mathbf{k}_i \cdot \boldsymbol{r}} \tag{2.4}$$

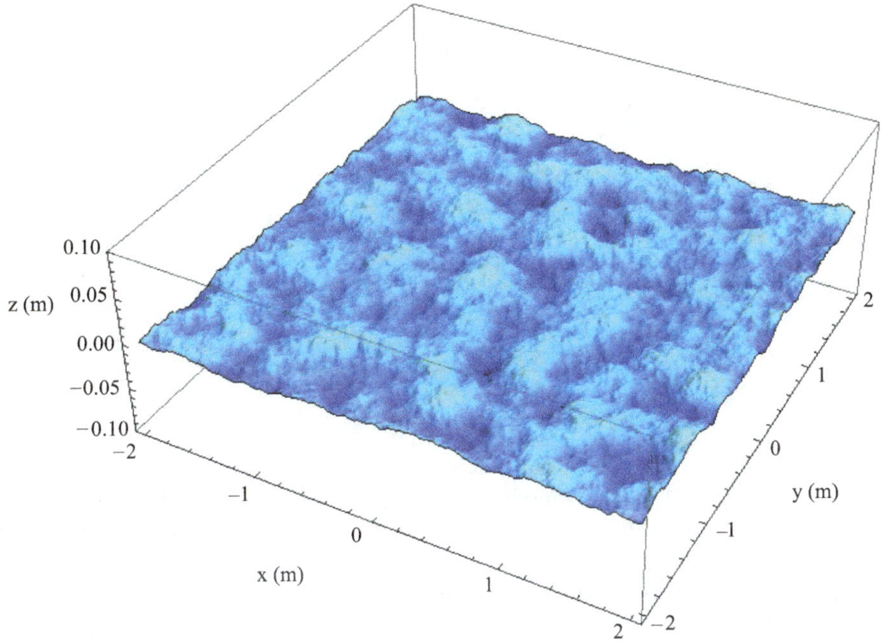

Figure 2.1 2D representation of a WM process, with $H = 0.75$

In (2.4), the unit vector $\widehat{\boldsymbol{p}}$ describes the polarization of the field and $\boldsymbol{r}$ is the position vector. A periodic time dependence $\exp(j\omega t)$ is understood and suppressed.

The geometry of the problem is shown in Figure 2.2, where ϑ_1 is the incidence angle. The electromagnetic scattered field in a direction identified by the angles ϑ_2 and ϑ_3 is characterized by the vector $\mathbf{k}_s$ and can be evaluated in integral form from Maxwell's equations. In the Fraunhofer region, it can be expressed in terms of the fields imposed on the surface by the primary source:

$$\boldsymbol{E}_s(\boldsymbol{r}) = \frac{\text{jkexp}(-\text{jkr})}{4\pi r}\left(\boldsymbol{I} - \widehat{\boldsymbol{k}}_s\widehat{\boldsymbol{k}}_s\right)$$

$$\int_A dS' \left\{\widehat{\boldsymbol{k}}_s \times [\widehat{\boldsymbol{n}} \times \boldsymbol{E}(\boldsymbol{r}')] + \zeta[\widehat{\boldsymbol{n}} \times \boldsymbol{H}(\boldsymbol{r}')]\right\}\exp(-j\boldsymbol{k}_s \cdot \boldsymbol{r}') \tag{2.5}$$

where k is the electromagnetic wavenumber, A is the area illuminated by the sensor, r is the distance from its center to the receiver, ζ the characteristic impedance of the medium, $\boldsymbol{E}$ and $\boldsymbol{H}$ are the electric and magnetic fields, $\widehat{\boldsymbol{n}}$ is the surface normal vector, and $\boldsymbol{I}$ is the identity matrix.

Equation (2.5) highlights that the scattered field evaluation requires the knowledge of the surface fields that are related to the sea surface shape. Its analytical evaluation in closed form is not available. Assuming a classical Gaussian surface description and moving from different hypotheses, several methods for describing the surface currents

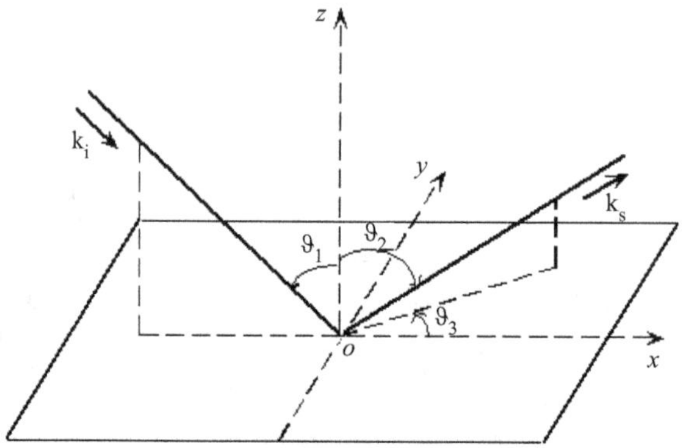

Figure 2.2 Geometry of the problem

have been developed. The methods commonly considered more reliable are the Kirchhoff approximation (KA), the small perturbation method (SPM), and the integral equation method (IEM) [18].

The basic hypothesis of the KA is that the radius of curvature is much higher than the incident wavelength. Then, the surface fields are evaluated in each point as if the surface could be confused with the tangent plane. This assumption brings to a local evaluation of the surface fields in terms of the local Fresnel coefficients [18].

With the SPM the surface fields are evaluated under the hypothesis that the surface standard deviation is much smaller than the incident wavelength and the surface currents are expressed in terms of a perturbative series, where the unperturbed currents are those on the mean plane surface.

A common limitation of all the classical methods is the inadequateness of the classical Gaussian model to describe the sea surface. A big improvement in this sense has been obtained with the introduction of the fractal description of the nature [20]. In the following, the most important electromagnetic methods based on fractal concepts are described.

In Section 2.2.2, we presented the fBM and the WM functions as suitable models to describe the sea surface. The rationale and the results related to the use of both the fBm and the WM in conjunction with the KA are here presented. If we apply the physical optics (PO) or the geometrical optics (GO) approximation, as described in [18], we can express the scattered field as

$$\begin{bmatrix} E_{ih} \\ E_{iv} \end{bmatrix} = \frac{jke^{-jkr}}{4\pi r}\mathbf{S}\begin{bmatrix} E_{ih} \\ E_{iv} \end{bmatrix} I_s \tag{2.6}$$

where

$$S_{pq} = \left[\left(\boldsymbol{I} - \widehat{\boldsymbol{k}}_s\widehat{\boldsymbol{k}}_s\right)\boldsymbol{F}(\widehat{\boldsymbol{k}}_i, \widehat{\boldsymbol{e}}_{ip}, \widehat{\boldsymbol{n}})\right] \cdot \widehat{\boldsymbol{e}}_{sq} \tag{2.7}$$

are the elements of the scattering matrix **S** and

$$I_s = \int e^{-j(\boldsymbol{k}_i - \boldsymbol{k}_s)\cdot \boldsymbol{r}'} d\boldsymbol{r}' \tag{2.8}$$

is the scattering integral I_s. The scattered field assumes the same expression in the PO and GO formulations, with the difference that in the PO case, the normal $\widehat{\boldsymbol{n}}$ in (2.7) refers to the mean scattering plane, in the GO formulation to the specular reflection points.

The use of the fBm and the WM models in (2.8) leads to obtain a second and a first-order solution, respectively, as recalled in the following.

By employing the fBm surface description in (2.8), after some manipulations [18], we obtain the following expression for the mean-square value of the scattered field:

$$\langle |E_{pq}|^2 \rangle = \frac{k^2 |E_{0p}|^2 |F_{pq}|^2}{4\pi r^2} 2\pi A \int_0^\infty J_0(\eta_{xy}\tau) e^{-\frac{1}{2}\eta_z^2 s^2 \tau^{2H}} \tau d\tau \tag{2.9}$$

where $\langle \cdot \rangle$ stands for statistical average, J_0 is the 0th order Bessel function, τ is the distance between $\boldsymbol{r}$ and $\boldsymbol{r}'$, $\eta_{xy} = \sqrt{(k_{ix} - k_{sx})^2 + (k_{iy} - k_{sy})^2}$, and $\eta_z = k_{iz} - k_{sz}$. The parameter s is the incremental standard deviation of the fBm process, that is, the standard deviation, measured in ($\mathrm{m}^{1-\mathrm{H}}$), of surface increments at unitary distance.

Two closed-form solutions can be derived for the integral (2.9), as reported in [18]; for the sea surface H is around 0.75 and the mean-square value of the scattered field can be written as

$$\langle |E_{pq}|^2 \rangle = \frac{|E_{0p}|^2 |F_{pq}|^2}{4\pi r^2} 2\pi A \frac{k^2 T^2}{2H} \sum_{n=0}^{\infty} \frac{(-1)^n}{2^{2n}(n!)^2} \Gamma\left(\frac{n+1}{H}\right) \frac{(\eta_{xy}\tau)^{2n}}{\left(\frac{\sqrt{2}}{2}|\eta_z| T\right)^{\frac{2n+2}{H}}} \tag{2.10}$$

where the topothesy T is used as an alternative to the s parameter to describe the fBm process [18]. Details on series convergence and validity limits are provided in [18].

The mean-square value of the scattered field is particularly useful in radar applications because it allows the evaluation of the radar cross section and the backscattering coefficient (σ^0), as defined in [21]. In Figure 2.3, we show a comparison between backscattering coefficients in dB scale, obtained by measured data and those evaluated with the KA-fBm and the KA-Gaussian model approaches, presented in [22]. For the Gaussian approach, two different correlation functions have been compared, the Gaussian (Ga-Ga) and the exponential (Ga-Exp). In Figure 2.3, we can note that the radar cross section evaluated with the fBm surface profile well matches the measured data. Good results can be obtained with the Ga-Exp approach but, it is important to note that, in order to obtain a good matching between the Gaussian model and experimental data, a tuning method is necessary to find the standard deviation and correlation length values, showing that the classical parameters depend on how we measure the surface properties, whilst fractal parameters are intrinsic features of the surface.

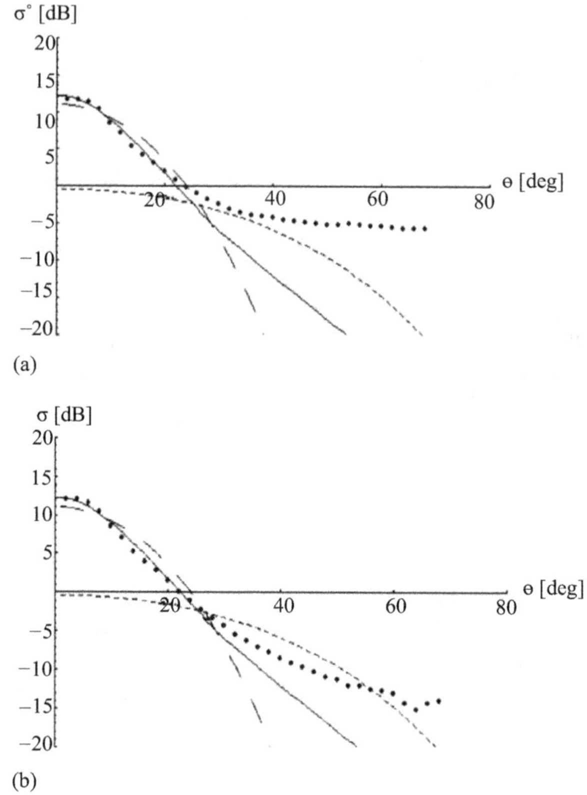

Figure 2.3 Backscattering coefficient comparison between X band experimental results (dots) and the KA results in conjunction with fBm (solid line), Ga-Ga (long dashed line), and Ga-Exp (short dashed line) for (a) VV polarization and (b) HH polarization

Note that the presented method allows only the retrieval of the second-order statistics of the scattered field, due to the fact that the fBm is a second-order statistic description of the surface. As presented in [23], WM functions act as a spectral sampler of the power spectrum typical of sea surface, hence they can be used to simulate sea surfaces when the evaluation of the first-order statistics of the scattered field is at hand.

In the KA procedure, after some mathematical manipulations, employing (2.5) as the surface profile, the scattered field turns out to be expressed as a superposition of generalized Floquet modes [18]:

$$E_{pq} = \frac{jkAe^{-jkr}}{4\pi r} E_p F_{pq}(\vartheta_1, \vartheta_2, \vartheta_3) \sum_{m_0=-\infty}^{+\infty} \cdots \sum_{m_{P-1}=-\infty}^{+\infty} \left\{ e^{jF} \prod_{p=0}^{P-1} A_p S_{2D} \right\} \tag{2.11a}$$

where

$$F = \sum_{n=0}^{P-1} m_n \phi_n \tag{2.11b}$$

$$A_p = \prod_{p=0}^{P-1} J_{m_p}\left(-\eta_z b C_p \nu^{Hp}\right) \tag{2.11c}$$

$$S_{2D} = \mathrm{sinc}\left[\left(-\eta_x + + k_0 \sum_{n=0}^{P-1} m_n \nu^n \cos(\psi_n)\right)\frac{X}{2}\right]$$

$$\mathrm{sinc}\left[\left(-\eta_y + k_0 \sum_{n=0}^{P-1} m_n \nu^n \sin(\psi_n)\right)\frac{Y}{2}\right] \tag{2.11d}$$

In (2.11), the WM parameters (b, H, k_0,ν, C_p, ϕ_p ψ_p, and P) defined within (2.3) are used; $A = XY$ is the illuminated surface; $\eta_x = k_{ix} - k_{sx,}$ and $\eta_y = k_{iy} - k_{sy}$. Directions of propagation of the modes are obtained by (2.11d), with η_x and η_y values such that the argument of the sinc function is null. The amplitudes A_p are related to a product of Bessel function, whose argument is proportional to the roughness parameter $\eta_z b$. The phases F are evaluated as a combination of the phases of the surface tones.

The great advantage of this formulation is the possibility to have a first-order description of the sea surface, which can be extremely advantageous in many fields as simulations, see [19] as an example.

It is interesting to note that the above-presented solutions assume simpler expressions in selected cases. As an example, in monostatic radars, only back-scattering directions are at hand. Therefore, the evaluation of the scattering integral can be simpler by considering that the main contributions to the backscattered field are provided by the ripple spectral components whose wavelength is of the same order of magnitude of the incident electromagnetic wavelength. Since the height of sea waves is much smaller than their wavelength (unless breaking waves are considered), then the height of ripple spectral components involved in the scattering mechanism must be much smaller than the electromagnetic wavelength, too. Under this hypothesis, the scattering integral can be evaluated at the first order as

$$I_s = j\frac{4\pi}{\lambda} Z_s(\overline{\kappa}_x, \overline{\kappa}_y)\cos\vartheta_1 \tag{2.12}$$

where

$$\begin{cases} \overline{\kappa}_x = \dfrac{4\pi}{\lambda}\delta_x \cos\vartheta_1 \\ \overline{\kappa}_y = \dfrac{4\pi}{\lambda}(\sin\vartheta_1 + \delta_y \cos\vartheta_1) \end{cases} \tag{2.13}$$

δ_x and δ_y are the facet's slopes and $Z_s(.)$ is the Fourier transform of $z_s(x,y)\mathrm{rect}[(x-x_0)/L_x]\mathrm{rect}\left[(y-y_0)/L_y\right]$, with $L_xL_y = A$ and x_0 and y_0 are the coordinates at the centre of the scene. In particular, $Z_s(.)$ is a complex circular Gaussian random variable with mean squared value:

$$\langle \left| Z_s(\bar{\kappa}_x, \bar{\kappa}_y) \right|^2 \rangle = A \cdot S(\bar{\kappa}_x, \bar{\kappa}_y) \tag{2.14}$$

where $S(.)$ is the wave spectrum evaluated at $(\bar{\kappa}_x, \bar{\kappa}_y)$. Equation (2.14) clearly shows that, according to the Bragg phenomenon, the synthetic aperture radar (SAR) sensor acts as a frequency selector.

In recent years, new refined methods have been proposed to describe the scattering from the sea surfaces. The two-scale approach (also known as composite model), whose principles can be found in [6,7], models the sea surface as a superposition of long and short waves. It was widely used to describe the scattering from the sea and recently enriched with the possibility to account for the inherent anisotropy of the sea spectrum [17]. The small slope approximation (SSA) was introduced in 1994 by Voronovich [24] to provide bridge between the KA and small perturbation method and has been refined also in recent years to provide valuable methods describing the scattering from the sea surface [25–27]. The Integral Equation Method [28] has been also used, as an example, for investigating the polarization effect in the GPS signal scattered by sea surfaces in bistatic configuration [29,30].

2.4 Scattering models for a ship

A recurrent application, gaining momentum in the last decade, is exactly that of ship detection with the CFAR algorithm [8,9,31] (see also Chapter 6) widely spread for the simplicity of its thresholding concept and fast processing times.

The CFAR detector is an example of maritime applications where models of the electromagnetic scattering from the sea surface play an important role in the determination of the threshold which separates noisy sea background from potential ship signatures labeled as detections.

The limit of such a rationale consists in considering and modeling only one aspect of the problem represented by the sea with its intrinsic properties of roughness and dielectric characteristics with no consideration, instead, going to the real feature of interest in the detection, which is the ship.

With the aim of having a better representation of the complex sea scenario, and possibly improving also ship detection rates, in [32] scattering models for the ship were for the first time considered in conjunction with scattering models from the sea and brought to the concept of a novel ship detector, the GLRT, which outperforms the classical CFAR at the expense of computational time.

The following describes which EM contributions it is likely to expect from a standard ship geometry illuminated by a radar. Discussion will consider the most relevant contributions and how to simplify the geometric model of the ship in order to have only them represented.

In [21], the radar cross section (RCS) of simple and complex targets is thoroughly discussed with a variety of examples and backscattering patterns at different working frequencies. In commenting the RCS pattern of a large naval auxiliary ship (shape details of which are not provided) detected from a radar deployed on the shores of the Chesapeake Bay, the authors generally conclude the relevance of ships RCS (despite the frequency). More than discussing single scintillations from the ship (which are many and fast-varying for different reasons, including the very large dimensions of the ship compared to microwave wavelengths and the variability of waves and sea surface), their statistics are there analyzed. The most interesting consideration is certainly the generally very large RCS despite which part of the ship is illuminated (whether the bow, the stern, or the broadside) with the broadside presenting a peak of RCS (at 64.1 dBm^2) and all other values always within a few decibels (and in any case never dropping below 50 dBm^2).

In [33], an attempt has been made in progressing the modeling and understanding of the ship signature by labeling the scattering contribution most responsible for the ship's RCS as a double reflection mechanism mainly caused by a dihedral geometry. Taking inspiration from a study modeling the scattering of canonical buildings [34], Iervolino *et al.* adapted the concept there conceived to fit the geometry of a simple ship.

A rigorous approach would need to consider a multitude of contributions. As the picture in Figure 2.4 shows, discontinuities and changes in the curvature radii are likely to arise new contributions to the overall scattering of different nature (single and multiple scattering or diffractions), but not all are of the same magnitude.

This justifies, especially from spaceborne SAR platforms, the possibility to drop the inclusion of minor contributions in favor of the main one which is equivalent to simplify the ship's geometrical model and reduce it to a parallelepiped over a sea surface.

The modeling of the scattering from this kind of target can be accomplished by using an approach similar to the one in [34], where now the canonical building is replaced by a canonical ship, and both are represented as parallelepipeds with smooth surfaces, while the ground on which the building is based is replaced by the sea surface in which the ship is sailing. In the following of this chapter, the ship will be considered still, in order to neglect any modification of the scattering due to Doppler effects. In [33] and [34], the consequences of a simplified geometrical model were discussed. For the ship, it is equivalent to disregard topside structures which are the main responsible for the asymmetry of the RCS pattern from port side to starboard side [21]. Another approximation implicitly made is that the freeboard of the ship (a portion of the ship emerging from the water) is perpendicular to the water surface which becomes less correct when moving away from the broadside. It is not a case that in [21], as previously mentioned, the peak of RCS is exactly at the broadside where this approximation holds better.

When such geometrical approximation is accepted and the ship's dimensions are much bigger than the working frequencies (as it is always the case in the microwave range), a scattering model can be derived as done in [33] to represent, in closed form, the main contributions to the radar receiver which are worth to be considered.

Before looking into any specific closed-form expression, it is useful to stress how the SAR image of a ship target looks like. The picture in Figure 2.5 represents a section

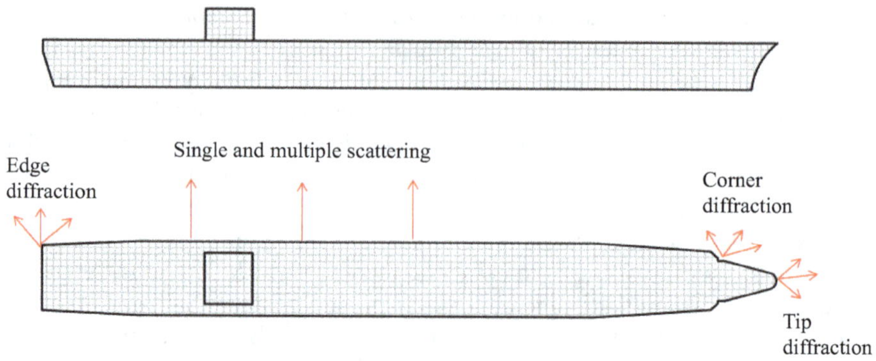

Figure 2.4 Different scattering and diffraction contributions arising from a canonical ship

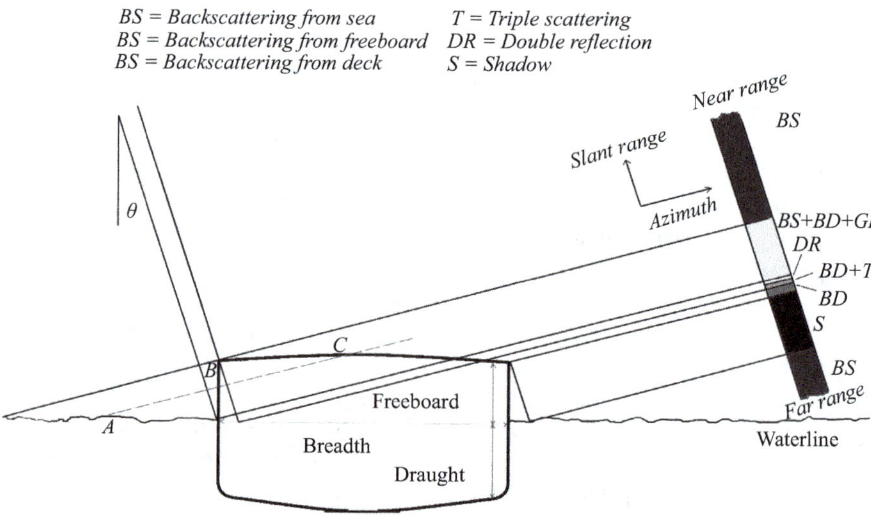

Figure 2.5 Cut at constant azimuth of the canonical ship with the composition of the different backscattering contributions arising in the SAR image where θ is the radar look angle

of the breadth of the ship (the longest length from port to starboard) and the freeboard. The draught of the ship is also indicated for clarity but, being this immersed in the sea, it will not contribute to the ship's signature in the image. Being the portion represented in Figure 2.5 only a section of a ship, the corresponding signature on the right side is indicative of how the scattering contributions would overlap if we ignored the overall length of the ship on waterline. If the radar wavefront arrives from the left side of the figure, and the center of the radar antenna beam forms an angle θ with the nadir, the echo from the left corner of the deck is the first to return to the radar soon after the surface scattering from the sea. Then, the wavefront continues propagating toward the scene and, by interacting with points of surfaces at the same range distance from the radar (such as points A, B, and C in the figure from respectively the sea, the freeboard, and the deck), it generates echoes in correspondence of these points that reach the radar receiver at the same time and therefore are collocated in range. For this reason, an entire region soon after the backscattering from the sea is actually described by the in-phase (for each specific range distance) sum of the backscattering contributions from the sea, the freeboard, and the deck. Continuing from near range to far range, the double scattering contribution follows. As well described in many references including [33] and [34], this mechanism caused by the dihedral surface, formed in this case by the ship freeboard and the sea surface, is responsible for the very bright appearance of ships. The reflection from the freeboard toward the sea surface and from there back to the freeboard and then to the radar (or the other way around) covers the same optical distance, irrespective of which point of the freeboard is considered, thus causing an in-phase summation of all echoes undergoing the double bounce previously described. After the double reflection, a contribution of triple scattering [34] together with backscattering from deck, and then backscattering from the deck only, follow immediately before a shadow area. Backscattering from sea surface starts again soon after.

It is worth to mention that, for the particular model adopted, where roughness is considered only for the sea surface but not for the port or starboard side of the ship (and neither the bow nor the stern or the deck surfaces), multiple reflections above the third order do not arise. Moreover, the particular overlapping of contributions depicted in Figure 2.5 is simply the result of the specific radar look angle selected and the freeboard/breadth ratio of the ship. Different values of those parameters/dimensions would produce a different combination of scattering, but would not affect the main conclusion that this analysis can help in drawing about the relevance of the double reflection mechanism, particularly for isolated ship targets on the sea surface (ships at anchor in the proximity of the coast are not considered here).

As already mentioned, the importance of the double reflection mechanism pushed researchers to find closed-form expressions of this contribution in which radar and scene parameters could be linked analytically [34]. These expressions have proved fundamental in the improvement of current ship detectors in terms, for example, of detection and false alarm rate [32].

The first part of this chapter has thoroughly analyzed the scattering models for sea surfaces. The assumption of a simple model is required also here and, for simplicity, a Gaussian stochastic process with Gaussian autocorrelation function is suggested for the sea surface. It is fundamental also to assume an infinite depth (in terms of working wavelength) and constant dielectric properties for the sea layer beneath the surface so that discontinuities do not exist and multiple bounces cannot arise from beneath the surface.

With these hypotheses, [33] used the Kirchhoff Approximation and evaluated the scattered field at each bounce with PO or GO solutions according to the sea roughness as originally done in [34] for the canonical building case.

In particular, PO approximation is applied if $k\sigma \ll 1$, where k is the radar wavenumber and σ is the standard deviation describing the stochastic process of the sea surface. GO approximation holds in the other case of $k\sigma \gg 1$.

For every bounce occurring on the ship freeboard only the GO approximation is instead considered.

In these hypotheses, the RCS for GO-PO double reflection contribution is given by

$$\sigma^o = h\left|S_{pq}\right|^2 l \tan\vartheta \cos\varphi \exp\left(-4k^2\sigma^2\cos^2\vartheta\right) \sum_{m=1}^{\infty} \frac{(2k\sigma\cos\vartheta)^{2m}}{m!} \frac{k^2L^2}{4m} \times \exp\left[-\frac{(2kL\sin\varphi\sin\vartheta)^2}{4m}\right] \tag{2.15}$$

while, for the GO-GO, the following expression for the RCS holds:

$$\sigma^o = \frac{h\left|S_{pq}\right|^2 l \tan\vartheta \cos\varphi\left(1+\tan^2\vartheta\sin^2\varphi\right)\exp\left[-\frac{\tan^2\vartheta\sin^2\varphi}{2\sigma^2(2/L^2)}\right]}{8\pi^2\sigma^2(2/L^2)\cos^2\vartheta} \tag{2.16}$$

In (2.15)–(2.16), σ^o represents the RCS relevant to the double reflection contribution; $S_{pq}$is the generic element of the scattering matrix with p and q standing for horizontal H, or vertical, V polarization, respectively; l is the length of the portion of the ship belonging to the resolution cell, assuming the ship length larger than the SAR spatial resolution, σ and L are the standard deviation and the correlation length, respectively, of the stochastic process representing the sea surface; φ is the angle between the sensor line of flight and the ship hull to the water surface; and $\vartheta$is the SAR look angle and h is the height of the freeboard forming a dihedral surface with the sea surface.

In (2.15) and (2.16), S_{pq} depends on the dielectric constant of the sea (ε_{SW}), the dielectric constant of the hull (ε_{HULL}), φ, ϑ, k, and the Fresnel coefficient according to the polarization of the propagating wave. The equations to compute S_{pq}, for both GO-GO and GO-PO solutions and for each polarization, are reported in [34].

In Section 2.4.1, some assumptions on the parameters involved in the electromagnetic model are made in order to model the distribution of the RCS values for the double reflection contribution.

2.4.1 RCS estimation of a canonical ship

In order to compute the RCS relevant to the double reflection contribution, all the parameters involved in (2.15) and (2.16) have to be known. Unfortunately, this a priori knowledge is not always available and just some parameters are generally known (i.e., the radar look angle and the wavelength from the ancillary data of the SAR dataset). Other parameters, instead, can be directly computed either from the SAR image (the sea roughness parameters) or from the literature (dielectric constant of the sea); while for the remaining ones (the orientation angle, the dielectric constant of the hull and the freeboard height), suitable probability distribution functions have to be taken in account. Due to the wind, the sea surface is never completely smooth and presents several capillary waves [33,35,36]. For this reason, the GO-GO approximation has been chosen and is the unique solution being considered in the following.

Summarising, it is possible to group the input parameters of (2.16) into four different categories [32]:

1. Vector **a**: parameters retrievable from the ancillary data of the SAR sensor. They are the radar look angle (ϑ), the spatial resolution (x set equal to the ship length), and the radar wavelength (λ).
2. Dielectric constant of the saline water (ε_{SW}) computable from models presented in literature for the particular radar parameters selected [34].
3. Roughness ratio (σ/L) computable directly from the SAR image. It is evaluated by minimizing the absolute error between the RCS relevant to the single scattering of the sea surface and the RCS measured on the SAR image [36].
4. Vector **b**: unknown parameters. They are the orientation angle (φ), the freeboard height (h), and the dielectric constant of the hull (ε_{HULL}).

In order to evaluate the RCS of the double reflection contribution, suitable distribution functions are considered for each parameter of vector **b**. In this way, the analytical model is turned into a statistical model, where the histogram of values relative to the double reflection contribution can be computed. In Sections 2.4.1.1–2.4.1.3, an insight about the computation of the input parameters is provided.

2.4.1.1 Roughness parameters computation

The roughness parameters (the ratio σ/L) is computed by minimizing the absolute error between the RCS (relevant to the single scattering from the sea) of the sea surface measured on the SAR images and the expected RCS within the GO solution [36]. For the sake of simplicity, the analytical expression provided by Franceschetti *et al.* in [34] for the single backscattering from a rough surface is reported for the GO approximation:

$$\sigma^0 = \frac{\left|S_{pq}\right|^2 ab}{64\pi^2 \cos^2\theta \sigma^2/L^2} \exp\left(-\frac{tg^2\theta}{4\sigma/L^2}\right) \tag{2.17}$$

where the symbols have the same meaning as in (2.15) and (2.16) and a and b represent the dimensions of the rectangular portion of sea where the RCS is evaluated. Equation (2.17) does not depend on the roughness parameters σ and L separately, but σ^0 is a function of the ratio σ/L; however, since (2.17) is not linear in respect to this ratio, it is not possible to invert the expressions to retrieve the closed-form equation. The ratio σ/L is estimated as the ratio which minimizes the absolute error E:

$$E = |\widehat{\sigma}_0 - \sigma_0| \tag{2.18}$$

where $\widehat{\sigma}_0$ is the RCS of the sea surface measured on the SAR images and σ^0 is the RCS computed according to (2.17).

The procedure has been tested over meaningful datasets acquired by TerraSAR-X over the Solent area in the United Kingdom [36]. Roughness estimation results are consistent with GO theory and with the ground truth (wind speed and direction) retrieved by buoy stations. Consequently, this procedure can be used for the estimation of the roughness parameters of (2.16).

2.4.1.2 Retrieval of the dielectric constant of the sea water

The dielectric constant of the sea water can be computed according to the *double-Debye dielectric model* reported in [37]. According to this model, the real ($\varepsilon'_{\mathrm{SW}}$) and the imaginary part ($\varepsilon''_{\mathrm{SW}}$) of the saline water are function of the radar wavelength, water conductivity, water salinity, and temperature. Figure 2.6 shows the frequency response of the saline water at a temperature of 20 °C and with a water salinity of 33‰. It is possible to demonstrate that the saline water model reduces to the pure water model when the salinity is 0‰.

2.4.1.3 Unknown parameters estimation

In this section, some assumptions on the parameters included in vector **b** are made in order to retrieve suitable distribution functions. With regard to the angle φ, the authors assume to work in the worst-case scenario, where the orientation angle can be retrieved neither from the ship signature nor from a visible wake. This consideration drives the choice of describing the angle φ statistically as uniformly distributed between 0° and 45° ($\varphi \sim U(0;45)$). Obviously, when the heading angle is greater than 45°, one side of

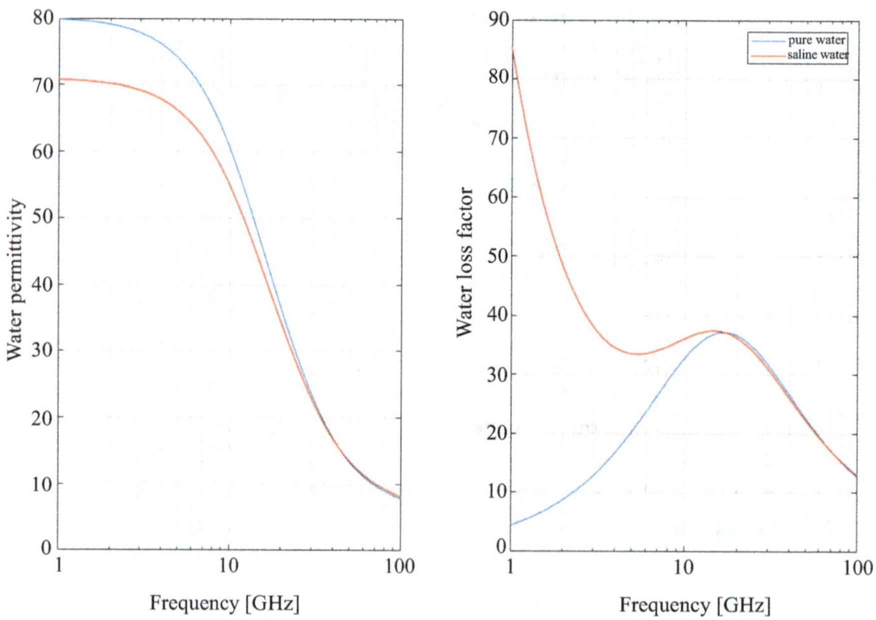

Figure 2.6 Microwave spectra of the permittivity (a) and the relative loss factor (b) of pure water (ε_W) and saline water ($\varepsilon_{\mathrm{SW}}$) at 20 °C and 33‰

the hull will always present an angle with the projection of the sensor on the sea smaller than 45° and, consequently, $\varphi \sim U(0; 45)$ takes into account all the possible scenarios.

The range of values for the freeboard height is selected according to the amendments of the 1974 SOLAS Convention which regulates the freeboard of the ships of 24 m length or more [27]. Ships are divided into two categories (Type A and Type B) and for each ship length (from 24 m to 365 m), the freeboard is provided. Freeboard values are included between 0.2 m and 5.3 m [34]. Having the latter as the only available information about the size of the ship, h is consequently selected uniformly distributed between 0.2 m and 5.3 m $[h \sim U(0.2; 5.3)m]$.

The dielectric constant of the hull is chosen by performing a weighted average of several dielectric constants of materials which mainly compose the structure of a ship. First of all, it is assumed that the canonical ship is made mostly of steel (with a percentage uniformly distributed between 60% and 90%, values based on authors' empirical evaluations) and for the remaining part of a mixture (equally distributed) of glass, aluminum, and fused silica. Within these hypotheses, $\varepsilon_{\text{HULL}}$ can be computed as already done in [38,39]:

$$\varepsilon_{\text{HULL}} = p\varepsilon_{st} + \frac{q}{3}(\varepsilon_a + \varepsilon_g + \varepsilon_{si}) \tag{2.19}$$

where $p \sim U(0.6; 0.9)$, $q = 1-p$ and ε_{st}, ε_a, ε_g, and ε_{si} are the complex relative dielectric constant of steel, aluminum, glass, and fused silica, respectively. Their values are reported in [39,40] and, $\varepsilon_{\text{HULL}}$ is computed for S-, C-, and X-band and values are listed in Table 2.1 in [33]. Finally, in Table 2.1, the way to compute or retrieve all the input parameters involved in the GOGO model is summarized.

2.4.2 RCS distribution

Once all input parameters are estimated or statistically modeled, the RCS relevant to the double reflection contribution of the canonical ship can be modeled too. First of all, a parametric study of (2.16) is carried out. One hundred values are generated for the freeboard height, the orientation angle, and the dielectric constant of the hull at C-band (5.4 GHz) according to the distribution functions reported in Table 2.1. Then, the square modulus of scattering matrix S is computed for VV polarization

Table 2.1 Estimation of the parameters needed to compute the RCS of the ship

Parameters	**Estimation**
Radar look angle ϑ (deg)	Radar ancillary data
Radar working frequency f (Hz)	Radar ancillary data
Roughness parameters σ/L	Minimizing the absolute error of the RCS [36]
Length of the portion of the ship l (m)	Set equal to the SAR spatial resolution
Dielectric constant of the sea ε_{SW}	See Figure 2.6 [37]
Dielectric constant of the ship hull $\varepsilon_{\text{HULL}}$	Equation (2.19)
Orientation angle φ (deg)	$U(0; 45)$
Freeboard height h (m)	$U(0.2; 5.3)$

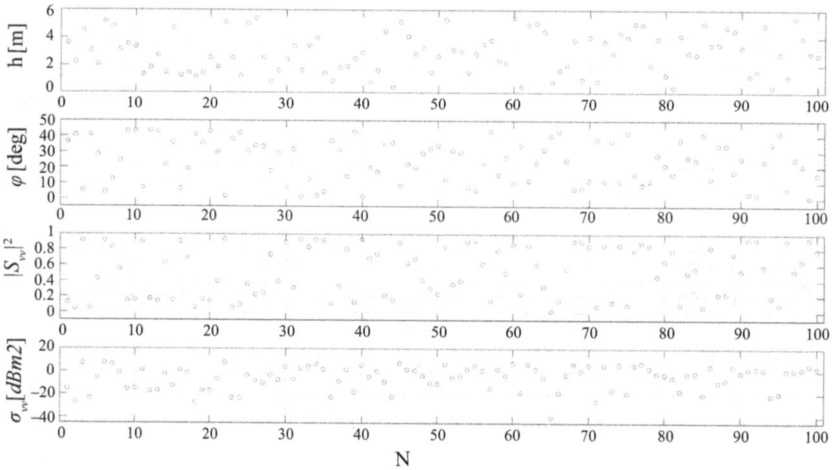

Figure 2.7 Freeboard height, orientation angle, square modulus of Scattering matrix at VV polarization and RCS relative to the double reflection contribution at C-band for 100 samples

and, finally, the RCS relative to the double reflection within the GOGO approximation is computed according to (2.16). The input parameters h, φ, and $S_{pq}{}^2$ along with σ^0_{VV} in dBm^2 scale are shown in Figure 2.7.

The greatest values of the RCS are obtained when the freeboard height is high and when the ship is parallel or near parallel to the SAR flight direction (low φ angle). Vice versa, when the freeboard is small or the angle φ great, the RCS is severely reduced resulting in missing targets in ship-detection algorithms. All the assumptions made on the input parameters make the canonical ship a heterogeneous target and this is witnessed by the large standard deviation compared to the mean RCS value.

Figure 2.8 shows the RCS histogram relative to the double reflection contribution at C-band for all the polarizations (VV, HH, and VH) when one million samples are used. HH and VV distributions are quite similar in shape, mean, and standard deviation while the HV distribution presents values three orders of magnitude smaller. This is why, according to the model presented in the literature [34], the RCS relevant to the double reflection line of the cross-polarized channel is much weaker due to the scattering coefficient S_{VH} being proportional to sin2φ. As a consequence, there is no cross-polarized double reflection component in the case of an ideal dihedral perfectly aligned with the sensor azimuth direction ($\varphi = 0$) and a very weak contribution for small values of the φ angle.

Finally, in [33], the histograms relative to the RCS contributions have been fitted with known distributions and it has been demonstrated that the *Gamma* and the *Weibull* probability distribution function best fit the histograms of copolarized (VV and HH) and cross-polarized (VH) channels, respectively.

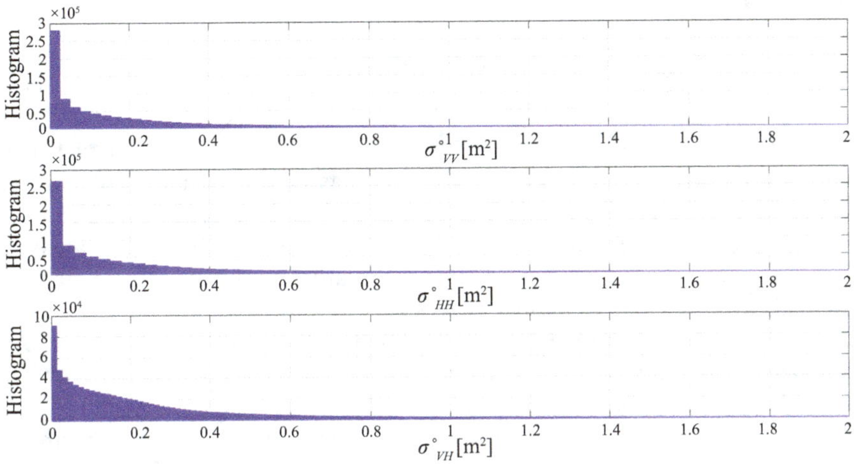

Figure 2.8 Histogram of the RCS values relevant to the double reflection contribution at C-band for VV polarization at the top (mean = 0.53 m² and SD = 1.40 m²), HH polarization in the middle (mean = 0.62 m² and SD = 1.63 m²), and VH polarization at the bottom (mean = 4.47 10⁻⁴ m² and SD = 6.94 10⁻⁴ m²)

2.4.3 Uncertainty budget analysis

In this section, the accuracy of the RCS relevant to the double reflection contribution from (2.16) is analyzed. According to the proposed model, the uncertainty sources are due to a nonperfect knowledge of the input parameters and to the inaccuracy of the model itself in describing all the details of a complex reflecting object as a ship.

The parameters that are *a priori* known (ϑ and k) and retrieved from the literature (ϵ_{SW}) are not considered as sources of uncertainty in the following. Instead, the uncertainty on the estimated value σ^0 of the unknown parameters (h, φ, and ϵ_{HULL}) and the parameters that are measured directly on the SAR images (σ/L) are considered, where each source of uncertainty is regarded separately from the other ones (sources of uncertainty are assumed to be uncorrelated to each other). It is therefore possible to write down an equation for the total uncertainty σ^0_{tot} on the estimated value σ^0 for all the sources of error:

$$\Delta\sigma^0_{\mathrm{tot}} = \left|\frac{\partial\sigma^0}{\partial h}\right|\Delta h + \left|\frac{\partial\sigma^0}{\partial\varphi}\right|\varphi + \left|\frac{\partial\sigma^0}{\partial\varepsilon'_{\mathrm{HULL}}}\right|\Delta\varepsilon'_{\mathrm{HULL}} + \left|\frac{\partial\sigma^0}{\partial\varepsilon''_{\mathrm{HULL}}}\right|\Delta\varepsilon''_{\mathrm{HULL}} + \left|\frac{\partial\sigma^0}{\partial(\sigma/L)}\right|\Delta\frac{\sigma}{L} \tag{2.20}$$

It is possible to separately analyze the terms on the second side of (2.20). Let us first consider the uncertainty σ^0 on the estimated value σ^0, caused by an uncertainty h on h:

$$\Delta\sigma^0 = \left|\frac{\partial\sigma^0}{\partial h}\right|\Delta h = \frac{\sigma^0}{h}\Delta h \Rightarrow \frac{\Delta\sigma^0}{\sigma^0} = \frac{\Delta h}{h} \tag{2.21}$$

Equation (2.21) suggests that the relative uncertainty on σ^0 is equal to the relative uncertainty on h; in other words, if h has been estimated with a certain error, the computed σ^0 will present an error of the same order.

Regarding the uncertainty on the remaining parameters, deriving their analytical expression is less useful. Precisely, even if the relative derivatives can be obviously computed, the retrieved analytical expression would be so involved that useful consideration about the influence on σ^0 estimation could not be carried on. As a consequence, the mathematical closed forms have not been computed and only their graphical representations are shown in Figure 2.9 for VV polarization and C-band.

$|\frac{\partial\sigma^0}{\partial\varphi}|$ presents the worst case when φ is about 15°, where even a minimum error on the knowledge of the orientation angle results in a completely wrong estimation of the RCS. The best range of value, instead, is included between $\varphi = 35°$ and $\varphi = 45°$ where a nonperfect knowledge of the orientation angle does not affect the estimation of the RCS. It is important to underline that, in this same range, the performances of the ship-detection algorithm are worse because most of the incidence radiation from SAR is reflected in the specular direction and consequently the ship could appear as dark as the sea clutter in the final SAR image [36].

The uncertainty $|\frac{\partial\sigma^0}{\partial(\sigma/L)}|$ tends to zero when the sea surface is smooth ($\sigma/L \to 0$) and when the sea surface is extremely rough ($\sigma/L \to \infty$). The worst case occurs when $\sigma_{\mathrm{dev}}/L = 0.10$, while the best case occurs when $\sigma/L = 0.16$.

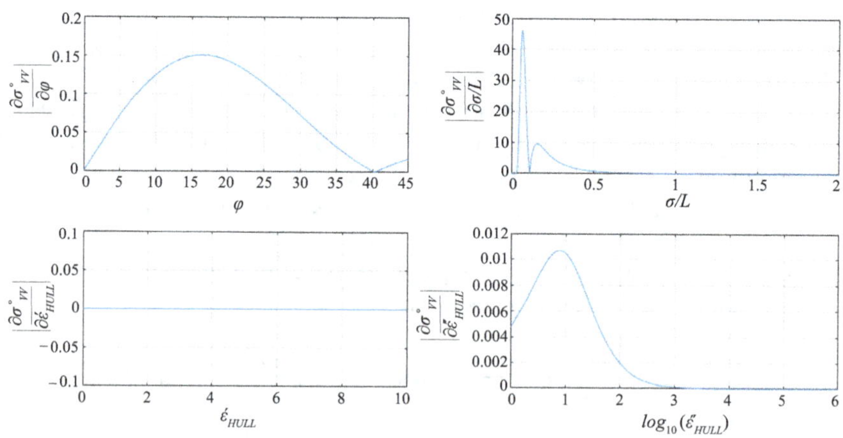

Figure 2.9 Plots of uncertainties computed at C-band and VV polarization relative to the angle φ (top left), roughness parameters (top right), permittivity of the hull (bottom left) and conductivity of the hull (bottom left)

The analysis concerning the dielectric constant of the hull is divided into two parts in order to consider separately the permittivity and conductivity effects. From the analysis of the real part of the dielectric constant, the range of variability of $\left|\frac{\partial\sigma^0}{\partial\varepsilon'_{\mathrm{HULL}}}\right|$ is several orders of magnitude smaller than the mean value of σ^0. Consequently, the influence from a nonperfect knowledge of the hull permittivity is negligible for any ship. Moving to the imaginary part of the dielectric constant, similar considerations can be drawn. The term $\left|\frac{\partial\sigma^0}{\partial\varepsilon''_{\mathrm{HULL}}}\right|$ presents remarkable variations for small values of the imaginary part of ϵ_{HULL}, but it approaches 0 for $Im(\epsilon_{\mathrm{HULL}}) > 10^3$. As a consequence, since the $Im(\epsilon_{\mathrm{HULL}})$ of the metals is much greater than 10^3 [36,38,40], the uncertainty relative to an imperfect knowledge of the conductivity is null if the ship is mostly made by metal, and this hypothesis is certainly verified in many real cases.

2.4.4 Model inaccuracy and validation

In this section, the errors on the RCS due to approximations on the shape of the canonical ship are analyzed. The simple basic parallelepiped model assumed for the ship is certainly a valuable starting basis, but it is not able to describe all the scattering mechanisms which occur in a real scenario.

Neglecting the superimposed structures of a ship (ship upper decks and masts), for example, leads to an underestimation of the dihedral surface, which contributes to the double reflection mechanism, with a consequent underestimation of the final RCS. Depending on the dimensions (length and heights) of masts and decks and the orientation angle of the ships, these contributions may be more or less relevant. In addition, the same superimposed structures may originate also strong trihedral reflection mechanisms with an even worse estimation [33]. An error budget and validation analysis is carried out in [33] for a couple of TerraSAR-X datasets acquired with Stripmap mode over the Isle of Wight in the UK. Several ship signatures were analyzed by measuring the RCS contribution relative to the double reflection. The measured RCS was then compared with the theoretical RCS contribution of the GOGO model and, finally, the absolute and the relative error of measurement was computed for each ship signature. Results show that the electromagnetic model always underestimates the measured RCS on real SAR images [33]. In particular, the average absolute error of measurement is $-0.0646\ \mathrm{m}^2$ while the average relative error of measurement is −41.37%, so in other words, the model underestimates the measured RCS of 1.5 dB on average. The discrepancy between the model and the measured RCSs may be caused by the simplified geometry of the canonical ship where no superimposed structure is taken into account.

References

[1] Goldstein H. "Frequency dependence of the properties of sea echo." *Phys. Rev.* 1946, vol. 70, no. 11, pp. 938–946.

[2] Davies H. "The reflection of electromagnetic waves from a rough surface." *Proc. IEEE.* 1954, vol. 101, no. 90, pp. 209–214.

[3] Cox C. and Munk W. "Measurement of the roughness of the sea surface from photographs of the sun's glitter." *J. Opt. Soc. Am.* 1954, vol. 44, no. 11, p. 838.

[4] Cox C. and Munk W. "Slopes of the sea surface deduced from photographs of sun glitter." UC San Diego Bulletin of the Scripps Institution of Oceanography, 1956.

[5] Phillips O. M. "The equilibrium range in the spectrum of wind generated waves." *J. Fluid Mech.* 1958, vol. 4, pp. 426–434.

[6] Valenzuela G. R. "Scattering of electromagnetic waves from a tilted slightly rough surface." *Radio Sci.* 1968, vol. 3, no. 11, pp. 1057–1066.

[7] Wright J. W. "A new model for sea clutter." *IEEE Trans. Antennas Propag.* 1968, vol. AP-16, no. 2, pp. 217–223.

[8] Cui Y., Yang J., Yamaguchi Y., Singh G., Park S. E., and Kobayashi H. "On semiparametric clutter estimation for ship detection in synthetic aperture radar images." *IEEE Trans. Geosci. Remote Sens.* 2013, vol. 51, no. 5, pp. 3170–3180.

[9] Leng X., Ji K., Yang K., and Zou H. "A bilateral CFAR algorithm for ship detection in SAR images." *IEEE Geosci. Remote Sens. Lett.* 2015, vol. 12, no. 7, pp. 1536–1540.

[10] Ott R. H. "A line-of-sight propagation model for calculating atmospheric, sea, terrain multipath, and clutter at microwave frequencies." *Radio Sci.* 1997, vol. 32, no. 2, pp. 369–386.

[11] Cho S. H. and Wait J. R. "Analysis of microwave ducting in an inhomogeneous troposphere." *Pure Appl. Geophys. PAGEOPH.* 1978, vol. 116, no. 6, pp. 1118–1142.

[12] Apel J. R. *Principles of ocean physics*. San Diego, CA: Academic Press, 1987.

[13] Harger R. O. "The SAR image of short gravity waves on a long gravity wave." In O. M. Phillips and K. Hasselmann (eds.), *Wave dynamics and radio probing of the ocean surface*. New York, NY: Plenum Press, 1986, pp. 371–392.

[14] Pierson W. J. and Moskowitz L. A proposed spectral form for fully developed wind seas based on the similarity theory of S.A. Kitaigorodskii. *J. Geophys. Res.* 1964, vol. 69, pp. 5181–5190.

[15] Elfouhaily T., Chapron B., Katsaros K., and Vandemark D. "A unified directional spectrum for long and short wind-driven waves." *J. Geophys. Res.* 1997, vol. 102, pp. 15781–15796.

[16] Hasselmann K., Barnett T. P., Bouws E., *et al.* "Measurements of wind-wave growth and swell decay during the joint sea wave project (JONSWAP)." *Dtsch. Hydrogr. Inst.* 1973, vol. 46.

[17] Di Martino G., Iodice A., and Riccio D. "Closed-form anisotropic polarimetric two-scale model for fast evaluation of sea surface backscattering." *IEEE Trans. Geosci. Remote Sens.* 2019, vol. 57, no. 8, pp. 6182–6194.

[18] Franceschetti G. and Riccio D. *Scattering, natural surfaces and fractals.* Burlington, MA: Academic Press, 2007.

[19] Riccio D. and Ruello G. "Synthesis of fractal surfaces for remote-sensing applications." *IEEE Trans. Geosci. Remote Sens.* 2015, vol. 53, no. 7, pp. 3803–3814.

[20] Mandelbrot B. B. *The fractal geometry of nature*. W. H. Freeman and Co., New York, NY, 1983.

[21] Knott E. F., Shaeffer J. F., and Tuley M. T. *Radar cross section*, 2nd dn. Boston, MA: Artech House, 1993.

[22] Ruello G., Blanco-Sánchez P., Iodice A., *et al.* "Measurement of the electromagnetic field backscattered by a fractal surface for the verification of electromagnetic scattering models." *IEEE Trans. Geosci. Remote Sens.* 2010, vol. 48, no. 4, pp. 1777–1787.

[23] Riccio D. and Ruello G. "Fractal surface synthesis for remote sensing." In *Proc. of the IEEE Int. Geosci. Remote Sens. Symp.*, 2012, no. 1, pp. 5778–5781.

[24] Voronovich A. "Small-slope approximation for electromagnetic wave scattering at a rough interface of two dielectric half-spaces." *Waves in Random Media*. 1994, vol. 4, pp. 337–367.

[25] Guerin C. A. and Johnson J. "A simplified formulation for rough surface cross-polarized backscattering under the second-order small-slope approximation." *IEEE Trans. Geosci. Remote Sens.* 2015, vol. 53, no. 11, pp. 6308–6314.

[26] Zavorotny V. U. and Voronovich A. G. "Bistatic radar scattering from an ocean surface in the small-slope approximation." IEEE 1999 Int Geosci. Remote Sens. Symp. IGARSS'99, no. 1, pp. 0–2, 1999.

[27] Li J., Zhang M., Wei P., and Jiang W. "An improvement on SSA Method for EM scattering from electrically large rough sea surface." *IEEE Geosci. Remote Sens. Lett.* 2016, vol. 13, no. 8, pp. 1144–1148.

[28] Fung A. K., *Microwave scattering and emission models and their applications.* Norwood, MA: Artech House, 1994.

[29] Fung A. K., Zuffada C., Member S., and Hsieh C. Y. "Incoherent bistatic scattering from the sea surface at L-band." 2001, vol. 39, no. 5, pp. 1006–1012.

[30] Zuffada C., Fung A., Parker J., Okolicanyi M., and Huang E. "Polarization properties of the GPS signal scattered off a wind-driven ocean." *IEEE Trans. Antennas Propag.* 2004, vol. 52, no. 1, pp. 172–188.

[31] Crisp D. J. The state-of-the-art in ship detection in synthetic aperture radar imagery. Canberra, Australia: Defence Science and Technology Group (DST), 2004.

[32] Iervolino P. and Guida R. "A novel ship detector based on the generalized-likelihood ratio test for SAR imagery." *IEEE J. Sel. Top. Appl. Earth Obs. Remote Sens.* 2017, vol. 10, no. 8, pp. 3616–3630.

[33] Iervolino P., Guida R., and Whittaker P. "A model for the backscattering from a canonical ship in SAR imagery." *IEEE J. Sel. Top. Appl. Earth Obs. Remote Sens.* 2016, vol. 9, no. 3, pp. 1163–1175.

[34] Franceschetti G., Iodice A., and Riccio D. "A canonical problem in electromagnetic backscattering from buildings." 2002, vol. 40, no. 8, pp. 1787–1801.

[35] Iervolino P., Guida R., Iodice A., and Riccio D. "Flooding water depth estimation with high-resolution SAR." *IEEE Trans. Geosci. Remote Sens.* 2015, vol. 53, no. 5.

[36] Iervolino P., Guida R., and Whittaker P. "Roughness parameters estimation of sea surface from SAR images." *Int. Geosci. Remote Sens. Symp.* 2014, pp. 5013–5016.

[37] Ulaby F. T. and Long D. G. *Microwave Radar and Radiometric Remote Sensing.* Ann Arbor, MI: The University of Michigan Press, 2014.
[38] Watson D. G. M., *Practical ship design.* Oxford, UK: Elsevier Science Ltd, 1998.
[39] Guida R., Iodice A., Riccio D., and Stilla U. "Model-based interpretation of high-resolution SAR images of buildings." *IEEE J. Sel. Top. Appl. Earth Obs. Remote Sens.* 2008, vol. 1, no. 2.
[40] Bussey H. E., Gray J. E., Bamberger E. C., *et al.* "International Comparison of Dielectric Measurements." *IEEE Trans. Instrum. Meas.* 2008, vol. IM–13, no. 4, pp. 305–311.
[41] Falconer K. *Fractal geometry*. Chichester, UK: John Wiley & Sons, 1990.

Chapter 3
Acquisition modes

Gerardo Di Martino[1], Antonio Iodice[1] and Andrea Monti-Guarnieri[2]

3.1 Introduction

The synthetic aperture radar (SAR) is a microwave remote sensing imaging system that consists of a pulsed radar mounted on a moving platform. Usually, the latter is an aeroplane (airborne SAR) or a satellite (spaceborne SAR) moving along an approximately straight-line trajectory (line of flight, or track) with an approximately constant velocity. Transmitted pulses are usually linearly frequency modulated (LFM), that is, they are chirp pulses, as they are called in radar jargon.

The SAR imaging system resolution along the range (or cross-track) direction is obtained by matched-filtering the received radar echoes, as in standard pulsed radars; therefore, range resolution depends on chirp bandwidth: it improves as the chirp bandwidth increases.

A high resolution in the azimuth (or along-track) direction is achieved by properly combining echoes of different pulses, so to generate a synthetic antenna much longer than the real one. The length of the synthetic antenna is the length of the part of platform trajectory from which a ground point is illuminated by the radar beam. It depends on the real antenna beam width, on the employed acquisition mode, that is, on the way (if any) in which the antenna beam is steered during the platform flight, and, possibly, on the employed pulse transmission timing strategy. The range and azimuth sizes of the imaged scene depend on the acquisition mode, too.

In fact, different operational modes can be used by SAR systems to acquire data from a ground area under survey. The standard and simplest one is the *stripmap* mode [1,2]: a fixed side-looking radar antenna is pointed toward the ground so that the antenna beam footprint draws a "strip" on the ground as the platform moves, see Figure 3.1. Accordingly, the illuminated area is theoretically unlimited in the along-track direction. Conversely, the range size of the imaged scene (i.e., the swath size) is related to the range size of the ground footprint of the radar antenna radiation pattern,

[1]Dipartimento di Ingegneria Elettrica e delle Tecnologie dell'Informazione, Università di Napoli Federico II, Naples, Italy
[2]Dipartimento di Elettronica Informazione e Bioingegneria,Politecnico di Milano, Milan, Italy

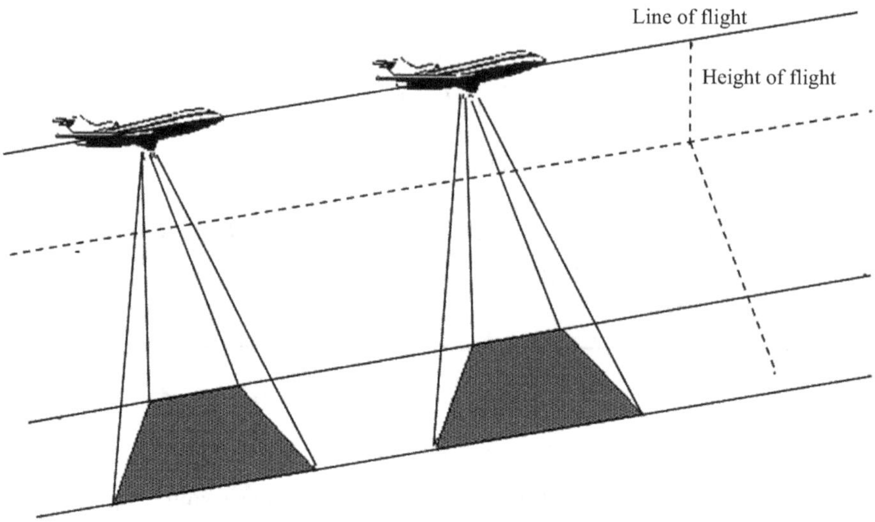

Figure 3.1 Geometry of the stripmap mode

and it is limited by the pulse repetition interval (PRI), that is, the time between transmission of two consecutive pulses. In fact, each inter-pulse interval must accommodate the reception of the entire echo of the illuminated swath, otherwise range ambiguity problems appear (see Chapter 5). In turn, the PRI cannot exceed half the real antenna azimuth length divided by the sensor velocity, in order to avoid azimuth ambiguity problems (see Chapter 5). With regard to the azimuth resolution, it is easy to realize that in this case, the synthetic antenna length is equal to the azimuth size of the ground footprint of the radar antenna radiation pattern (azimuth footprint). Accordingly, azimuth resolution turns out to be independent of the sensor-to-target distance, and equal to half the real antenna azimuth length, see Section 3.2. To improve this resolution it is necessary to increase the synthetic aperture length. This is the case in the *spotlight*, or *staring spotlight*, mode [1,2], for which the radar antenna beam is steered from forward to backward, during the acquisition time, see Figure 3.2, in order to point always in the same area over the ground. Obviously, in this way the achieved azimuth resolution improvement is traded off by the loss of ground azimuth coverage. An intermediate solution is the so-called *hybrid stripmap/spotlight*, or *sliding spotlight*, mode [3–5]: the radar antenna beam is steered about a point farther away from the radar than the illuminated ground, see Figure 3.3 so that the antenna beam footprint on the ground moves more slowly than the sensor; this allows the generation of microwave images with an azimuth resolution better than that achieved in the stripmap configuration and a ground coverage better than the one of the staring spotlight configuration.

While staring and sliding spotlight modes provide an improvement of azimuth resolution at the expense of azimuth coverage, *scanSAR* [1,2] and *TOPSAR* [6] modes aim at improving range coverage (i.e., at extending the range swath) at the expense of azimuth resolution. In the scanSAR mode, the antenna beam elevation is

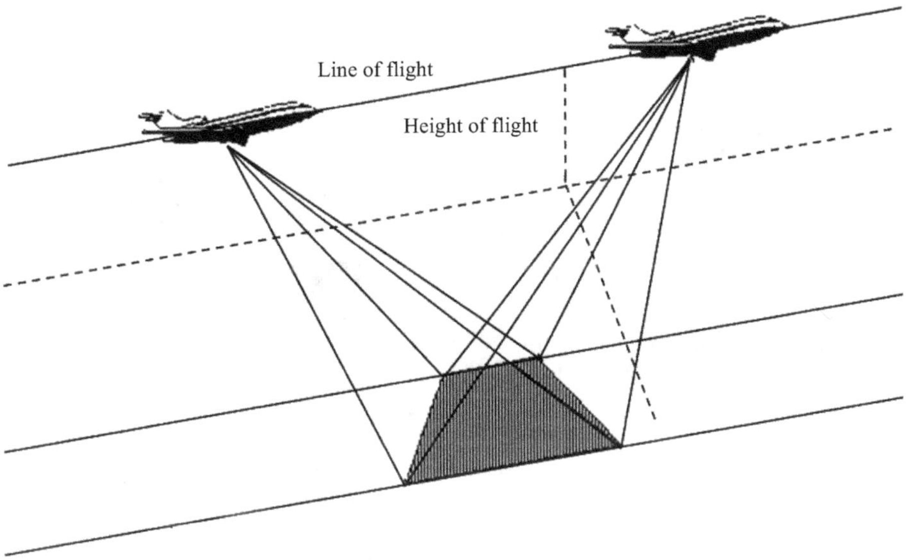

Figure 3.2 Geometry of the staring spotlight mode

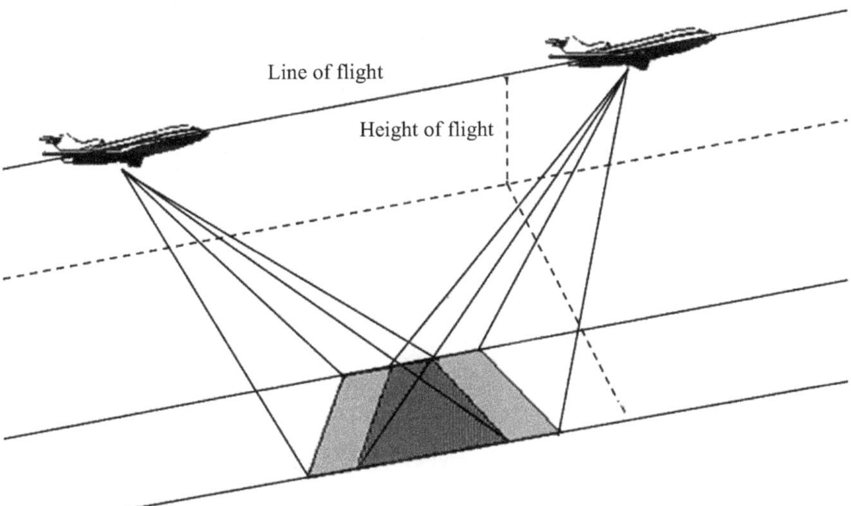

Figure 3.3 Geometry of the sliding spotlight mode

periodically switched to illuminate different range subswaths, so that a wide range swath is obtained, see Figure 3.4. Therefore, each subswath is periodically illuminated by *bursts* shorter than the stripmap integration time, so that the synthetic antenna is shortened, and azimuth resolution is worsened. A drawback of the

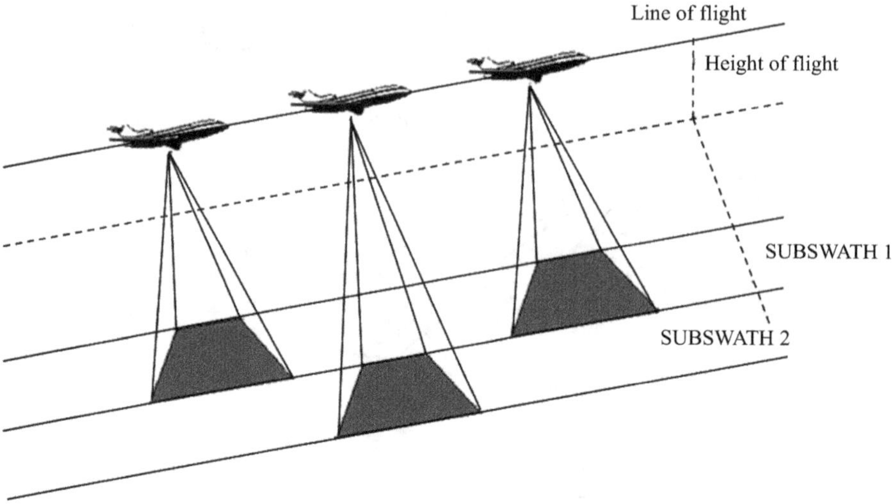

Figure 3.4 Geometry of the scanSAR mode

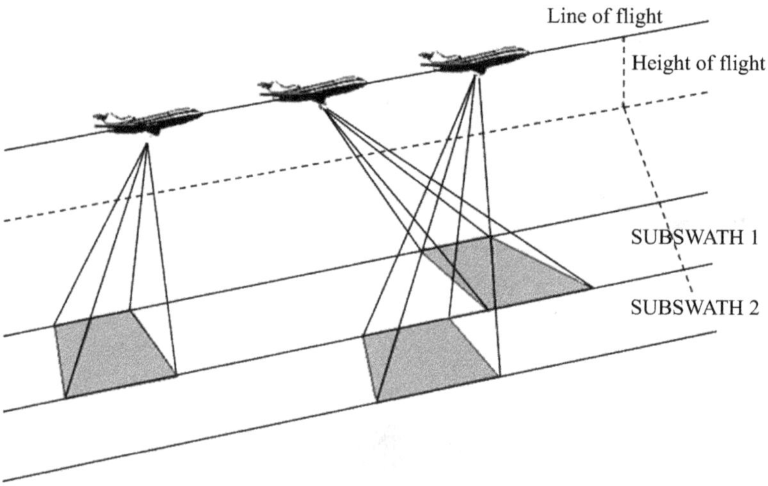

Figure 3.5 Geometry of the TOPSAR mode

scanSAR mode is that ground points at different azimuth coordinates are illuminated by different parts of the antenna azimuth pattern so that an inconvenient periodical modulation of image amplitude (*scalloping*) is obtained. To avoid this drawback, the TOPSAR mode has been proposed: along with the periodical switch of the antenna beam elevation among different range subswaths, within each burst, the beam is also steered from backward to forward (i.e., in the opposite way with respect to the staring or sliding spotlight cases), see Figure 3.5, so that the antenna footprint on the ground moves faster than the sensor. In this way, all points are illuminated by the whole antenna azimuth pattern, and no scalloping is obtained.

An acquisition modality specifically developed for ocean monitoring, and aimed at reducing the data rate and the required data storage, is the wave mode [7]. In this modality, the radar is turned on only periodically for a rather short time, so to obtain small images of the ocean (imagettes), each with an area of the order of 5 km × 5 km, placed at distances of the order of the hundreds of kilometers one from another. These imagettes are employed to retrieve the ocean surface spectra, and hence the wind velocity field [8].

All the acquisition modes described above are by now well established, and they have been used, or are currently used, by several spaceborne missions. In fact, while the first spaceborne SAR systems, such as ERS and ENVISAT, used to employ stripmap, wave, and scanSAR modes, most recent ones, namely, Cosmo/SkyMed, TerraSAR-X, and Sentinel-1, are also using sliding spotlight and/or TOPSAR modes.

More recently, other modalities, often specifically developed for ship detection applications, have been devised to reduce the necessary data rate and/or increase coverage while preserving resolution: low-PRF mode [9], compressive sensing SAR (CS-SAR) [10], staggered SAR [11], and Coprime SAR (CopSAR) [12]. They are very promising, but they are still at an experimental stage.

In the rest of this chapter, the acquisition modes cited above will be described in greater detail, their advantages and drawbacks will be highlighted, and their applicative domains will be illustrated.

3.2 Stripmap mode

In a SAR system, the radar moves at constant velocity v and at constant height H with respect to ground, along a fixed direction that we take as the x-axis of a cylindrical coordinate system, see Figure 3.6, in which r is the distance from the x-axis, that is, from the line of flight, and ϑ is the off-nadir angle. The radar transmits chirp pulses at times $t_n = n\Delta t$, where Δt is the PRI and n is integer:

$$p_n(t) = \cos\left[2\pi f(t - t_n) \pm \tfrac{\pi\Delta f}{\tau}(t - t_n)^2\right]\mathrm{rect}\left(\frac{t - t_n}{\tau}\right) \tag{3.1}$$

where f is the carrier frequency, Δf the pulse bandwidth, τ the pulse duration, and "rect" is the rectangular window function, equal to 1 when the absolute value of its argument is smaller than ½, and zero otherwise. If in (3.1) the + signum is chosen, then an *up-chirp* (i.e., a chirp with instantaneous frequency increasing with time) is obtained; otherwise, a *down-chirp* (i.e., a chirp with instantaneous frequency decreasing with time) is generated. In the following, we will always consider the transmission of an up-chirp, but substantially the same results are obtained for a down-chirp.

The sensor position at the time of transmission of the nth pulse is $x_n = vt_n$. If we consider the stripmap mode, the radar antenna is constantly pointed along the cross-track direction, or, more in general, along a fixed direction forming a *squint angle* ψ with the cross-track direction. Accordingly, the antenna beam footprint on

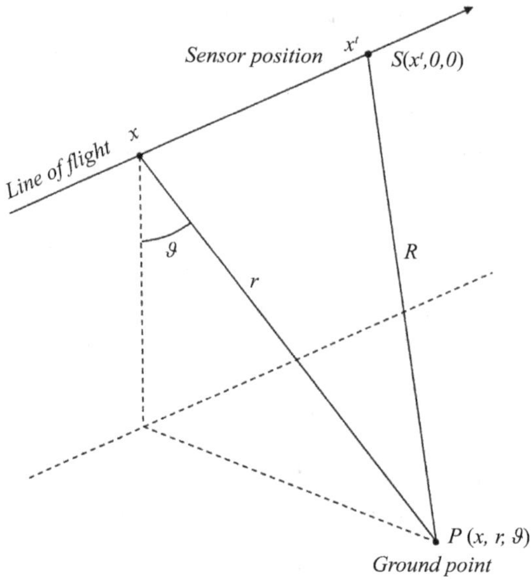

Figure 3.6 Geometry of the problem and reference coordinate system

the ground moves with the same velocity of the sensor (actually, for spaceborne systems this is an approximation). Each transmitted pulse is backscattered by the ground points within the antenna beam footprint and its echo is received by the sensor. The contribution from each ground point of coordinates (x, r, ϑ) is delayed by the round-trip time $2R/c$, where R is the distance from the ground point to the sensor and c is the speed of light. Here we make the "stop-and-go" approximation, that is, we assume that the position of the sensor during reception of the echo of the nth pulse is still x_n: this is acceptable for most practical uses [1,2]. Accordingly, the received signal at the sensor (the *raw signal*) after coherent demodulation (i.e., the baseband raw signal) can be written as

$$h(x_n, r') = \iint \gamma(x, r) g(x_n - x, r' - r; r) dx dr \tag{3.2}$$

where

$$\begin{aligned} g(x_n - x, r' - r; r) = \exp\left[-j\frac{4\pi}{\lambda}\Delta R\right] \exp\left[j\frac{4\pi}{\lambda}\frac{\Delta f/f}{c\tau}(r' - r - \Delta R)^2\right] \\ \times \operatorname{rect}\left[\frac{(r' - r - \Delta R)}{c\tau/2}\right] w^2\left(\frac{x_n - x - r\tan\psi}{X}\right) \end{aligned} \tag{3.3}$$

is the two-dimensional (2D) SAR system impulse response, and

$$\Delta R = \Delta R(x_n - x; r) = R - r = \sqrt{r^2 + (x_n - x)^2} - r \tag{3.4}$$

In (3.2)–(3.4), λ is the carrier wavelength, $\lambda = c/f$; r' is $c/2$ times the time elapsed from each pulse transmission, $2r' = c(t\text{-}t_n)$; $\gamma(x, r)$ is the scene reflectivity pattern including the phase factor $\exp\left[-j\frac{4\pi}{\lambda}r\right]$ and the range antenna pattern $w_r\left(\frac{r-r_0}{S_r}\right)$, S_r being the range size of the antenna beam footprint and r_0 the distance from the line of flight to the center of the range swath; $w(\cdot)$ is the azimuth antenna pattern; and $X = \lambda r_0/L$ is the azimuth size of the antenna beam footprint (we assume that $w(\cdot)$ is negligible when the absolute value of its argument is larger than $1/2$), L being the real antenna azimuth length.

Note that in (3.2), an inessential multiplicative constant has been neglected. In addition, the received signal is actually sampled at frequency f_s, so that, with respect to r', it is actually sampled at spatial frequency $2f_s/c$; however, if f_s is larger than or equal to the pulse bandwidth Δf, the Nyquist criterion is satisfied, so that we can neglect the effect of sampling and consider r' as a continuous variable. Similarly, if the pulse repetition frequency (PRF), that is, the inverse of the PRI, is larger than or equal to $2v/L$ (the azimuth, or Doppler, bandwidth), then we can consider x_n as a continuous variable and let $x_n = x'$. The effect of azimuth sampling, that is, of a finite PRF, will be discussed in Chapter 5.

The final SAR image can be recovered from the raw signal in (3.2) by a proper processing (*focusing*) that consists in compensating the phase terms in (3.3) to recover the scene reflectivity:

$$\hat{\gamma}(\hat{x}, \hat{r}) = \iint h(x', r')g^*(x' - \hat{x}, r' - \hat{r}; \hat{r})dx'dr' \tag{3.5}$$

where * stands for complex conjugate. Processing performed directly in time domain via (3.5) is often referred to as *direct backprojection*, and has a general validity, but a rather high computational cost, since it implies the numerical evaluation of a twofold integral.

Note that, due to the r-dependence of ΔR in (3.4), the SAR system impulse response in (3.3) also depends separately on r (*range variation*, or *focus-depth variation*). If this range variation can be neglected, the integral in (3.2) becomes a convolution, and processing in (3.5) coincides with 2D matched-filtering the raw signal. In addition, by using the convolution theorem, the Fourier transform (FT) of the raw signal is expressed as

$$H(\xi, \eta) = \Gamma(\xi, \eta)G_0(\xi, \eta) \tag{3.6}$$

where $\Gamma(\xi, \eta)$ is the FT of $\gamma(x, y)$ and $G_0(\xi, \eta)$ is the FT of $g(x', r'; r = r_0)$, that is, the 2D SAR system transfer function. The latter can be analytically evaluated by using the stationary phase method as [2]

$$G_0(\xi, \eta) = \exp\left[-j\frac{\eta^2}{4b}\right]\exp\left[j\frac{\xi^2}{4a(1 + \eta\lambda/(4\pi))}\right]\text{rect}\left[\frac{\eta}{bc\tau}\right]w^2\left[\frac{\xi - \xi_{DC}}{2aX}\right] \tag{3.7}$$

where

$$a = \frac{2\pi}{\lambda r_0}, b = \frac{4\pi}{\lambda}\frac{\Delta f/f}{c\tau}, \quad \xi_{DC} = 2ar_0 \tan\psi \tag{3.8}$$

Actually, the argument of the second exponential as expressed in (3.7) implies an unnecessary narrow-beam approximation, that is adopted here for the sake of simplicity. A more complex and general expression can be found in [2]. According to (3.6)–(3.7), the SAR focusing can be more efficiently performed in the Fourier domain, due to the availability of fast Fourier transform (FFT) algorithms:

$$\begin{aligned}\widehat{\Gamma}(\xi,\eta) &= H(\xi,\eta)\exp\left[j\frac{\eta^2}{4b}\right]\exp\left[-j\frac{\xi^2}{4a(1+\eta\lambda/(4\pi))}\right]\mathrm{rect}\left[\frac{\eta}{bc\tau}\right]\mathrm{rect}\left[\frac{\xi-\xi_{DC}}{2aX}\right]\\ &= \Gamma(\xi,\eta)w^2\left[\frac{\xi-\xi_{DC}}{2aX}\right]\mathrm{rect}\left[\frac{\eta}{bc\tau}\right]\mathrm{rect}\left[\frac{\xi-\xi_{DC}}{2aX}\right]\end{aligned} \tag{3.9}$$

The inverse FT of (3.9) allows to recover the focused SAR image:

$$\widehat{\gamma}(\widehat{x},\widehat{r}) \cong \iint \gamma(x,r)\ \mathrm{sinc}\left[\frac{\pi}{\Delta x}(x-\widehat{x})\right]\mathrm{sinc}\left[\frac{\pi}{\Delta r}(r-\widehat{r})\right]dxdr \tag{3.10}$$

where $\mathrm{sinc}(t) = \sin(t)/t$ and

$$\Delta x = \frac{2\pi}{2aX} = \frac{L}{2} \text{ and } \Delta r = \frac{2\pi}{bc\tau} = \frac{c}{2\Delta f} \tag{3.11}$$

are the azimuth and range spatial resolutions, respectively.

If range variation cannot be neglected, focusing can still be efficiently performed via FFT by using the processing defined by (3.9), followed by a proper interpolation in the Fourier domain [2]. The latter can be in turn efficiently implemented by using a *chirp-scaling* algorithm [2].

From the discussion above, it is clear that the main advantages of the stripmap mode are the simplicity of required hardware and software and the along-track coverage. In fact, the employed antenna can be very simple, since no beam steering is required, and processing can be efficiently performed by using standard FFT algorithms; in addition, the azimuth size of the imaged scene is in principle unlimited. However, the synthetic aperture length is limited to the azimuth footprint size X, so that the azimuth resolution is $L/2$, see (3.11). Unfortunately, the real antenna azimuth length L cannot be arbitrarily reduced, since, as already underlined, it limits the PRI (to avoid azimuth ambiguity) that in turn limits the range swath size (to avoid range ambiguity). In fact, we have

$$\frac{2S_r}{c} + \tau \leq PRI - \tau \leq \frac{L}{2v} - \tau \tag{3.12}$$

Equation (3.12) also accounts for the fact that the sensor cannot receive during the transmission of a pulse. For a spaceborne SAR system in the low Earth orbit (LEO), whose velocity v is about 7 km/s, an azimuth resolution of a few meters implies a swath width of no more than some tens of kilometers.

Accordingly, the main limitation of the stripmap mode is that, at least for spaceborne SAR systems (for airborne ones, velocity v is much smaller), azimuth resolution cannot be improved without reducing the range swath. Note that a smaller range swath implies a longer revisit time: a larger number of orbits are needed for a sun-synchronous orbit satellite to cover the entire globe. Typical spaceborne stripmap SAR resolutions span from 3 to about 20 meters. Here, and in the entire chapter, we always mean single-look resolution.

Accordingly, we can conclude that the stripmap mode is a general-purpose operational modality, suitable for all the applications that do not require a very high resolution and/or a very short revisit time. In particular, with regard to maritime applications, it can be used for monitoring the sea state, sea ice, and also for ship detection, unless it is required that very small boats are detected. An example of stripmap SAR image is shown in Figure 3.7.

3.3 Staring spotlight mode

To increase the length of the synthetic aperture, in the staring spotlight mode, the radar beam is steered in such a way to always illuminate the same spot on the ground, see Figure 3.2. Therefore, the center of rotation of the beam is the illuminated scene center. By considering the same transmitted pulse as in the stripmap case, that is, (3.1), the baseband SAR raw signal can be written as

$$h(x', r') = \text{rect}\left[\frac{x'}{X_1}\right]\tilde{h}(x', r') \tag{3.13}$$

where

$$\tilde{h}(x', r') = \iint \gamma(x, r) w^2 \left[\frac{x}{X}\right] g(x' - x, r' - r; r) dx dr \tag{3.14}$$

$$\begin{aligned} g(x' - x, r' - r; r) = \exp\left[-j\frac{4\pi}{\lambda}\Delta R\right] \exp\left[j\frac{4\pi}{\lambda}\frac{\Delta f/f}{c\tau}(r' - r - \Delta R)^2\right] \\ \times \text{rect}\left[\frac{(r' - r - \Delta R)}{c\tau/2}\right] \end{aligned} \tag{3.15}$$

X_1 is the length of the trajectory flight portion used to acquire the raw data (of course, $X_1 > X$), and all other symbols are defined in Section 3.2.

To recover the final image from the raw signal, a time-domain focusing can be performed via direct backprojection, that is, by implementing (3.5). However, note that, apart from the range variation that can be treated as in the stripmap case, the integral in (3.14) is still a convolution, and we can write the FT of $\tilde{h}(x', r')$ as

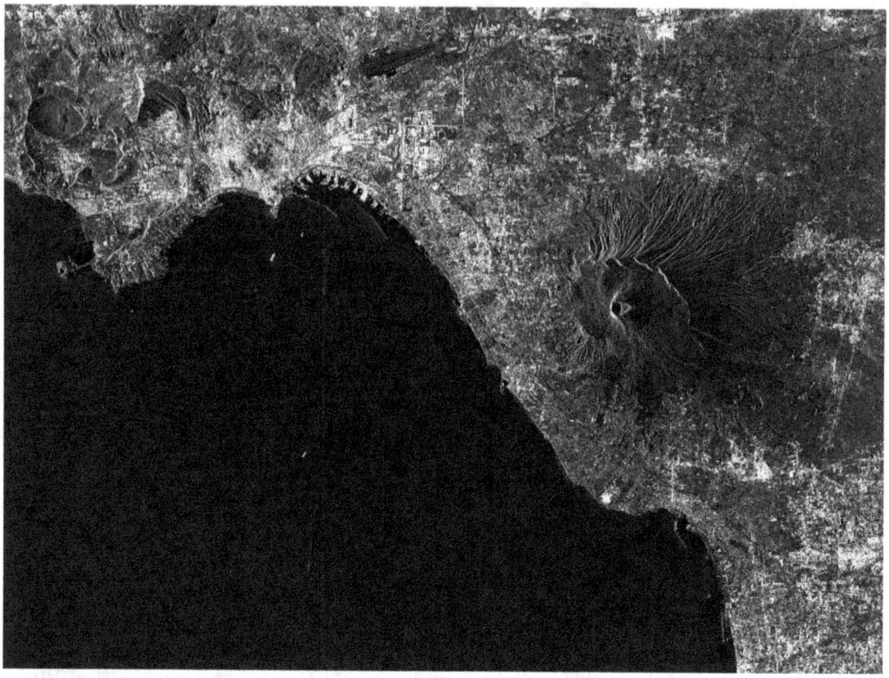

(a)

(b)

Figure 3.7 *(a) TerraSAR-X (DLR) stripmap mode SAR image of the Gulf of Naples (Italy). Imaged area is about 30 km × 30 km. (b) 1428 × 909 pixel full-resolution excerpt from the same image. Resolution is about 3 m × 3 m, pixel spacing is 1.9 m × 1.9 m*

$$\tilde{H}(\xi,\eta) = \tilde{\Gamma}(\xi,\eta)G_0(\xi,\eta) \tag{3.16}$$

where $\tilde{\Gamma}(\xi,\eta)$ is the FT of $\gamma(x,y)w^2(x/X)$ and $G_0(\xi,\eta)$ is the FT of $g(x',r';r=r_0)$. The latter can be analytically evaluated by using the stationary phase method as [13]

$$G_0(\xi,\eta) = \exp\left[-j\frac{\eta^2}{4b}\right]\exp\left[j\frac{\xi^2}{4a(1+\eta\lambda/(4\pi))}\right]\mathrm{rect}\left[\frac{\eta}{bc\tau}\right]\mathrm{rect}\left[\frac{\xi}{2a(X+X_1)}\right] \tag{3.17}$$

Equations (3.16)–(3.17) are in the same form of (3.6)–(3.7), so that, in principle, an efficient Fourier-domain processing, similar to that employed for the stripmap case, can be used. However, the last factor in (3.17) shows that the spotlight raw signal azimuth bandwidth is much larger than the stripmap one (which allows to obtain a higher resolution). If we want to maintain the same PRF as in the stripmap case, either we have to use time domain processing (at least for azimuth focusing), or stripmap processing must be preceded by application of a "spectral analysis" (SPECAN) algorithm [4,14], that is, multiplication by a chirp signal ("deramping") followed by an FFT. This first processing step performs a preliminary bulk azimuth processing and allows to recover a sufficient azimuth sampling rate to deal with the increased azimuth bandwidth [4,14]. The final focused image can still be expressed as (3.10), with Δr given by the second of (3.11), and

$$\Delta x = \frac{2\pi}{2aX_1} = \frac{L}{2}\frac{X}{X_1} \tag{3.18}$$

Accordingly, the azimuth resolution is improved by a factor X_1/X with respect to the stripmap one, in agreement with the fact that the synthetic antenna length is X_1, instead of X. This significant improvement is paid with a complication of hardware and software, and, above all, with a drastic reduction of the illuminated scene azimuth size. In fact, the antenna must be electronically steered during acquisition, so that a phased array must be used; and processing cannot be fully efficiently performed in the Fourier domain. Finally, the azimuth size of the illuminated scene is limited to the azimuth footprint size. Accordingly, a spaceborne SAR system in the spotlight acquisition mode can achieve even submetric resolution, but only for an azimuth scene size of a few kilometers. Limitations on the range swath size are the same as in the stripmap case, that is, (3.12).

We can conclude that the spotlight mode is suitable for applications in which a very high resolution is needed for a specific small area under survey. In particular, with regard to maritime applications, it can be used for ship detection, when it is required that very small boats are detected in a specific limited sea area.

3.4 Sliding spotlight mode

An intermediate solution between stripmap and staring spotlight modes is the hybrid, or sliding spotlight, acquisition mode. In this mode, the antenna beam is still steered during the acquisition flight, but the beam rotation center is farther from the antenna than the illuminated ground, so that the antenna beam footprint

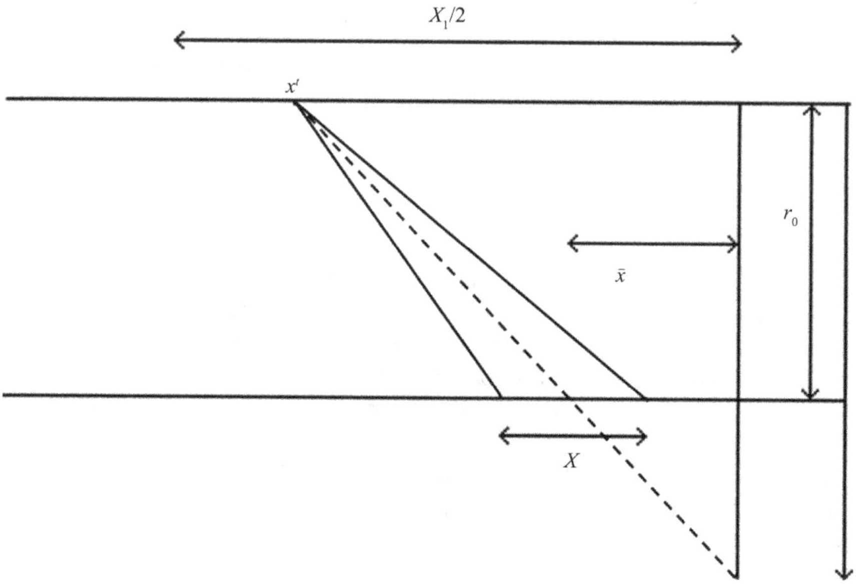

Figure 3.8 Sliding spotlight mode: azimuth footprint size X, azimuth footprint center $\bar{x}$, and synthetic antenna length X_1

on the ground moves more slowly than the sensor, see Figures 3.3 and 3.8. Accordingly, when the sensor is at position x', the center of the antenna azimuth footprint is at $\bar{x} = Ax'$, where

$$A = \frac{r_1 - r_0}{r_1} \tag{3.19}$$

and r_1 is the distance from the line of flight to the beam rotation center, see Figure 3.8. If a flight trajectory portion of length X_1 is used to acquire the raw data, then the azimuth size of the fully focused area (i.e., the area that is illuminated for the maximum possible flight time) is $X_F = |AX_1 - X|$, so that an increase of illuminated area with respect to the spotlight is obtained if $A > 2X/X_1$. The length of the synthetic antenna is then X/A, and the baseband raw signal can be expressed as

$$h(x', r') = \text{rect}\left[\frac{x'}{X_1}\right]\tilde{h}(x', r') \tag{3.20}$$

where

$$\tilde{h}(x', r') = \iint \gamma(x, r) g(x' - x, r' - r; x, r) dx dr \tag{3.21}$$

$$g(x' - x, r' - r; x, r) = \exp\left[-j\frac{4\pi}{\lambda}\Delta R\right]\exp\left[j\frac{4\pi}{\lambda}\frac{\Delta f/f}{c\tau}(r' - r - \Delta R)^2\right] \times \text{rect}\left[\frac{(r' - r - \Delta R)}{c\tau/2}\right] w^2\left(\frac{Ax' - x}{X}\right) \tag{3.22}$$

As usual, to recover the final image from the raw signal, a time-domain focusing can be performed via direct backprojection, implementing (3.5).

At variance with the previous cases, the integral in (3.21) cannot be treated as a convolution, and its FT is still in integral form:

$$\widetilde{H}(\xi,\eta) = \iint \gamma(x,r)G(\xi,\eta;x,r)\exp[-j\xi x]\exp[-j\eta r]dxdr \tag{3.23}$$

where

$$G(\xi,\eta;x,r) = \exp\left[-j\frac{\eta^2}{4b}\right]\exp\left[j\frac{\xi^2(r/r_0)}{4a(1+\eta\lambda/(4\pi))}\right]\mathrm{rect}\left[\frac{\eta}{bc\tau}\right] \times w^2\left[\frac{A\xi - 2a(A-1)x}{2aX}\right] \tag{3.24}$$

By treating range variation similarly to the stripmap case, in principle, an efficient Fourier-domain processing, similar to that employed for the stripmap case, can be used. However, the last factor in (3.24) shows that, as in the staring spotlight case, the sliding spotlight raw signal azimuth bandwidth is much larger than the stripmap one (it is $1/A$ times larger). Again, if we want to maintain the same PRF as in the stripmap case, either we have to use time domain processing (at least for azimuth focusing), or stripmap processing must be preceded by application of a SPECAN algorithm, as shown in [4]. Several specific algorithms, with different levels of efficiency and/or accuracy, have been proposed to focus sliding spotlight raw data (see, e.g., [3–5]). In all cases, the final focused image can still be approximated by (3.10), with Δr given by the second of (3.11), and

$$\Delta x = \frac{2\pi A}{2aX} = \frac{L}{2}A \tag{3.25}$$

Accordingly, the azimuth resolution is improved by a factor $1/A$ with respect to the stripmap one, in agreement with the fact that the synthetic antenna length is X/A. In addition, the azimuth size of the imaged scene is increased by a factor $AX_1/X - 1$ with respect to the staring spotlight case. Hardware and software complexities are substantially the same as the ones of the staring spotlight. A spaceborne sliding-spotlight SAR system can achieve a metric resolution for a scene size of the order of 10 to 20 km. Limitations on the range swath size are the same as in the stripmap case, that is, (3.12).

The applications of the sliding spotlight mode are substantially the same as the staring spotlight ones, in cases when the resolution requirement can be slightly relaxed in favor of a larger imaged scene. An example of sliding spotlight SAR image is shown in Figure 3.9.

3.5 ScanSAR mode

The ScanSAR mode aims to achieve a ground coverage in range much larger than in STRIPMAP mode, trading it by a coarser azimuth resolution. The principle can

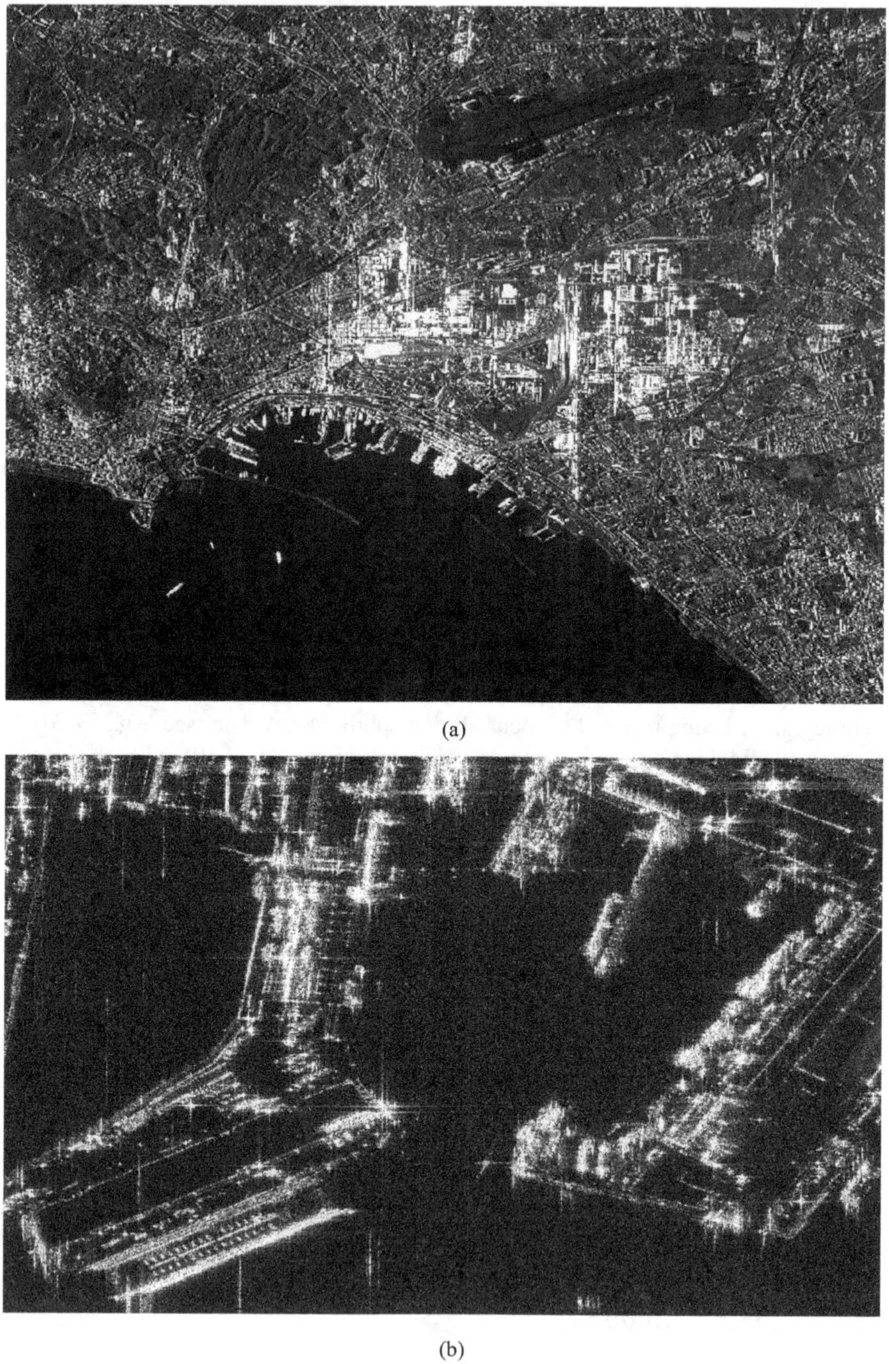

Figure 3.9 *(a) Cosmo/SkyMed (ASI) sliding-spotlight mode SAR image of the Gulf of Naples (Italy). Imaged area is about 10 km × 10 km. (b) 1428 × 909 pixel full-resolution excerpt from the same image. Resolution is about 1 m × 1 m, pixel spacing is 0.5 m × 0.7 m*

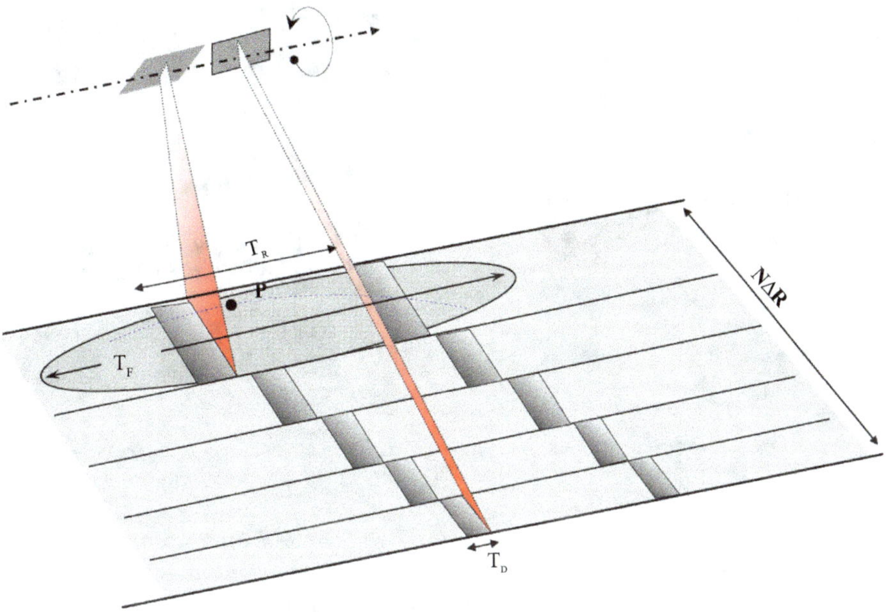

Figure 3.10 Geometry of a five subswath, two looks ScanSAR

be guessed from Figure 3.10, that represents an $N = 5$ subswaths ScanSAR like ESA EnviSAR wide swath mode. The antenna is electronically steered to cyclically scan the five subswaths, with a cycle time T_R (also named "return time"). T_R should be lower than the footprint time, T_F (the time a target is observed with the azimuth antenna footprint, that is, $T_F = X/v$) to achieve the full coverage of each subswath without gaps. This in turn implies that for each subswath, the number or pulses used for imaging is reduced to a "burst" lasting less than T_F/N, that is, N times less than the synthetic aperture time in a STRIPMAP mode. The shortening of the observation time results in a coarsening of the azimuth resolution that scales inversely to the number of subswaths, hence to the total swath depth.

ScanSAR time-division principle can be exploited to switch over different polarizations instead of range swath, as for Envisat Alternating Polarization mode or Cosmo-Skymed "Ping Pong" mode. In both cases, a multipolarimetric acquisition is achieved without need to halve the PRF, though the two sets of images, say HH/HV and VV/VH, cannot be coherently combined.

The properties of ScanSAR data can be inferred in the time-frequency-domain (TFD) from Figure 3.11. On the left, the antenna illumination pattern is represented for two consecutive bursts in the same swaths. The Doppler histories of several targets are superposed in the TFD diagram, appearing as a slanted line is the TFD domain, whose slope is the Doppler rate, k_D. The antenna main lobe is filled in red, while green and blue represent sidelobes. The target whose zero-Doppler position is in P is the one that should be imaged both in burst n and in

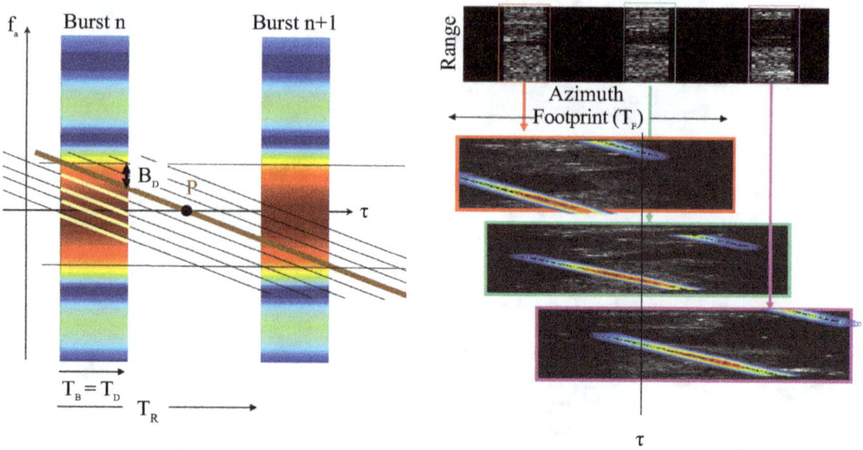

Figure 3.11 Time-frequency diagrams. (a) Two ScanSAR bursts: color bands represent the azimuth antenna pattern and slant lines the target Doppler histories. (b) Three focused subswaths: colors represent the spectrograms

burst $n+1$ to achieve continuous coverage along-track. This bounds the maximum return time to

$$T_R \leq T_F - T_B T_F \simeq \frac{B_F}{|k_D|} \tag{3.26}$$

T_B is the target dwell time or burst time and B_F is the Doppler bandwidth imaged for each target, that is, reciprocal to the azimuth resolution. The same figure, on the right, shows a set of three bursts before (above) and after (below) focusing.

The focusing of each of the bursts results in an imagette that extends for the whole return time so that the mosaic of the focused bursts create a continuous strip. In the imagette, each target contributes to a TF spectral support that depends upon the azimuth of the target itself. The target in P, for example, contributes nearly to PRF/2 in burst n and -PRF/2 in burst $n+1$. Consequently, different Doppler contributions are weighted by different portions of the antenna pattern (shades of red).

This target-dependent azimuth antenna pattern weighting can cause—if not properly compensated—a radiometric artifact that appears as cyclical modulation of the amplitude, known as "scalloping." This is quite evident in Figure 3.11, where many Doppler histories are shown as straight lines in the time-frequency plot, each shifted by a different central frequency and weighted by a different portion of the antenna pattern. This is made clear in Figure 3.12 by representing the real part of

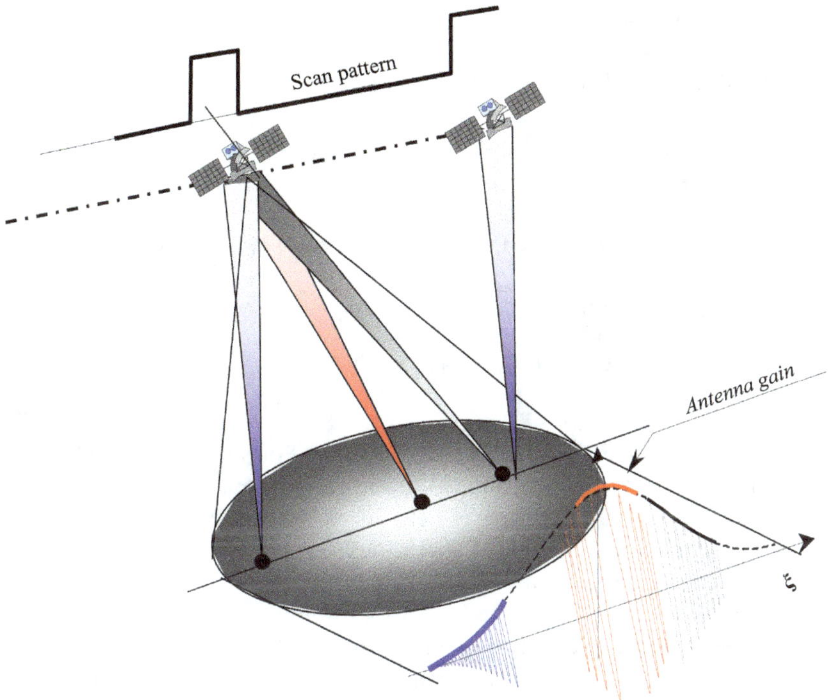

Figure 3.12 Scalloping effects of three different targets

impulse responses of three targets located at different azimuth with respect to the burst center.

The ScanSAR impulse response there shown, after range compression, can be derived just by windowing the STRIPMAP one in (3.2)–(3.3) and assuming impulsive behavior in range. We assume Zero-Doppler imaging, that is, squint $\psi = 0$:

$$h(x_n;x) = \exp\left(j\frac{4\pi}{\lambda}\sqrt{r^2 + (x_n - x)^2}\right) \times w^2(x_n - x) \times rect\left(\frac{x_n - x_c}{v \cdot T_B}\right) \tag{3.27}$$

where v is the sensor velocity and x_c is the position of the burst center. Due to the "rect" burst windowing, the last term in (3.27), the impulse response is nonstationary, since it is no longer dependent on $(x_n - x)$. Such nonstationarity is cause of the many drawbacks of the ScanSAR mode.

We can approximate azimuth focusing by (3.1) superposing a window to the burst (to mitigate sidelobes), $w_b(x)$ and (3.2) correlating with a matched reference that is as long as the full synthetic aperture (related to the footprint): $\Delta = v \cdot T_F$. The resulting impulse response can be computed by approximating the first phase term in (3.27) to be quadratic, with phase center on the actual target location, x and assuming x_n to be continuous:

$$\begin{aligned}
\widehat{\gamma}(x) &= w_b(x_n) \times h(x_n; x) * \exp\left(-j\pi k(x_n - x_c)^2\right) \\
&= \int_{-\frac{\Delta}{2}}^{\frac{\Delta}{2}} w_b(\xi - x_c) \times w(\xi - x) \times \exp\left(j\pi k(\xi - x)^2\right) \\
&\quad \cdot \exp\left(-j\pi k(\xi - x_c)^2\right) d\xi = \int_{-\frac{\Delta}{2}-x_c}^{\frac{\Delta}{2}+x_c} w_b(\rho) \times w(\rho + x_c - x) \\
&\quad \times \exp\left(j\pi k(\rho + x_c - x)^2\right) \exp\left(-j\pi k_D \rho^2\right) d\rho \simeq exp\left(j\pi k(x_c - x)^2\right) \\
&\quad \times w(x_c - x) \times W_b(k \times x)
\end{aligned} \tag{3.28}$$

where the asterisk stands for convolution, W_b is the Fourier transform of the burst window and k is the sweep rate, related to the Doppler rate k_D by $k = k_D/v$. The focused impulse response function (FIRF) in (3.28) has been approximated by assuming that the azimuth antenna pattern, $w(x_n\text{-}x)$, is almost constant within the FIRF. If a rectangular burst window is adopted, the term W_b in (3.28) becomes a sinc-like pattern, whose width is the ScanSAR resolution, centered on the azimuth position of the target. The first and second terms in (3.28) account for a quadratic phase, that depends upon the displacement of the target with respect to the burst center (the Zero Doppler position) and an amplitude factor proportional to the azimuth antenna pattern, also shifted by the target position. The first term introduces an approximately linear modulation of the local spectrum, which is shown in the TFD in Figure 3.11(b), whereas the azimuth varying amplitude, shown in Figure 3.12, is responsible for scalloping.

Scalloping is basically a radiometric artifact but it can be removed by scaling the focused image times by the reciprocal of the antenna pattern: $1/w(x_c\text{-}x)$ in (3.28). This is quite effective in most cases, though it introduces a modulation of the possible noise term, which behaves like an artifact. Furthermore, scalloping is not perfectly compensated when the Zero-Doppler position is estimated with error, usually in the presence of sea currents and moving targets in general. Another artifact that results in ScanSAR focusing is a nonstationary amplitude of ambiguities that also can be deduced from Figure 3.11(a).

The applications of the ScanSAR mode are substantially the same as the stripmap ones, in cases when the resolution requirement can be relaxed in favor of a larger swath.

3.6 TOPSAR mode

TOPSAR mode has been introduced to get the same goal of ScanSAR, that is, an extended range coverage achieved by coarsening azimuth resolution. However, in doing that, it avoids the nonstationary spectral weighing of targets and the resulting artifacts just mentioned.

TOPSAR is a burst mode acquisition, like ScanSAR, and the difference is that, during the acquisition, the antenna pattern is electronically swept from aft to fore, in a way opposite to spotlight, as shown in Figure 3.13(a). The effect is that each target is illuminated by the whole antenna beam pattern, as shown in Figure 3.13 (b), though the beam center of each target, hence the Doppler Centroid, sweeps from negative to positive Doppler. This fact is represented in Figure 3.14, that is, the correspondent of Figure 3.11 for the ScanSAR case. The antenna pattern shifts in the time-frequency domain in direction opposite to the Doppler rate so that the

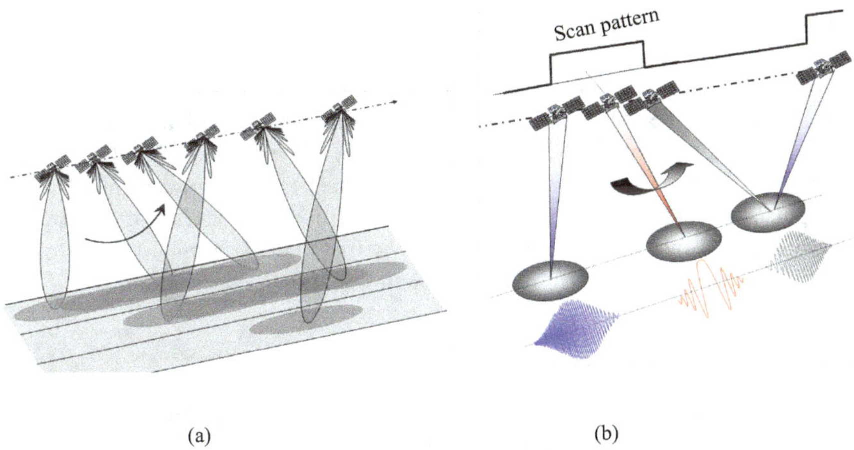

Figure 3.13 (a) TOPSAR geometry and scanning and (b) real part of raw data contributions of three targets, to be compared with Figure 3.13

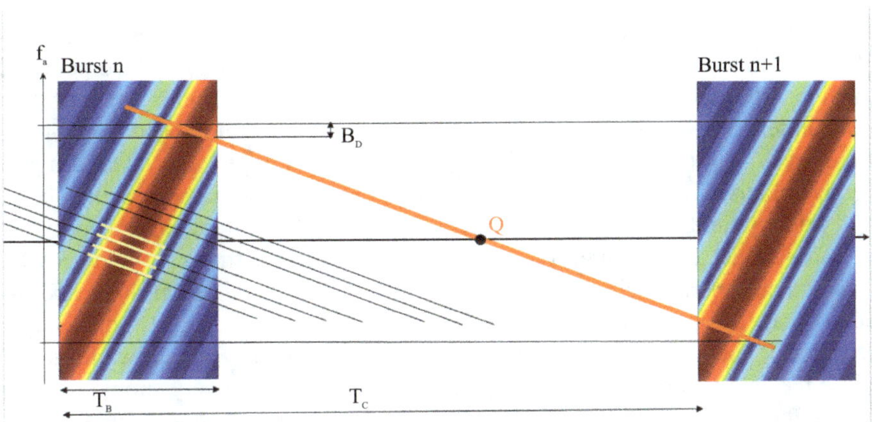

Figure 3.14 Time-frequency diagrams for TOPSAR, to be compares with Figure 3.11(a)

Doppler histories of the targets in the same subswath are weighted by the full antenna pattern (white lines in the figure).

In order to achieve a continuous coverage, the target observed in the foremost antenna pattern, named "Q" in the sketch of Figure 3.14, should be observed again after the cycle time, T_c, and the corresponding of ScanSAR repeat time, T_R. This condition is achieved by imposing that the Doppler shift due to target Doppler rate in the return time is the same as the shift during the burst:

$$T_C \times |k_D| \leq (T_B + T_D) \times |k_a| \rightarrow T_c \leq (T_B + T_D)\frac{|k_a|}{|k_d|} \tag{3.29}$$

where k_a is the antenna frequency sweep rate.

Equation (3.29) compares with (3.26) related to ScanSAR, but one observes that a fast antenna sweep rate makes TOPSAR bursts significantly longer than ScanSAR, while leaving a degree of freedom in k_a to optimize the burst length in each subswath, and that should be tuned to the local (range-variant) Doppler rate. The longer burst duration results also in a better efficiency since, at the beginning of each bursts, some "rank" pulses are lost before an echo is effectively received from the ground.

One consequence of the longer bursts is that the variation of the Doppler Centroid with the target position is larger than the antenna bandwidth:

$$(T_B + T_D) \times k_a \gg \frac{2 \times v}{L} \tag{3.30}$$

resulting in a wider Doppler shift. This shift is sometimes exploited in repeat-pass interferometric acquisitions to sense along-track displacements.

An example of TOPSAR mode SAR image is shown in Figure 3.15.

3.7 Wave mode

Monitoring sea state and wind velocity field requires a rather high resolution to be able to accurately estimate the sea surface directional spectrum. On the other hand, global monitoring of the oceans requires a wide coverage. The combination of these two requirements would lead to a huge data volume to be acquired and stored if one of the previously described acquisition modes were employed. However, sea state does not spatially change very rapidly, so that a rather low "sampling rate" of the wind velocity field or sea state is sufficient to accurately monitor them. Accordingly, the wave mode can be employed: a standard spaceborne stripmap-mode SAR is turned on only periodically for a rather short time, so to obtain small images of the ocean (imagettes), each with a linear size of a few kilometers, placed at distances of the order of the hundreds of kilometers one from another. This allows to greatly reduce the data volume, and, as a side advantage, allows for a very simple focusing (in fact, due to the small size of the imagettes, range variation is negligible).

For instance, Sentinel-1 wave mode operates with an elevation beam alternating between 23° and 36.5° incidence angles and a fixed on/off duty cycle which leads to generate 20 km × 20 km imagettes with a 5 m × 5 m resolution in regular

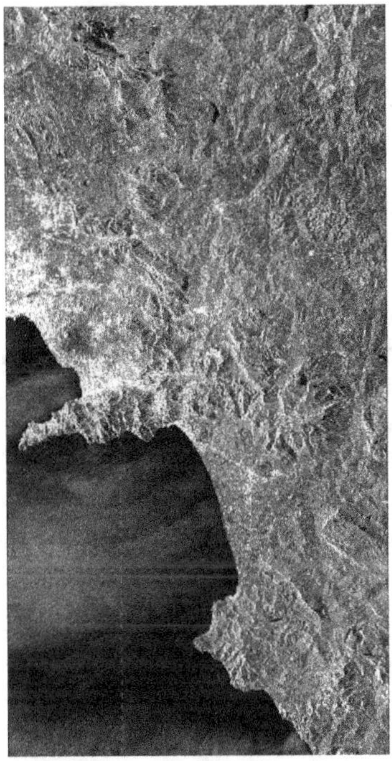

(a)

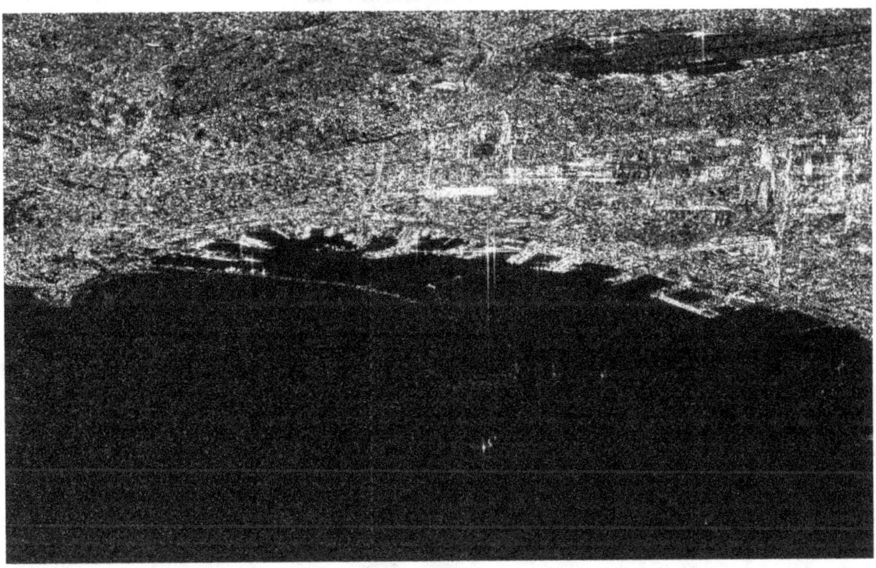

(b)

Figure 3.15 *(a) Sentinel-1 (ESA) TOPSAR mode SAR image of the Gulf of Naples (Italy). Imaged area is about 80 km in range (one of the three subswaths) × 160 km in azimuth and (b) 1428 × 909 pixel full-resolution excerpt from the same image. Resolution is about 5 m × 20 m, pixel spacing is 2.3 m × 13.9 m* intervals of 100 km [15]. The sea state can be then evaluated in correspondence of each imagette. In particular, significant wave height, dominant wavelength, and dominant direction of propagation can be retrieved with uncertainties of 0.3 m, 40 m, and 12°, respectively [15].

3.8 Experimental modes

All the acquisition modes described in Sections 3.2 to 3.7 have been and are already operatively employed in past and current spaceborne SAR missions. However, recently, a number of new acquisition modes have been proposed, mainly aimed at reducing the necessary data rate and/or increasing coverage while preserving resolution. They are at different stages of experimentation and all seem promising, with their own advantages and drawbacks. A nonexhaustive list includes Low-PRF mode [9,16], CopSAR [12,17], CS-SAR [10,18,19], Staggered SAR [11,20], Digital Beamforming (DB) [21,22], multiple-input-multiple-output SAR (MIMO-SAR) [23,24], and Forward-Looking SAR (FLoSAR) [25]. In the following, we describe some of these experimental modes; in particular, we choose the modes whose performance can be predicted by using available SAR data acquired with traditional modes. For the others, the reader is referred to available literature [21–25].

3.8.1 Low-PRF mode

As it will be better shown in Chapter 5, due to the finite value of the PRF, replicas of the scene reflectivity pattern (*ghosts*) appear on SAR images (azimuth ambiguity effect). These replicas are shifted both in azimuth and, although in a smaller extent, in range:

$$\Delta x_i = i\frac{\text{PRF}\lambda r_0}{2v} \text{ and } \Delta r_i = \frac{\Delta x_i^2}{2r_0} \tag{3.31}$$

where Δx_i and Δr_i are the azimuth and range shifts of the *i*th replica. However, the intensities of such replicas are weighted by the antenna azimuth pattern, so that replicas shifted in azimuth by more than the azimuth footprint (i.e., such that $|\Delta x_i| > X$) are strongly attenuated and are usually negligible. Therefore, if PRF $>$ $2v/L$, all replicas are suppressed by the antenna azimuth pattern and the effect of the finite PRF can be ignored, as we did in Sections 3.2.to 3.7. Otherwise, replicas and "true" reflectivity are mixed up and this prevents the correct image interpretation if an extended scene is imaged. However, if the scene is made of a few isolated bright targets over a dark background, based on relations (3.31), the replicas of "true" targets can be identified and discarded. This is the case of ship detection in ocean scenes: on SAR images, ships appear as bright target, whereas the sea appears as a dark background, whose brightness depends on the sea state: calm sea appears as very dark in SAR images, whereas very rough sea may appear rather bright. Therefore, in order to perform ship monitoring at global scale, the low-PRF mode has been proposed as a way to reduce the necessary amount of data and to enlarge the range swath without losing spatial resolution. In fact, in this mode, pulses are transmitted at PRF = PRF_0/N, where PRF_0 satisfies the Nyquist condition, that is, $\text{PRF}_0 > 2v/L$, and N (the *undersampling ratio*) is an integer. Accordingly, the necessary data rate and amount of data are reduced by N, and the range swath can be widened by a factor equal to about N.

A low-PRF mode is active on the recently launched (September 2018) NovaSAR-1 sensor: by combining scanSAR operation and low PRF, a 400 to 750 km wide range swath can be achieved by this sensor, with a 30 m spatial resolution [9]. A drawback of the low-PRF mode is that proper algorithms to select the "true target" (the ship) among its "ghosts" must be employed [16]. In addition, the reduced number of transmitted pulses causes a reduction of the focusing gain, so that a reduction of signal-to-background ratio (SBR) and of signal-to-noise ratio (SNR) with respect to full-PRF modes has to be accepted.

3.8.2 Coprime SAR

The CopSAR mode can be employed in the same kind of scenario of the low-PRF mode, that is, a few isolated bright targets over a dark background, and in this case it allows to automatically suppress the ghosts, at variance with the low-PRF mode.

In the CopSAR mode, the system transmits two interlaced sequences of pulses: one at $\text{PRF}_1 = \text{PRF}_0/N_1$, and the other at $\text{PRF}_2 = \text{PRF}_0/N_2$, where PRF_0 satisfies the Nyquist condition, that is, $\text{PRF}_0 > 2v/L$, and N_1 and N_2 are two coprime integers (i.e., their greatest common divisor is 1). The two sequences can be separately processed to obtain two SAR images. Geometric resolution of these images is not changed with respect to the standard SAR case, because the synthetic array length is unchanged, but these images are affected by the presence of "ghosts" (they can be considered as two low-PRF mode images with different undersampling ratios). However, if the scene consists of bright targets on a dark background (as in the case of boats or ships on the ocean), then only the true targets will be present on both images at the same location, whereas replicas will be at different locations on the two images. In fact, on the first image, each target gives rise to replicas displaced as

$$\Delta x_{i_1} = i_1 \frac{\text{PRF}_0 \lambda r_0}{N_1 2v}, \Delta r_{i_1} = \frac{\Delta x_{i_1}^2}{2r_0} \tag{3.32}$$

whereas on the second image, each target gives rise to replicas displaced as

$$\Delta x_{i_2} = i_2 \frac{\text{PRF}_0 \lambda r_0}{N_2 2v}, \Delta r_{i_2} = \frac{\Delta x_{i_2}^2}{2r_0} \tag{3.33}$$

with i_1 and i_2 integers, so that replicas on the two images do not appear at the same location unless $i_1/i_2 = N_1/N_2$. But, since N_1 and N_2 are coprime, this only happens if $i_1 = i\,N_1$ and $i_2 = i\,N_2$, so that

$$\Delta x_{i1} = \Delta x_{i2} = i \frac{\text{PRF}_0 \lambda r_0}{2v} \tag{3.34}$$

that is, only at the positions of the replicas in the image that would be obtained in the standard acquisition mode. And, since $\mathrm{PRF}_0 > 2v/L$, these replicas are suppressed by the antenna azimuth pattern. Therefore, if the two images are combined by choosing, at each pixel, the image value from the image in which its modulus is smaller, then a final image is obtained in which no replicas are present [12].

In this mode, the necessary data rate and amount of data are reduced by about $\frac{N_1+N_2}{N_1 N_2}$, and the range swath can be widened by a factor equal to about the smaller between N_1 and N_2 if up-chirps are transmitted in the first sequence and down-chirps are transmitted in the second sequence (*Orthogonal CopSAR*, [17]).

As in the low-PRF mode, also in CopSAR, the reduced number of transmitted pulses causes a reduction of the focusing gain so that a reduction of signal-to-background ratio (SBR) and of signal-to-noise ratio (SNR) with respect to full-PRF modes has to be accepted.

Promising results have been obtained both on simulated data and on equivalent CopSAR data generated by undersampling ERS-2 SAR real raw signals [12]. However, at the time of writing this book, no mission is implementing this approach.

3.8.3 Compressive sensing SAR

Low-PRF mode and CopSAR exploit the fact that the scene is known to be composed of a few *sparse* targets to recover a full-resolution image from a limited amount of observed data (the low-PRF raw signal). However, standard SAR focusing techniques are used in these modes. Recently, *Compressive Sensing* (CS), or *Compressed Sensing*, has emerged as a new paradigm to retrieve sparse signals from a limited number of observations. Therefore, application of this approach to SAR systems has been proposed (see, e.g., [10,18,19]). To illustrate application of CS to SAR, let us rewrite (3.2) in vector form as

$$h_k = \sum_{i=1}^{M} g_{ik}\gamma_i \qquad \text{with} \qquad k = 1 \ldots K \tag{3.35}$$

where γ_i are the elements of the vector $\underline{\gamma}$ obtained by sampling the (unknown) reflectivity at full resolution, h_k are the elements of the vector $\underline{h}$ obtained by sampling the observed raw signal at PRF and (possibly) f_s lower than the Nyquist limit, and g_{ik} are the elements of the matrix $\underline{\underline{g}}$ obtained by sampling the SAR impulse response according to the sampling rates of reflectivity and raw signal. Due to undersampling of the raw signal, $K < M$ and the system of (3.35) are underdetermined. Now, if the reflectivity vector $\underline{\gamma}$ is sparse or compressible, that is, a basis exists in which $\underline{\gamma}$ can be represented by a limited number of non-negligible coefficients, and if the matrix $\underline{\underline{g}}$ obeys a restricted isometry condition [10], then the estimate $\widehat{\underline{\gamma}}$ of the full-resolution reflectivity (i.e., the focused SAR image) can be retrieved as

$$\widehat{\underline{\gamma}} = \underset{\underline{\gamma}}{\operatorname{argmin}} \|\underline{\gamma}\|_{l_1} \text{ subject to } \underline{h} = \underline{\underline{g}}\,\underline{\gamma} \tag{3.36}$$

where the l_1 norm of $\underline{\gamma}$ is

$$\|\underline{\gamma}\|_{l_1} = \sum_{i=1}^{M} |\gamma_k| \tag{3.37}$$

This means that among all the reflectivities that produce the observed raw signal, the sparsest one is selected. With regard to conditions required by CS SAR, compressibility of the reflectivity depends on the scene. Examples of compressible reflectivity maps are those corresponding to a few ships in an ocean scene, but also those corresponding to some urban area scenes seem to satisfy this condition. Regarding the restricted isometry condition on the matrix $\underline{\underline{g}}$, it depends on the SAR system properties and on the employed undersampling strategy (i.e., the timing of the pulse transmission sequence). The advantage of CS SAR with respect to low-PRF mode and CopSAR is its applicability to a wider set of scenarios. However, a main drawback is that solution of the optimization problem defined by (3.36) requires time-consuming algorithms, and this at present seems to prevent the operational use of this approach in SAR missions. In addition, the data reduction rate is not easily predictable, and it is usually smaller than the one of previous methods.

3.8.4 Staggered SAR

While the experimental modes considered in subsections 3.8.1 to 3.8.3 were mainly aimed at reducing data volume, the Staggered SAR is aimed at significantly widening the range swath. We recall that swath width S_r is limited by (3.12), that can be rewritten as

$$S_r \leq \frac{c}{2}\mathrm{PRI} - c\tau \tag{3.38}$$

A possibility to obtain a wide swath is to use multiple antenna beams (each with a separate receiver), with different elevations, so that different parallel swaths are illuminated, each satisfying (3.38). However, since the sensor cannot receive during pulse transmission, such swaths must be separated by gaps of width $c\tau$. If the PRI is constant, such gaps form blind strips parallel to the line of flight, from which no echo pulse is available. However, if the PRI is periodically linearly varied from a maximum to a minimum, then the blind strips are obliquus, and on each given equi-range line only a small fraction of the echo pulses forming the synthetic antenna are missing. Accordingly, azimuth focusing is still possible also in correspondence of the blind strips, and no gap is present on the final wide-swath image. This is the main concept of Staggered SAR [11]. However, due to the regular pattern of the missing pulses, grating lobes will appear in the synthetic antenna pattern, which can produce ghosts on the final image similar to those produced by azimuth ambiguity. To avoid this, either the maximum PRI should be small enough to allow for recovering missing pulses by interpolation, or some irregularity must be introduced in the pulse transmission timing. In the former case, complexity is added to the processing algorithms, and a higher data rate is needed; in the latter, hardware complexity is increased.

Table 3.1 Summary of the main SAR acquisition modes

Acquisition mode	**Resolution**	**Coverage**	**Hardware complexity**	**Sensors**
Stripmap	Intermediate (~ 5–10 m)	Intermediate (~ 40 km)	Low	Cosmo/SkyMed (ASI), TerraSAR-X (DLR), Sentinel-1 (ESA), Radarsat-2 (CSA), ALOS-PALSAR 2 (JAXA)
Staring spotlight	Very high (~ 0.5 m)	Very low (~ 5 km)	High	TerraSAR-X (DLR), Radarsat-2 (CSA), ALOS-PALSAR 2 (JAXA)
Sliding spotlight	High (~ 1 m)	Low (~ 10 km)	High	Cosmo/SkyMed (ASI), TerraSAR-X (DLR)
ScanSAR	Low (~ 50 m)	High (~ 300 km)	Intermediate	Cosmo/SkyMed (ASI), TerraSAR-X (DLR), Radarsat-2 (CSA), ALOS-PALSAR 2 (JAXA)
TOPSAR	Low (~ 15–30 m)	High (~ 250–400 km)	High	Sentinel-1 (ESA)
Wave	Intermediate (~ 5–10 m)	Low (~ 10–20 km)	Low	Sentinel-1 (ESA)

Promising results have been obtained on simulated data and on equivalent staggered SAR data generated by properly undersampling very high PRF airborne SAR real raw signals [20]. However, at the time of writing this book, no mission is implementing this approach.

3.9 Summary

In this chapter, different SAR acquisition modes have been described, their advantages and drawbacks have been highlighted, and their applicative domains have been illustrated.

The information provided in this chapter is summarized in Table 3.1, where the currently (2019) operative satellite SAR sensors implementing different modes are also reported.

References

[1] Curlander J. C. and McDonough R. N. *Synthetic aperture radar: Systems and signal processing.* New York: Wiley; 1991.

[2] Franceschetti G. and Lanari R. *Synthetic aperture radar processing.* New York: CRC Press; 1999.

[3] Belcher D. P., and Baker C. J. "High resolution processing of hybrid strip-map/spotlight mode SAR." *IEE Proc., Radar Sonar Navig.* 1996, vol.143, pp. 366–374.

[4] Lanari R., Zoffoli S., Sansosti E., Fornaro G., and Serafino. "New approach for hybrid strip-map/spotlight SAR data focusing." *IEE Proc. Radar Sonar Navig.* 2001, vol.148, pp. 363–372.

[5] Mittermayer J., Lord R. T., and Boerner E. "Sliding Spotlight SAR Processing for TerraSAR-X using a new formulation of the extended chirp scaling algorithm." *Proceedings of IGARSS* 2003, Toulose, France, 2003, pp. 1462–1464.

[6] De Zan F. and Monti Guarnieri A. "TOPSAR: Terrain Observation by Progressive Scans." *IEEE Trans. Geosci. Remote Sens.* 2006, vol. 44, no. 9, pp. 2352–2360.

[7] *Envisat ASAR Product Handbook*, European Space Agency, Noordwijk, The Netherlands, Issue 2.2, 2007.

[8] Lehner S., Schulz-Stellenfleth J., Schättler J. B., Breit H., and Horstmann J. "Wind and wave measurements using complex ERS-2 wave mode data." *IEEE Trans. Geosci. Remote Sens.*, vol. 38, no. 5, pp. 2246– 2257, 2000.

[9] Iervolino P., Guida R., and Whittaker P. "NovaSAR-S and maritime surveillance." *Proceedings of IGARSS* 2013, Melbourne, Australia, 2013, pp. 1282–1285.

[10] Patel V. M., Easley G. R., Healy D. M., and Chellappa R. "Compressed synthetic aperture radar." *IEEE J. Sel. Topics Signal Process.* 2010, vol. 4, no. 2, pp. 244–254.

[11] Villano M., Krieger G., and Moreira A. "Staggered SAR: High-resolution wide-swath imaging by continuous PRI variation." *IEEE Trans. Geosci. Remote Sens.* 2014, vol. 52, no. 7, pp. 4462–4479.

[12] Di Martino G. and Iodice A. "Coprime synthetic aperture radar (CopSAR): A new acquisition mode for maritime surveillance." *IEEE Trans. Geosci. Remote Sens.* 2015, vol. 53, no. 6, pp. 3110–3123.

[13] Cimmino S., Franceschetti G., Iodice A., Riccio D., and Ruello G. "Efficient spotlight SAR raw signal simulation of extended scenes." *IEEE Trans. Geosci. Remote Sens.* 2003, vol. 41, no. 10, pp. 2329–2337.

[14] Sack M., Ito M. R., and Cumming I. G. "Application of efficient linear FM matched filtering algorithms to synthetic aperture radar processing." *Proceedings of the Institution of Electrical Engineers*, Part F, vol. 132, 1985, pp. 45–57.

[15] Torres R., Snoeij P., Geudtner D., *et al.* "GMES Sentinel-1 mission." *Remote Sens. Environ.* 2012, vol. 120, pp. 9–24.

[16] Iervolino P., Cohen M., Guida R., and Whittaker P. "Ship-detection in SAR imagery using low pulse repetition frequency radar." In *Proceedings of EuSAR*, Berlin, Germany, 2014, pp. 1–4.

[17] Di Martino G. and Iodice A. "Orthogonal coprime synthetic aperture radar." *IEEE Trans. Geosci. Remote Sens.* 2017, vol. 55, no. 1, pp. 432–440.

[18] Tello Alonso M., López-Dekker P., and Mallorquí J. J. "A novel strategy for radar imaging based on compressive sensing." *IEEE Trans. Geosci. Remote Sens.* 2010, vol. 48, no. 12, pp. 4285–4295.

[19] Fang J., Xu Z., Zhang B., Hong W., and Wu Y. "Fast compressed sensing SAR imaging based on approximated observation." *IEEE J. Sel. Topics Appl. Earth Observ. Remote Sens.* 2014, vol. 7, no. 1, pp. 352–363.

[20] Villano M., Krieger G., Jäger M., and Moreira A. "Staggered SAR: Performance analysis and experiments with real data." *IEEE Trans. Geosci. Remote Sens.* 2017, vol. 55, no. 11, pp. 6617–6638.

[21] Younis M., Fischer C., and Wiesbeck W. "Digital beamforming in SAR systems." *IEEE Trans. Geosci. Remote Sens.* 2003, vol. 41, no. 7, pp. 1735–1739.

[22] Gebert N., Krieger G., and Moreira A. "Digital beamforming on receive: Techniques and optimization strategies for high-resolution wideswath SAR imaging." *IEEE Trans. Aerosp. Electron. Syst.* 2009, vol. 45, no. 2, pp. 564–592.

[23] Wang W.-Q. "Space-time coding MIMO-OFDM SAR for high-resolution imaging." *IEEE Trans. Geosci. Remote Sens.* 2011, vol. 49, no. 8, pp. 3094–3104.

[24] Krieger G. "MIMO-SAR: Opportunities and pitfalls." *IEEE Trans. Geosci. Remote Sens.* 2014, vol. 52, no. 5, pp. 2628–2645.

[25] Franceschetti G., Iodice A., and Riccio D. "Forward-looking synthetic aperture radar (FLoSAR): The array approach." *IEEE Geosci. Remote Sens. Lett.* 2014, vol.11, no.1, pp. 303–307.

Chapter 4
SAR polarimetry

Maurizio Migliaccio[1,2], *Ferdinando Nunziata*[1] *and Andrea Buono*[1]

4.1 Introduction

This chapter has a twofold objective: on one side, it provides the basics of synthetic aperture radar (SAR) polarimetry; on the other side, SAR polarimetry is specialized with respect to the sea surface observation for marine and maritime applications.

Polarimetric SAR represents an advanced imaging radar system that plays a key role in radar remote sensing. In fact, by means of polarimetric SARs, an unprecedented amount of information on the scattering mechanisms that characterize the observed scene can be obtained if compared to conventional single-polarization SAR architectures. Nowadays, there is a wide and general consensus that SAR polarimetry offers extra information that can be exploited for a large set of environmental applications [1], including ocean ones. The most up-to-date state of the art related to earth observation applications and added-value products developed starting from polarimetric SAR measurements, alone or in combination with complementary techniques as across-track/differential SAR interferometry, includes but is not limited to the following domains/products:

- Forest, that includes land use/cover classification, detection of vertical structures, estimation of stand height and above-ground biomass, and fire scan by means of change detection [2,3];
- Agriculture, including phenology, crop type mapping, estimation of soil moisture, and flood monitoring by means of change detection [4,5];
- Urban, that includes 3D rendering, land use/cover classification, landslides and subsidence estimation, and change [6,7];
- Cryosphere, including estimation of snow water equivalent, estimation of sea and land ice extent, permafrost, and iceberg detection [8,9];
- Ocean, that includes ship and critical infrastructure detection, coastline extraction, oil pollution, estimation of wind speed, and ocean waves [10,11].

[1]Dipartimento di Ingegneria, Università degli Studi di Napoli Parthenope, Napoli, Italy
[2]Istituto Nazionale di Geofisica e Vulcanologia (INGV), Roma, Italy

Hence, SAR polarimetry is a quite mature and advanced approach for a large set of environmental applications.

In this chapter, after a brief overview of the main features that characterize operational polarimetric SAR satellite sensors (see Section 4.2), the mathematical formalism adopted to describe polarimetric SAR observables is introduced together with a set of polarimetric features (see Section 4.3). Then, target scattering decompositions are provided in Section 4.4, while in Section 4.5 polarimetric sea surface scattering is addressed and explained using showcases related to actual polarimetric SAR data sets, while conclusions are drawn in Section 4.6.

4.2 Polarimetric SARs

In this section, an overview of the different polarimetric SAR satellites is presented (see Table 4.1). In the last decades, the technological development of the space industry boosted the launch of numerous SAR satellites where different polarimetric modes are implemented. With reference to the polarimetric information content, polarimetric SAR can be classified as follows:

- Full-polarimetric (FP) SAR: it provides the complete scattering matrix related to the observed scene transmitting and receiving according to an orthogonal linear polarization basis, see (4.2);
- Compact-polarimetric (CP) SAR: it provides information on the observed scene transmitting a single circular or slant linear polarization while receiving according to a linear orthogonal basis;
- Dual-polarimetric (DP) SAR: it provides a single row/column of the scene's scattering matrix (see (4.2)) transmitting a single linear polarization while receiving coherently (i.e., in both amplitude and phase) in the corresponding orthogonal linear basis.

Within this context, the FP SAR architecture represents the polarimetric imaging mode that offers the largest information content on the observed scene, even though this costs in terms of significant hardware complexity, limited area coverage, and narrow range of usable angles of incidence. Nevertheless, it was shown that for a given target/application, DP SARs are enough to retrieve geophysical parameters that characterize the imaged scenario. In recent years, tailored DP SAR configurations, that is, CP SAR architectures, were proposed as a trade-off solution between the amount of polarimetric information and the area coverage.

Table 4.1 presents a sketch of the recent history of polarimetric spaceborne SAR architectures, starting from the first FP SAR, that is, the Spaceborne Imaging Radar (SIR) C/X, launched by National Aeronautics and Space Administration (NASA) in 1994 and now dismissed, up to the polarimetric SAR missions already planned for the near future, that is, the European Space Agency (ESA) Biomass equipped with a P-band FP SAR sensor. Nowadays, a virtual constellation of multiplatform and multifrequency polarimetric SARs that includes the Japanese Aerospace Exploration Agency (JAXA) Advanced Land Observing Satellite (Alos) Phased Array type

Table 4.1. Overview of the main satellite polarimetric SAR missions. Features refer to the finest spatial resolution imaging mode providing the largest polarimetric information content

Space agency SAR mission	Launch (year)	Polarimetric imaging mode	Frequency (band)	NESZ (dB)	Spatial resolution (slant range × azimuth) (m)	Area coverage (range × azimuth) (km)
NASA SIR C/X	1994	FP	L, C	–28[a]	20 × 8	15 × 15
JAXA Alos PalSAR-1	2006	FP	L	–30	9 × 5	30 × 30
CSA Radarsat-2	2007	FP	C	–36.5[b]	5 × 8	25 × 25
CAST Gaofen-3	2016	FP	C	N/A	8 × 8	40 × 40
ASI Cosmo-SkyMed 2	2019	FP	X	N/A	3 × 3	15 × 250
KARI Kompsat-5	2013	FP	X	≤ –17	3 × 3[d]	30 × 30[d]
ESA Biomass	2021[c]	FP/CP	P	≤ –27	50 × 50	100 × 100
NASA NISAR	2019	FP/CP	L	–25	7 × 3–48	> 240
CONAE/ASI SAOCOM	2018	FP/CP	L	N/A	10 × 10	15 × 15
JAXA Alos PalSAR-2	2014	FP/CP	L	–28	10 × 10	30 × 30
CSA RCM	2019	FP/CP	C	≤ –22	15 × 9	25 × 25
ISRO Risat-1	2012	FP/CP	C	≤ –19	6 × 4	25 × 25
DLR TerraSAR-X	2007	FP/DP	X	–19 to –26	3 × 2	10 × 10
DLR TanDEM-X	2010	FP/DP	X	–25	3 × 3	30 × 50[d]
ESA Sentinel-1	2014	DP	C	≤ –22	3 × 22	250 × 250
INTA PAZ	2018	DP	X	–21	6 × 6[e]	15–2000 × 15

[a]At C-band.
[b]±3 standard deviation.
[c]Planned.
[d]At AOI = 45°.
[e]Stripmap mode.

L-band SAR (PalSAR)-2, the C-band Canadian Space Agency (CSA) Radarsat-2, the ESA Sentinel-1 and the China Academy of Space Technology (CAST) Gaofen-3, and the X-band German Aerospace Center (DLR) TerraSAR-X and TanDEM-X SARs operating in both DP and FP modes (see Table 4.1). In addition, the L-band JAXA Alos PalSAR-2 and the Argentina National Space Activities Commission/ Italian Space Agency (CONAE/ASI) Microwave Observation Argentinian Satellite (SAOCOM), and the C-band Indian Risat-1 and CSA Radarsat Constellation Mission (RCM) SAR satellites are equipped with CP imaging modes.

As it can be noted in Table 4.1, the exploitation of spaceborne polarimetric SAR data is strictly limited in terms of noise floor, spatial resolution, and area coverage. When accurate estimation of polarimetric features is needed, the underlying polarimetric measurements must be reliable, that is, high-quality SAR data should be considered (see Radarsat-2, characterized by a nominal noise equivalent sigma zero—NESZ—of about −36.5 dB). When operational services must be provided based on polarimetric SAR data, a wide area coverage is a key point. Hence, DP SARs should be preferred, for example, Sentinel-1 SAR imagery that calls for a nominal swath width of about 250 km or the new CP architectures that aim at providing an information content that tends to the one available when FP SAR is considered but without its severe swath limitation. In fact, the DP swath width can be obtained. When the detection of small targets must be addressed, fine resolution polarimetric SAR measurements are needed. Accordingly, stripmap or even spotlight imaging modes characterized by spatial resolution finer than very few meters (see the nominal 3 m spatial resolution of TerraSAR-X) should be exploited. However, even in this case, a trade-off between area coverage and achievable spatial resolution must be considered according to the operational needs, for example, maritime surveillance, ship classification, etc.

Please note that, in practical cases, a trade-off between the different SAR imaging parameters that affect polarimetric feature extraction must be accounted for, and that they are often interconnected, for example, NESZ, incident wavelength and angle of incidence are related to each other.

4.3 Radar polarimetry

In this section, the basics of radar polarimetry are introduced. The different polarimetric observables that aim at describing the scattering process of the observed surface are presented together with some of the most common features used to characterize its polarimetric properties.

When dealing with remote sensing of the Erath's surface by monostatic radar instruments, a microwave pulse is transmitted and the electromagnetic wave scattered back to the transmitting antenna is received. The latter, resulting from the interaction between the transmitted wave and the scattering one, inherently carries on information about the scene that can be extracted by considering the received and the transmitted waves. A single-polarization radar is only able to perform measurements according to a fixed transmitting/receiving basis, while a multipolarization radar implements polarization diversity in both the transmitting and receiving sides. Hence, a multipolarization radar provides a larger amount of information on the scattering scene since it also accounts for the modification of the polarization of the transmitted wave induced by the scene [12].

The transformation of the incident electromagnetic wave into the one scattered off a target can be described, in the far zone of the scatterer, that is, when the plane wave assumption is considered for both the incident and the scattered fields, according to the Jones formalism [13]:

$$\boldsymbol{E}^{\mathrm{s}} = \frac{e^{-\mathrm{j}kr}}{r}\mathbf{S}\boldsymbol{E}^{\mathrm{i}} \tag{4.1}$$

where j is the imaginary unit, k is the electromagnetic wavenumber, r is the distance, and $\boldsymbol{E}^{\mathrm{s}}$ and $\boldsymbol{E}^{\mathrm{i}}$ are the complex Jones vectors describing the scattered and incident waves, respectively, and **S** is the scattering matrix.

The scattering matrix **S**, in the monostatic backscattering case, that is, when the transmitting and receiving SAR antennas are placed in the same location, and under the Backscatter Alignment Convention (BSA), is also known as Sinclair matrix and, in the linear horizontal (h)—vertical (v) polarization basis, is given by

$$\begin{aligned} \mathbf{S} &= \begin{pmatrix} S_{\mathrm{hh}} & S_{\mathrm{hv}} \\ S_{\mathrm{hv}} & S_{\mathrm{vv}} \end{pmatrix} = \begin{pmatrix} |S_{\mathrm{hh}}|e^{\mathrm{j}\varphi_{\mathrm{hh}}} & |S_{\mathrm{hv}}|e^{\mathrm{j}\varphi_{\mathrm{hv}}} \\ |S_{\mathrm{hv}}|e^{\mathrm{j}\varphi_{\mathrm{hv}}} & |S_{\mathrm{vv}}|e^{\mathrm{j}\varphi_{\mathrm{vv}}} \end{pmatrix} \\ &= e^{\mathrm{j}\varphi_{\mathrm{hh}}} \begin{pmatrix} |S_{\mathrm{hh}}| & |S_{\mathrm{hv}}|e^{\mathrm{j}(\varphi_{\mathrm{hv}}-\varphi_{\mathrm{hh}})} \\ |S_{\mathrm{hv}}|e^{\mathrm{j}(\varphi_{\mathrm{hv}}-\varphi_{\mathrm{hh}})} & |S_{\mathrm{vv}}|e^{\mathrm{j}(\varphi_{\mathrm{vv}}-\varphi_{\mathrm{hh}})} \end{pmatrix} \end{aligned} \tag{4.2}$$

where invoking reciprocity $S_{\mathrm{hv}} = S_{\mathrm{vh}}$. The elements $\mathrm{S}_{pq} = |\mathrm{S}_{pq}|e^{\mathrm{j}\varphi_{pq}}$ with $\{p,q\} = \{\mathrm{h,v}\}$ are termed complex scattering amplitudes, with the diagonal (off-diagonal) terms of the scattering matrix being termed copolar (cross-polar) terms since they involve the same (orthogonal) polarization for the incidence and the scattered waves.

When fixing the frequency and the viewing geometry, **S** carries on information about the scattering properties of the observed scene [12]. Hence, in this particular case, **S** is completely characterized by five independent parameters, that is, the modulus of the three scattering amplitudes and their relative phases, see (4.2). In fact, the absolute phase term φ_{hh}, which carries on information about the target-to-radar distance, is not considered as an independent parameter since it presents an arbitrary value. It must be pointed out that, in addition to the BSA convention, the forward scatter alignment (FSA) can be used to define the scattering matrix [12].

A SAR instrument which is able to measure the complete scattering matrix for each resolution cell is termed full-polarimetric or FP. This means that the transmitting antenna is able to observe the scene according to two orthogonal polarizations, that is, h and v, while receiving coherently in the same linear polarization basis. This is the case, for example, of the Canadian Radarsat-2 and the German TerraSAR-X SAR systems, see Table 4.1. However, simpler SAR architectures are also available that are not characterized by transmitting polarization diversity, that is, they transmit a single linear polarization (h or v), while receiving coherently in an orthogonal basis [14]. Those SAR architectures are termed dual-polarimetric or DP since they measure a single row/columns of the scattering matrix (see (4.2)), that is, a copolar and a cross-polar channel. This is the case, for example, of the European Sentinel-1 and the Spanish PAZ SAR systems, see Table 4.1. Recently, hybrid DP configurations have been proposed and implemented on-board of different SAR satellites that aim at providing polarimetric measurements whose information content tends to the one provided by FP configurations while being simpler and more performing in terms of swath coverage [14]. This is the case, for example, of the Indian Risat-1, for which a right-hand circularly polarized wave is transmitted while a standard h-v basis is used for receiving, and the Japanese Alos PalSAR-2, for which a slant linearly polarized wave is transmitted while a standard h-v basis is used for receiving, see Table 4.1.

Once **S** is measured, some key features on the observed target can be obtained, for example, the total scattered power, also known as SPAN:

$$\mathrm{SPAN} = \mathrm{Tr}\left(\mathbf{S}\mathbf{S}^{\dagger}\right) = |S_{\mathrm{hh}}|^2 + 2|S_{\mathrm{hv}}|^2 + |S_{\mathrm{vv}}|^2 \tag{4.3}$$

where Tr(·) is the matrix trace operator and † stands for complex conjugate transpose. Note that in (4.3) monostatic backscattering case and reciprocity assumption are considered. The SPAN is a basis-invariant parameter, that is, the total received power does not depend on the polarization basis adopted to represent the backscattered electromagnetic wave. If the wave scattered off the illuminated target is completely polarized, that is, a deterministic target is in place, the polarization of the wave can be deduced from the polarization of the transmitting antenna using the Jones formalism (4.1). However, when dealing with a generic distributed or "partial" target, that is, a target that extends over an area larger than the sensor's resolution cell and composed by several scatterers, high-order descriptors are often used. Under the hypothesis of scattering amplitudes following a complex Gaussian distribution, second-order statistics are necessary and sufficient to fully characterize the scattering behavior of a distributed target. Hence, a powerful and effective way to deal with the most general scattering behavior consists of developing descriptors that are based on second-order products of the scattering matrix. Nevertheless, considering a distributed target in a dynamic environment, assumptions on stationarity, homogeneity, and ergodicity must be made. Three operators are typically used that stem from the scattering matrix **S** by introducing proper averaging. The three operators are the coherency (**T**), the covariance (**C**) and the Mueller (**M**) matrix.

The Mueller matrix can be considered as the straightforward generalization of (4.1) to the general case of depolarizing transformations. According to Stokes formalism, the Mueller matrix is a 4 × 4 real-value never symmetric matrix that rules the transformation of the incident ($\boldsymbol{g}^{\mathrm{i}}$) Stokes vector into the scattered ($\boldsymbol{g}^{\mathrm{s}}$) one [15,16]:

$$\langle \boldsymbol{g}^{\mathrm{s}} \rangle = (kr)^{-2} \langle \mathbf{M} \rangle \boldsymbol{g}^{\mathrm{i}} \tag{4.4}$$

where ⟨ · ⟩ stands for ensemble average that, assuming ergodicity, can be replaced with spatial averaging and the power term accounts for the wave propagation as the similar term in (4.1). The Stokes vector $\boldsymbol{g}$ is an alternative representation of the polarization properties of an electromagnetic plane waves based on noncoherent, that is, power measurements. It is derived by making the outer product of the electric field vector with its conjugate transpose expressed according to the Jones formalism and projecting the resulting Hermitian matrix onto the Pauli matrix set [17]. As a result, the four real Stokes parameters are obtained:

$$\boldsymbol{g} = \begin{pmatrix} g_0 \\ g_1 \\ g_2 \\ g_3 \end{pmatrix} = \begin{pmatrix} |E_\mathrm{h}|^2 + |E_\mathrm{v}|^2 \\ |E_\mathrm{h}|^2 - |E_\mathrm{v}|^2 \\ 2\Re(E_\mathrm{h}E_\mathrm{v}^*) \\ -2\Im(E_\mathrm{h}E_\mathrm{v}^*) \end{pmatrix} \tag{4.5}$$

where g_0 is the total power of the wave, g_1 is the part of power associated to linear horizontal or vertical polarizations, g_2 is the part of the power associated to the linear polarization tilted at 45° in the h-v plane and g_3 is the part of the power associated to the circular polarization. Note that in (4.4), no ensemble average is made on the incident Stokes vector since the incident wave is deterministic, that is, fully polarized (the total power of the wave is the sum of powers of the different standard polarization), while after the interaction with the distributed target, depolarization occurs and, therefore, ensemble average is needed for the scattered Stokes vector (i.e., part of the scattered power may be not polarized).

Hence, since the Mueller matrix contains second-order products of the scattering matrix and, the Stokes vector can describe both fully- and partially polarized waves, it can account for the most general depolarizing scattering mechanism. Considering the monostatic backscattering case and the BSA convention and assuming that reciprocity holds, the Mueller matrix can be expressed as follows:

$$\mathbf{M} = \begin{pmatrix} \frac{1}{2}\langle|S_\mathrm{hh}|^2 + 2|S_\mathrm{hv}|^2 + |S_\mathrm{vv}|^2\rangle & \frac{1}{2}\langle|S_\mathrm{hh}|^2|S_\mathrm{vv}|^2\rangle & \langle\Re(S_\mathrm{hh}S_\mathrm{hv}^* + S_\mathrm{hv}S_\mathrm{vv}^*)\rangle & \langle\Im(S_\mathrm{hh}S_\mathrm{hv}^* + S_\mathrm{hv}S_\mathrm{vv}^*)\rangle \\ \frac{1}{2}\langle|S_\mathrm{hh}|^2|S_\mathrm{vv}|^2\rangle & \frac{1}{2}\langle|S_\mathrm{hh}|^2 2|S_\mathrm{hv}|^2 + |S_\mathrm{vv}|^2\rangle & \langle\Re(S_\mathrm{hh}S_\mathrm{hv}^*S_\mathrm{hv}S_\mathrm{vv}^*)\rangle & \langle\Im(S_\mathrm{hh}S_\mathrm{hv}^*S_\mathrm{hv}S_\mathrm{vv}^*)\rangle \\ \langle\Re(S_\mathrm{hh}S_\mathrm{hv}^* + S_\mathrm{hv}S_\mathrm{vv}^*)\rangle & \langle\Re(S_\mathrm{hh}S_\mathrm{hv}^*S_\mathrm{hv}S_\mathrm{vv}^*)\rangle & \langle\Re(S_\mathrm{hh}S_\mathrm{vv}^* + |S_\mathrm{hv}|^2)\rangle & \langle\Im(S_\mathrm{hh}S_\mathrm{vv}^* + |S_\mathrm{hv}|^2)\rangle \\ \langle\Im(S_\mathrm{hh}S_\mathrm{hv}^* + S_\mathrm{hv}S_\mathrm{vv}^*)\rangle & \langle\Im(S_\mathrm{hh}S_\mathrm{hv}^*S_\mathrm{hv}S_\mathrm{vv}^*)\rangle & \langle\Im(S_\mathrm{hh}S_\mathrm{vv}^* + |S_\mathrm{hv}|^2)\rangle & \langle\Re(S_\mathrm{hh}S_\mathrm{vv}^*|S_\mathrm{hv}|^2)\rangle \end{pmatrix} \tag{4.6}$$

In (4.6), the symbol * stands for complex conjugate. According to the Stokes vector representation, additional key features on the imaged target can be derived. In fact, the partially polarized wave scattered off the observed scene can be always decomposed into its fully polarized, that is, characterized by a unitary degree of polarization (DoP) (see (4.12)), and fully unpolarized, that is, characterized by DoP = 0 components [18]:

$$\boldsymbol{g} = g_0 \begin{pmatrix} \mathrm{DoP} \\ \mathrm{DoP}\cos 2\chi \cos 2\psi \\ \mathrm{DoP}\cos 2\chi \sin 2\psi \\ \mathrm{DoP}\sin 2\chi \end{pmatrix} + g_0 \begin{pmatrix} 1 - \mathrm{DoP} \\ 0 \\ 0 \\ 0 \end{pmatrix} \tag{4.7}$$

The polarization of a fully polarized wave is elliptical in the most general case and the polarization ellipse is determined by two geometrical parameters: the ellipticity angle χ and the orientation angle ψ whose domains of variation are [−45°, 45°] and [0°, 180°], respectively. One of the great theoretical and practical advantages of polarimetric radars with respect to single-polarization ones lies in the fact that, once measurements collected according to an orthogonal basis are available, it is possible to synthesize the power that is received by any combination of transmitted and received antenna polarizations. This technique, called polarization synthesis [18], is based on the Kennaugh matrix—**K**—which, under the BSA convention is given by

$$\mathbf{K} = \mathbf{U}_4\mathbf{M} \tag{4.8}$$

where $\mathbf{U}_4$ is a 4 × 4 diagonal matrix whose elements are (1, 1, 1,−-1). Once **K** is introduced, the normalized radar cross section (NRCS) for a generic received—*p*—and transmitted—*q*—polarization can be synthesized using the following formula:

$$\sigma_{pq} = \frac{1}{2}\frac{4\pi}{k^2}(\boldsymbol{g}^{\mathrm{r}})^{\mathrm{T}}\langle\mathbf{K}\rangle\boldsymbol{g}^{\mathrm{t}} \tag{4.9}$$

where the superscript "T" means transpose operator, while the superscripts "r" and "t" of the Stokes vectors stand for the receiving and transmitting antenna polarizations, respectively. Generally, (4.9) is used to synthesize the copolarized (cross-polarized) NRCS signature, that is, a 3D figure representing σ relevant to identical (or orthogonal) transmitted and received polarizations. According to (4.7), unpolarized backscattered energy results in a pedestal on which the polarization signature is set, see Figure 4.1. Hence, the height of the normalized pedestal can be considered as a measure of the amount of unpolarized backscattered energy.

Although **M** contains all the information to characterize the polarimetric behavior of the observed scene, it does not present the attractive matrix properties that stem from **T**/**C**. The latter can be obtained from **S** by projecting it onto special basis. In particular, first a target scattering vector ($\boldsymbol{k}$) is constructed by projecting it on a complete orthogonal set (under the inner Hermitian product) of a 2 × 2 complex basis matrices [17]:

$$\boldsymbol{k} = \frac{1}{2}\mathrm{Tr}(\mathbf{S}\psi) \tag{4.10}$$

When the Pauli and lexicographic basis are used, the coherence (**T**) and covariance (**C**) matrices are obtained [13,14]. Under the hypothesis of reciprocity and using the BSA convention, the **T** matrix is obtained using the Pauli basis matrices augmented with the identity matrix:

$$\psi_{\mathrm{P}} = \sqrt{2}\left\{\begin{pmatrix}1 & 0\\0 & 1\end{pmatrix}\ \begin{pmatrix}1 & 0\\0 & -1\end{pmatrix}\ \begin{pmatrix}0 & 1\\1 & 0\end{pmatrix}\right\} \tag{4.11}$$

Hence, the **T** matrix is obtained as follows:

$$\mathbf{T} = \langle\boldsymbol{k}\boldsymbol{k}^{\dagger}\rangle$$
$$= \frac{1}{2}\begin{pmatrix}\langle|S_{\mathrm{hh}}+S_{\mathrm{vv}}|^2\rangle & \langle(S_{\mathrm{hh}}+S_{\mathrm{vv}})(S_{\mathrm{hh}}-S_{\mathrm{vv}})^*\rangle & 2\langle(S_{\mathrm{hh}}+S_{\mathrm{vv}})S_{\mathrm{hv}}^*\rangle\\ \langle(S_{\mathrm{hh}}-S_{\mathrm{vv}})(S_{\mathrm{hh}}+S_{\mathrm{vv}})^*\rangle & \langle|S_{\mathrm{hh}}-S_{\mathrm{vv}}|^2\rangle & 2\langle(S_{\mathrm{hh}}-S_{\mathrm{vv}})S_{\mathrm{hv}}^*\rangle\\ 2\langle S_{\mathrm{hv}}(S_{\mathrm{hh}}+S_{\mathrm{vv}})^*\rangle & 2\langle S_{\mathrm{hv}}(S_{\mathrm{hh}}-S_{\mathrm{vv}})^*\rangle & 4\langle|S_{\mathrm{hv}}|^2\rangle\end{pmatrix} \tag{4.12}$$

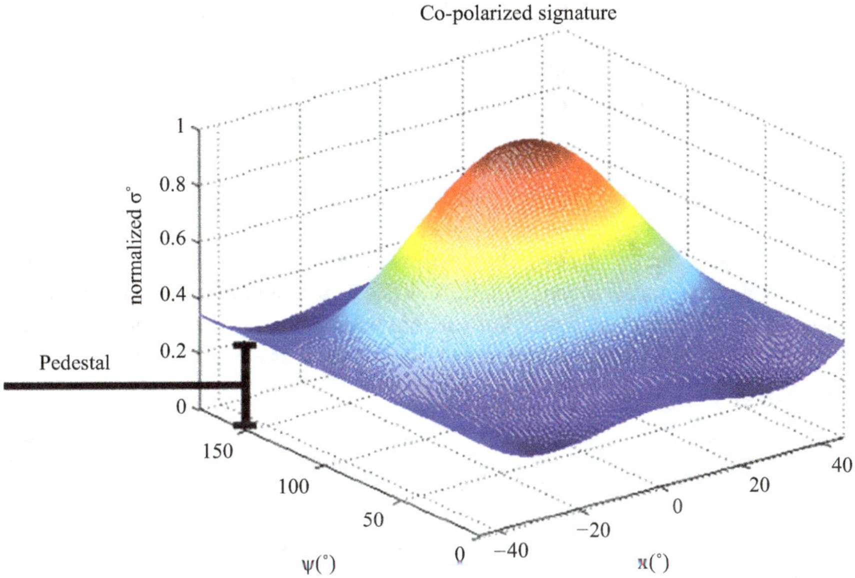

Figure 4.1. Pictorial view of the normalized pedestal height—NPH—on which the copolarization signature is set

When the lexicographic basis set is used:

$$\psi_1 = \sqrt{2}\left\{ \sqrt{2}\begin{pmatrix} 1 & 0 \\ 0 & 0 \end{pmatrix} \quad 2\begin{pmatrix} 0 & 1 \\ 0 & 0 \end{pmatrix} \quad \sqrt{2}\begin{pmatrix} 0 & 0 \\ 0 & 1 \end{pmatrix} \right\} \tag{4.13}$$

the **C** matrix can be constructed as follows:

$$\mathbf{C} = \langle \boldsymbol{k}\boldsymbol{k}^{\dagger} \rangle = \begin{pmatrix} \langle |S_{\mathrm{hh}}|^2 \rangle & \sqrt{2}\langle S_{\mathrm{hh}} S_{\mathrm{hv}}^* \rangle & \langle S_{\mathrm{hh}} S_{\mathrm{vv}}^* \rangle \\ \sqrt{2}\langle S_{\mathrm{hv}} S_{\mathrm{hh}}^* \rangle & 2\langle |S_{\mathrm{hv}}|^2 \rangle & \sqrt{2}\langle S_{\mathrm{hv}} S_{\mathrm{vv}}^* \rangle \\ \langle S_{\mathrm{vv}} S_{\mathrm{hh}}^* \rangle & \sqrt{2}\langle S_{\mathrm{vv}} S_{\mathrm{hv}}^* \rangle & \langle |S_{\mathrm{vv}}|^2 \rangle \end{pmatrix} \tag{4.14}$$

In (4.12) and (4.14), the coefficients 2 and $\sqrt{2}$ are needed to satisfy the total backscattered power invariance, that is, to keep the norm of the scattering vectors independent from the particular basis matrix set and equal to the SPAN of the scattering matrix [17].

It can be shown that **T**, **C**, and **M** are linked to each other by a 1-1 mapping and one can go from **T**/**C** to **M** and vice versa without the need to go via **S** [13,15]. This means that the most general depolarizing scattering mechanism, see (4.4), can be described using **T**/**C** that, unlike **M**, are Hermitian and positive semidefinite (PSD) matrices by construction, that is, they have real non-negative eigenvalues and orthogonal eigenvectors. Note that **C** and **T** share the same eigenvalues but they call for different eigenvectors [15].

In this chapter, the approach based on the covariance matrix **C** is advanced. This matrix can be decomposed into the incoherent sum of three fully deterministic scattering mechanisms, described by the eigenvectors $\boldsymbol{u}_n$, whose relative power is given by the eigenvalue λ_n [13,15]:

$$\mathbf{C} = \sum_{n=1}^{N} \lambda_n \boldsymbol{u}_n \boldsymbol{u}_n^{\dagger} \tag{4.15}$$

where $N = 3$ in the backscattering case. Following this decomposition, also termed to the Cloude–Pottier target decomposition (TD), some synthetic parameters can be introduced to describe the scattering mechanisms. Among them, the polarimetric entropy H can be defined to measure the depolarization associated with the scattering scenario [13,15]:

$$H = -\sum_{i=1}^{3} p_i \log_3 p_i,\; p_i = \frac{\lambda_i}{\sum_{i=1}^{3} \lambda_i},\; \lambda_1 \geq \lambda_2 \geq \lambda_3 \geq 0 \tag{4.16}$$

where p_i can be interpreted as the probability associated with the nth independent deterministic (i.e., polarized) state, described by its eigenvector $\boldsymbol{u}_n$ [13,15]. The spread of probabilities is represented by $0 \leq H \leq 1$ and it provides an effective and basis-invariant way to measure the depolarization. When H tends to 0, the system described by **C** is weakly depolarizing and the dominant scattering mechanism can be considered to be the one associated with the largest eigenvector. When H tends to 1, **C** corresponds to a high depolarizing system and, therefore, no dominant scattering mechanism can be extracted. A basis-invariant parameter that is complementary to H is the degree of polarization that can be defined as follows [19]:

$$\mathrm{DoP} = \sqrt{1 - 27\frac{\prod_{n=1}^{3} \lambda_n}{\left(\sum_{n=1}^{3} \lambda_n\right)^3}} \tag{4.17}$$

When DoP tends to 1 (0), the system described by **C** is weakly (highly) depolarizing.

Another feature that measures the depolarization associated to the scattering scene is the normalized pedestal height—NPH—see Figure 4.1, that can be mathematically defined as the ratio between the smallest and the largest eigenvalue of **C**:

$$\mathrm{NPH} = \frac{\lambda_3}{\lambda_1} \tag{4.18}$$

Although **C** contains all the scattering information related to the observed scene, partial information on the scattering scene may be extracted using specific features, for example, the standard deviation of the copolarized phase difference (CPD) [20,21] and the correlation between co- and cross-polarized scattering amplitudes X_c [13,22,23].

Those features are of paramount importance when partial polarimetric information is available, for example, when a SAR with dual-polarimetric capabilities is considered.

The CPD is defined as

$$\mathrm{CPD} = \angle C_{13} = \angle \langle S_{\mathrm{hh}} S_{\mathrm{vv}}^{*} \rangle \tag{4.19}$$

where $\angle$ means phase. It can be shown that the CPD probability density function (pdf) is related to the complex correlation coefficient between the copolarized channels ρ [21]:

$$\rho = |\rho| e^{j\overline{\varphi}} = \frac{\langle S_{\mathrm{hh}} S_{\mathrm{vv}}^{*} \rangle}{\sqrt{\langle |S_{\mathrm{hh}}|^2 \rangle \langle |S_{\mathrm{vv}}|^2 \rangle}} \tag{4.20}$$

For a given number of looks, the width of the CPD pdf depends on $|\rho|$. In particular, when $|\rho|$ tends to zero (total decorrelation between the copolarized channels), the pdf becomes uniformly distributed between $[-\pi, \pi)$, while for $|\rho|$ approaching one (hh and vv totally correlated), the pdf tends to a Dirac delta function. For $0 \leq |\rho| \leq 1$, the pdf resembles a Gaussian bell with a mean value and a standard deviation σ inversely related to $|\rho|$. Hence, σ can be considered as an indicator of the degree of correlation prevailing between copolarized channels [21]. In particular, in [24], it can be shown that σ is an unbiased indicator of $|\rho|$ even when a small number of samples is considered.

To define the correlation between like- and cross-polarized scattering amplitudes, the symmetry properties that characterize the observed scenario must be analyzed. In particular, no hypothesis on the scattering system except for the linearity and the reciprocity has been made to obtain the 3×3 covariance matrix (4.9) that, therefore, consists of nine independent parameters. However, the number of independent parameters is reduced when symmetry properties are satisfied. This is the case of naturally distributed scenarios that generally satisfy reflection symmetry [22,23,25,26]. A distributed scene that satisfies this symmetry property is characterized by the following covariance matrix [26]:

$$\mathbf{C} = \begin{pmatrix} \langle |S_{\mathrm{hh}}|^2 \rangle & 0 & \langle S_{\mathrm{hh}} S_{\mathrm{vv}}^{*} \rangle \\ 0 & \sqrt{2} \langle |S_{\mathrm{hv}}|^2 \rangle & 0 \\ \langle S_{\mathrm{vv}} S_{\mathrm{hh}}^{*} \rangle & 0 & \langle |S_{\mathrm{vv}}|^2 \rangle \end{pmatrix} \tag{4.21}$$

A direct consequence of reflection symmetry is that the correlation between like- and cross-polarized scattering amplitudes vanishes:

$$\langle S_{\mathrm{hh}} S_{\mathrm{hv}}^{*} \rangle = \langle S_{\mathrm{hv}} S_{\mathrm{vv}}^{*} \rangle = 0 \tag{4.22}$$

This occurs when the symmetry plane is orthogonal to the line-of-sight. When this is not the case, it is parametrized by a rotation angle that introduces an apparent correlation between the like- and cross-polarized scattering amplitudes. However, since this latter case is due to cross-polarization by the axis misalignment (and is in not due to depolarization), it can be always removed, for instance, by a singular value decomposition (SVD) of the scattering matrix [13].

In most naturally distributed targets, and in particular when dealing with the sea surface, reflection symmetry is so robust that (4.22) is considered as a reference for sea surface scattering and it is even used for calibration purposes [25,26].

In addition to reflection symmetry, other symmetry properties may apply [17,25]. When dealing with a distributed target characterized by rotation symmetry around the line-of-sight, that is, a rotational-invariant target, the corresponding averaged covariance matrix is unchanged under the special unitary rotation with respect to a given angle θ. This means that the covariance matrix is invariant under the rotation around an axis:

$$\mathbf{C} = \begin{pmatrix} \langle |S_{\mathrm{hh}}|^2 \rangle & \frac{\sqrt{2}}{2}\langle S_{\mathrm{hh}} S_{\mathrm{hv}}^* \rangle & \langle |S_{\mathrm{hh}}|^2 \rangle - 2\langle |S_{\mathrm{hv}}|^2 \rangle \\ -\frac{\sqrt{2}}{2}\langle S_{\mathrm{hh}} S_{\mathrm{hv}}^* \rangle & \sqrt{2}\langle |S_{\mathrm{hv}}|^2 \rangle & \frac{\sqrt{2}}{2}\langle S_{\mathrm{hh}} S_{\mathrm{hv}}^* \rangle \\ \langle |S_{\mathrm{hh}}|^2 \rangle - 2\langle |S_{\mathrm{hv}}|^2 \rangle & -\frac{\sqrt{2}}{2}\langle S_{\mathrm{hh}} S_{\mathrm{hv}}^* \rangle & \langle |S_{\mathrm{hh}}|^2 \rangle \end{pmatrix} \tag{4.23}$$

When a distributed target exhibits reflection symmetry in any plane that contains the rotation symmetry axis, it is termed to be azimuthal-symmetric [17,25]. In this case, by combining (4.22) and (4.23), $\mathbf{C}$ is given by

$$\mathbf{C} = \begin{pmatrix} \langle |S_{\mathrm{hh}}|^2 \rangle & 0 & \langle |S_{\mathrm{hh}}|^2 \rangle - 2\langle |S_{\mathrm{hv}}|^2 \rangle \\ 0 & \sqrt{2}\langle |S_{\mathrm{hv}}|^2 \rangle & 0 \\ \langle |S_{\mathrm{hh}}|^2 \rangle - 2\langle |S_{\mathrm{hv}}|^2 \rangle & 0 & \langle |S_{\mathrm{hh}}|^2 \rangle \end{pmatrix} \tag{4.24}$$

The modulus of the correlation between like- and cross-polarized scattering amplitudes r_{c}:

$$r_{\mathrm{c}} = \left| \langle S_{\mathrm{hh}} S_{\mathrm{hv}}^* \rangle \right| \tag{4.25}$$

can be considered as a consistent measure for the departure from the reflection symmetry case. When r_{c} tends to 0, the observed scene presents the reflection symmetry property and the scattering scenario is described by (4.24). Larger r_{c} values call for departure from reflection symmetry, implying that (4.14) applies [23].

The above-mentioned theoretical rationale can be exploited to give an understanding of polarimetric scattering from sea surface. Under moderate wind conditions and at intermediate incidence angles, sea surface scattering calls for Bragg or tilted-Bragg scattering, that is, a single-reflection mechanism that, being quasi-deterministic, is characterized by a low polarimetric entropy, a high correlation between like-polarized channels and a low unpolarized backscattered energy, as demonstrated in [18,19]. In addition, since sea surface is a natural distributed target, reflection symmetry is satisfied [23,25] and r_{c} values close to 0 are obtained. The slight departure from 0 is mainly due to a deterministic contribution related to the misalignment between sensor coordinates and the symmetry axis of the medium.

4.4 Target scattering decomposition

In this section, a primer on model-based polarimetric scattering decomposition is presented.

At the very root, TD methods aim at decomposing the polarimetric scattering information available from polarimetric observables (typically **T** or **C**) into elementary scattering mechanisms either using matrix algebra approaches (e.g., in the eigenvalue/eigenvector decompositions, see (4.15)) or approximating a scattering model (e.g., model-based decompositions) [17].

When looking at the nature of the polarimetric observables to be investigated in the decomposition process, TD methods can be grouped into two main classes [15]:

- Coherent decompositions, which are based on the knowledge of the scattering matrix and
- Incoherent decompositions, which rely on the second-order polarimetric descriptors **C**/**T**.

The class of incoherent TDs includes both "eigenvector-based" and "model-based" decomposition theorems. The former group consists of decomposing an actual covariance matrix into an incoherent and weighted sum of elementary independent covariance matrices associated to elementary scattering mechanisms, where the weights and the elementary scattering mechanisms are given by the eigenvalues and the eigenvectors of the covariance matrix, respectively. Well-known eigenvector-based TDs are the Cloude, Holm, and Van Zyl ones [15,17]. Model-based decompositions consist of representing the measured covariance matrix into an incoherent weighted linear combination of a pre-defined set of elementary scattering mechanisms. Well-established model-based TDs are the Freeman-Durden and the Yamaguchi decompositions [15,17]. It is important to point out that the validity of TDs is assessed according to the following criteria:

- The actual significance order of the considered scattering mechanism components must be preserved;
- The number of pixels characterized by negative power values must be reduced as much as possible; and
- The SPAN must be conserved.

In the following, the four-component Yamaguchi model-based TD is showcased [27]. The reference scattering model according to which an actual covariance matrix is decomposed consists of (1) a Bragg surface scattering component that models the backscattering from a slightly rough surface; (2) volume scattering from cloud of randomly oriented dipoles; (3) even/double-bounce scattering from a couple of orthogonal surfaces made by different materials; and (4) helix scattering from nonreflection symmetric structures characterized by complex geometry. Accordingly, the Yamaguchi polarimetric scattering model, assuming that the four components are uncorrelated, can be seen as the sum of the covariance matrices associated to each individual scattering mechanism [27]:

$$\mathbf{C} = \mathbf{C}_{\mathrm{S}} + \mathbf{C}_{\mathrm{D}} + \mathbf{C}_{\mathrm{H}} + \langle \mathbf{C}_{\mathrm{V}} \rangle_{\vartheta}$$

$$= \begin{pmatrix} f_{\mathrm{S}}|\beta|^2 + f_{\mathrm{D}}|\alpha|^2 + \dfrac{f_{\mathrm{H}}}{4} & \pm j\dfrac{\sqrt{2}f_{\mathrm{H}}}{4} & f_{\mathrm{S}}\beta + f_{\mathrm{D}}\alpha - \dfrac{f_{\mathrm{H}}}{4} \\ \mp j\dfrac{\sqrt{2}f_{\mathrm{H}}}{4} & \dfrac{f_{\mathrm{H}}}{2} & \pm j\dfrac{\sqrt{2}f_{\mathrm{H}}}{4} \\ f_{\mathrm{S}}\beta^* + f_{\mathrm{D}}\alpha^* - \dfrac{f_{\mathrm{H}}}{4} & \mp j\dfrac{\sqrt{2}f_{\mathrm{H}}}{4} & f_{\mathrm{S}} + f_{\mathrm{D}} + \dfrac{f_{\mathrm{H}}}{4} \end{pmatrix}$$

$$+ f_{\mathrm{V}} \begin{pmatrix} a & 0 & d \\ 0 & b & 0 \\ d & 0 & c \end{pmatrix} \tag{4.26}$$

where the subscripts "S," "D," "H," and "V" refer to surface, double-bounce, helix, and volume scattering components. Hence, the four-component Yamaguchi scattering model provides five independent equations in six scalar unknowns ($\alpha, \beta, f_{\mathrm{S}}, f_{\mathrm{H}}, f_{\mathrm{D}}$, and f_{V}), where the parameter a, b, c, and d can be set according to the considered volume scattering averaged covariance matrix $\langle \mathbf{C}_{\mathrm{V}} \rangle_{\vartheta}$. Note that the averaging is to be performed with respect to the orientation angle ϑ that is assumed to be uniformly distributed in the 0°–90° angular range when the orientation of the dipoles' cloud is completely random; or it is assumed to follow a nonuniform statistical distribution (i.e., $1/2\cos\vartheta$), when the dipoles' cloud exhibit an asymmetric orientation, that is, a dominant vertical/horizontal alignment. Note that the orientation angle is taken with respect to the horizontal axis as seen from the SAR [17]. In practical cases, [27] suggested to select volume scattering component according to the copolarization (vv-pol vs. hh-pol) power ratio: when it is in the range ±2 dB, a symmetric distribution for the orientation angle is more suitable, while asymmetric ϑ statistics should be adopted when the relative magnitude of copolar channels is larger than 2 dB.

With reference to the SPAN, the power contribution of each scattering mechanism is given by [27]:

$$\mathrm{SPAN} = P_{\mathrm{S}} + P_{\mathrm{D}} + P_{\mathrm{H}} + P_{\mathrm{V}} = f_{\mathrm{S}}\left(1 + |\beta|^2\right) + f_{\mathrm{D}}\left(1 + |\alpha|^2\right) + f_{\mathrm{H}} + f_{\mathrm{V}} \tag{4.27}$$

4.5 Polarimetric sea surface scattering

In this section, reference is made to sea surface scattering that is considered to be the reference scenario to discuss the polarimetric characterization of the observed scene according to the theoretical background presented in Section 4.3.

It is well-known that, under low-to-moderate wind conditions (i.e., approximately 2 ms^{-1}–15 ms^{-1}) and at intermediate incidence angles (i.e., about 20°–60°), microwave sea surface scattering is governed by Bragg or tilted-Bragg scattering that, at once, calls for the following properties:

- Large copolarized backscattering and negligible cross-polarized one;
- vv-pol backscattering larger than hh-pol one, in average;
- Large correlation between copolarized channels resulting in a low standard deviation of the copolarized phase difference;
- A dominant eigenvalue resulting in low polarimetric entropy;
- A significantly polarized scattering resulting in a large degree of polarization and low pedestal height;
- Reflection symmetry that results in like- and cross-polarized channels being uncorrelated; and
- A dominant surface scattering component.

Note that all the above-mentioned properties are due to the Bragg-like nature of the sea surface scattering, except for the last property which is related to the inherent symmetry properties of naturally distributed targets. Nonetheless, it is important to underline that, even though they are general properties of polarimetric sea surface microwave backscattering, the particular values assumed by the different polarimetric features significantly depend on both SAR imaging parameters (i.e., frequency, incidence angle, and noise floor) and sea state conditions (i.e., wind speed, presence of wave patterns, swell, etc.).

In this section, as a showcase, a full-polarimetric SAR acquisition collected on 30 October 2014 in the Gulf of Mexico by the L-band (1.24 GHz) PalSAR 2 SAR sensor on board of the JAXA space mission Alos, is considered. The SAR scene, collected in ascending pass with an incidence angle spanning from about 29.5° up to 32.6° (from near range to far range), covers an area of about 23 km × 66 km and is characterized by a very fine spatial resolution of 2.86 m × 3.13 m (range × azimuth direction, respectively). External ancillary information provided at the time of the SAR acquisition suggests that low-to-moderate wind conditions applied. The Single-Look Complex (SLC) level 1.1 product is considered for processing in order to preserve both amplitude and phase information. A pre-processing step that includes radiometric calibration and geocoding (producing a 5.5 m square pixel in the resulting SAR image) is applied. Note that, in this section, a 7 × 7 average moving window is used to perform spatial averaging in order to replace the ensemble average operators in the equations of Section 4.3.

A false-color RGB image relevant to an almost square excerpt of the Alos PalSAR-2 SAR scene is shown in Figure 4.2, where the following RGB color-coding is adopted (see (4.14)): $\langle |S_{\text{hv}}|^2 \rangle$, that is, $C_{22}/2$, is assigned to red color, $\langle |S_{\text{hh}}|^2 \rangle$, that is, C_{11}, is associated to green; while $\langle |S_{\text{vv}}|^2 \rangle$, that is, C_{33}, is assigned to blue. The excerpt refers to the Campeche Bay area in the southern part of the Gulf of Mexico. Bright spots at sea are clearly visible in Figure 4.2 that correspond to metallic targets related to the offshore infrastructures (rigs/platforms) owned by oil and gas companies. In fact, the Campeche Bay area is very rich in natural oil and gas seeps. However, it can be noted that, in this representation, the heterogeneous sea surface calls for a dominant bluish color, resulting from the main role played by the copolarized channels (hh and vv) in ruling sea surface backscattering. Sea targets, instead, appear as bright spots

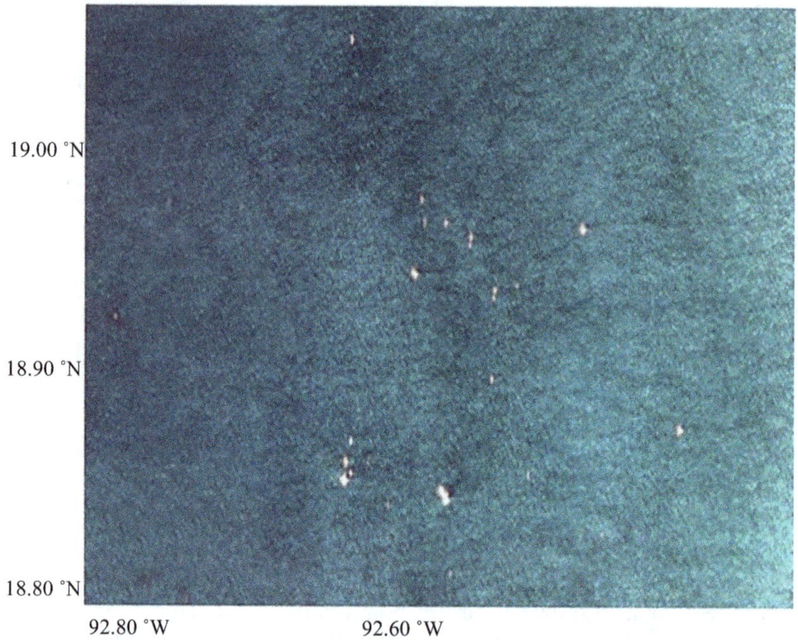

Figure 4.2 False-color RGB image relevant to the Alos PalSAR-2 SAR acquisition collected in the Campeche Bay (Gulf of Mexico). The color-coding Red = $\langle |S_{hv}|^2 \rangle$; Green = $\langle |S_{hh}|^2 \rangle$; Blue = $\langle |S_{vv}|^2 \rangle$ is adopted, see (4.9). Metallic targets at sea can be seen as bright spots

since their complex metallic structure results into very strong backscattering in both co- and cross-polarized channels.

According to Section 4.3, to characterize polarimetric scattering second-order descriptors are needed, that is, the covariance/coherency matrix. Hence, the modulus of the nine elements of the 3×3 covariance matrix **C** (reciprocity is assumed, see (4.14)) are evaluated and shown, in decibel (dB) scale, in Figure 4.3. The back-scattering power of each polarimetric channel are along the main diagonal of **C**, see (4.14). Again, it can be noted that over the heterogeneous sea surface, the mean backscattered power associated to the copolarized channels (see Figure 4.3(a) and (i)) is about more than 6 dB larger than the one associated to the cross-polarized channel (see Figure 4.3(e)). By visual inspecting Figure 4.3(a) and (i), it is clear that over sea surface the vv-pol backscattered power is slightly larger (about 2 dB), in average, than the corresponding hh-pol one, as predicted by the Bragg scattering theory. The structure of the sea surface **C** matrix shown in Figure 4.3 also points out that reflection symmetry property applies, see (4.22). In fact, the off-diagonal elements that represent the modulus of the complex correlation between like- and cross-polarized channels, see Figure 4.3(b), (d), (f) and (h), call for values very close to 0

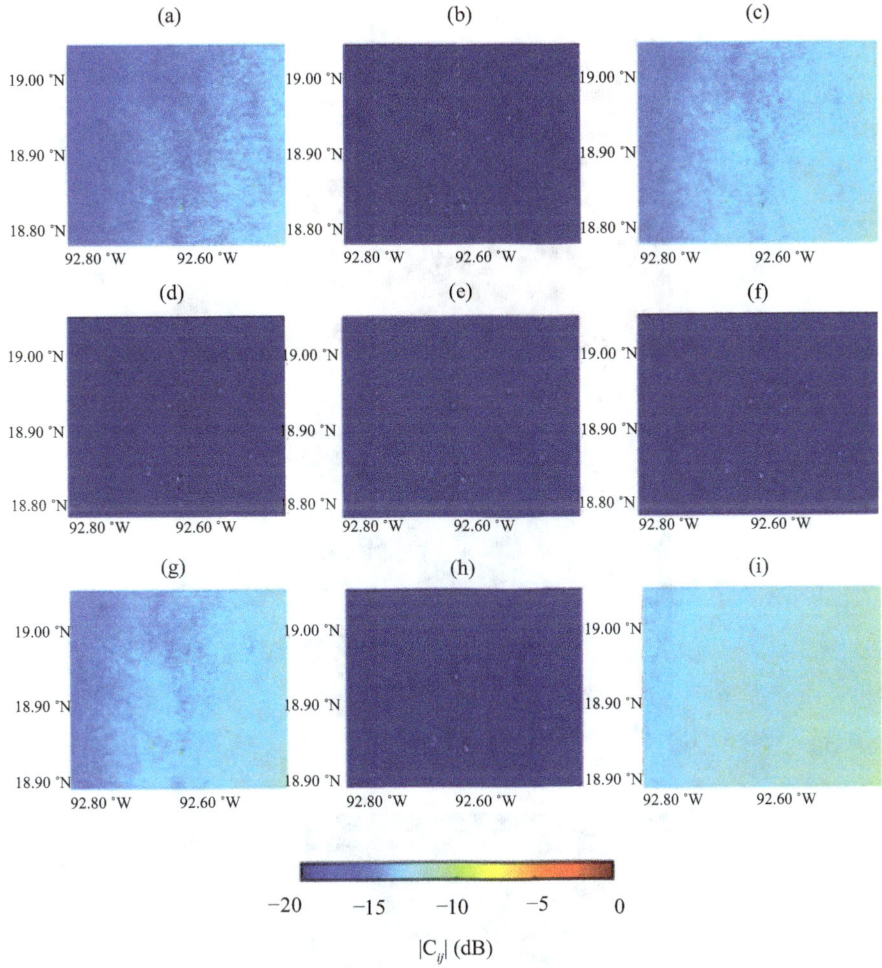

Figure 4.3 *Elements of the covariance matrix* **C***, see (4.16), relevant to the full-pol SAR image shown in Figure 4.2. (a): C_{11}; (b): $|C_{12}|$; (c): $|C_{13}|$; (d): $|C_{21}|$; (e): C_{22}; (f): $|C_{23}|$; (g): $|C_{31}|$; (h): $|C_{32}|$; (i): C_{33}. dB scale is adopted*

(lower than −20 dB). Note that, over metallic targets, those elements call for values larger than −10 dB.

Figure 4.4 shows the three trace-normalized eigenvalues of **C**. The latter, being Hermitian and PSD, results in three real and non-negative eigenvalues, see (4.15). As expected by the Bragg scattering theory, a dominant eigenvalue is in place over sea surface, that is, λ_1 is larger than 0.8 while λ_2 and λ_3 are both lower than 0.1. This means that the single-reflection scattering mechanism associated to the first eigenvalue rules sea surface backscattering, while the contribution of the secondary eigenvalues,

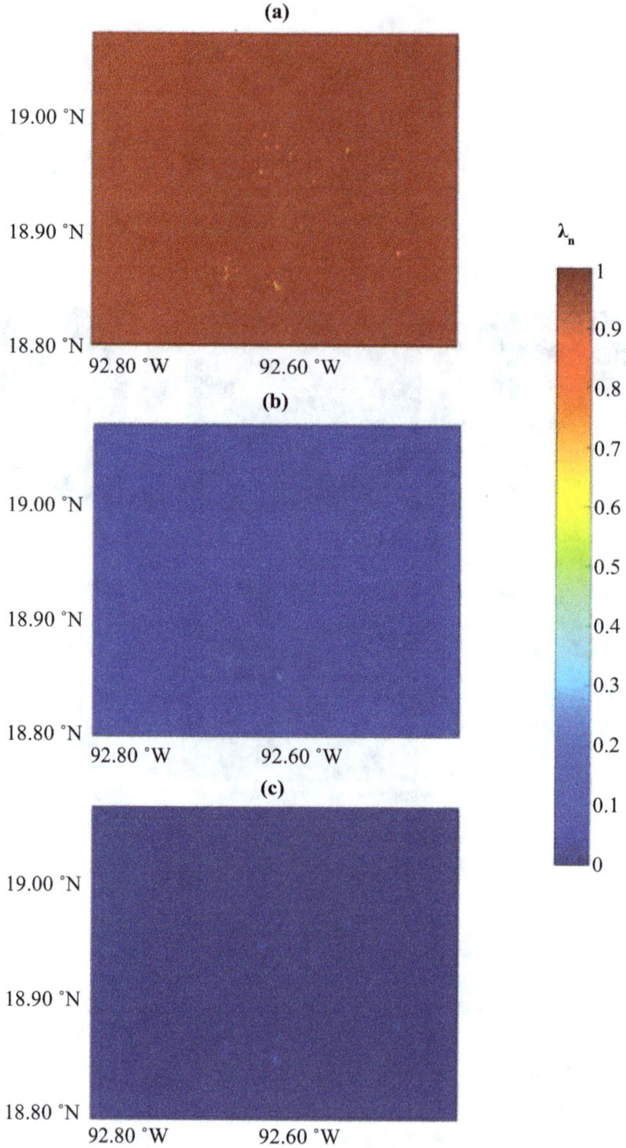

Figure 4.4 Trace-normalized eigenvalues of the covariance matrix **C***, see (4.16), relevant to the full-pol SAR image shown in Figure 4.2: (a): λ_1; (b): λ_2; and (c): λ_3 image*

associated to double-bounce and multiple scattering elementary scattering mechanisms, is marginal. Note that, over metallic sea targets, the difference between the eigenvalues is reduced due to the more complex and random scattering behavior that applies. In fact,

λ_1 assumes values around 0.5–0.6 while λ_2 and λ_3 increase up to about 0.3 and 0.2, respectively (see Figure 4.4).

The behavior of the eigen-decomposition of **C** directly reflects into the behavior of different eigenvalue-based polarimetric parameters, that is, the ones resulting from the Cloude–Pottier TD. The latter provides information on the physical characteristics of the scattering mechanisms ruling the observed scene. As an example, polarimetric entropy *H*, see (4.11), is shown in Figure 4.5. The dominant eigenvalue experienced over sea surface results in very low entropy values, that is, *H* is in the range 0.2–0.3, which are the typical values that characterize an almost deterministic scattering process. This is no longer true when dealing with sea metallic targets that result in *H* values significantly larger if compared to the surrounding sea, that is, larger than 0.7, due to the remarkably random nature of their scattering process which involves surface back-scattering from plane structures together with double-bounce and multiple scattering due to the interaction between the incident electromagnetic wave and the vertical/horizontal structures of the target.

An additional polarimetric parameter that allows estimating the degree of randomness of the scattering mechanisms characterizing the observed scene is the degree of polarization—DoP, see (4.17) —which is shown in Figure 4.6. It can be noted that sea surface calls for very large DoP values, that is, about 0.75. This is due to the fact that the Bragg/tilted-Bragg scattering that dominates over sea surface is characterized by a quasi-deterministic behavior. This is no longer the

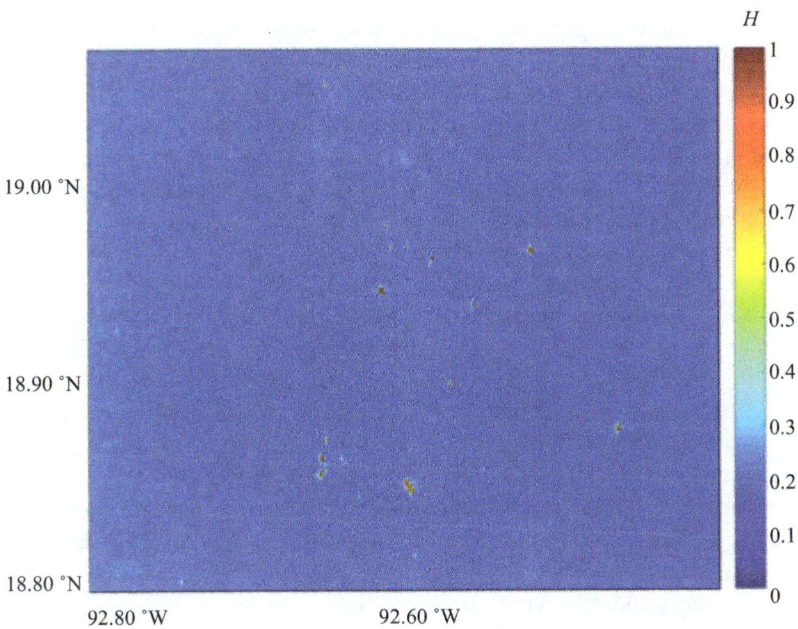

Figure 4.5 Polarimetric entropy H, derived from the Cloude–Pottier TD (see (4.11)), relevant to the full-pol SAR image shown in Figure 4.2

Figure 4.6 Degree of polarization—DoP (see (4.12))—image relevant to the full-pol SAR image shown in Figure 4.2

case for the metallic targets, whose random combination of different elementary scattering mechanisms results in lower DoP values, that is, approximately 0.5.

The degree of polarization of the scattering process is strictly reflected on the properties of the backscattered wave. Figure 4.7 shows the map of the pedestal height parameter normalized with respect to the SPAN—NPH, see (4.18)—that was shown to be a reliable estimator of the amount of unpolarized energy scattered off the observed target (see Section 4.3). As expected due to the low-depolarizing properties of the Bragg-like sea surface scattering, significantly low NPH values are observed over sea surface, that is, lower than 0.05. This means that the almost deterministic behavior of sea surface scattering predicted by the Bragg theory results in an energy associated to the backscattered wave and received by the SAR antenna which is almost completely polarized. In Figure 4.7, it can be also observed how the depolarizing characteristics of sea metallic targets result into much larger NPH values (i.e., about 0.3–0.4), which are more than about 10 times larger than the corresponding values measured over the surrounding sea.

Another key property that characterizes polarimetric sea surface scattering is the large degree of correlation exhibited by the copolarized channels. The latter can be estimated in a robust and effective way from actual polarimetric SAR measurements by evaluating the standard deviation of the CPD—σ_{CPD}, see (4.19)—which is shown in Figure 4.8. Over sea surface, σ_{CPD} values lower than 40° are observed, while metallic targets at sea call for significantly larger σ_{CPD} values, that is, about 70°–80°, due to the fact that they result in a more complex scattering process.

Figure 4.7 Normalized pedestal height—NPH (see (4.13))—image relevant to the full-pol SAR image shown in Figure 4.2

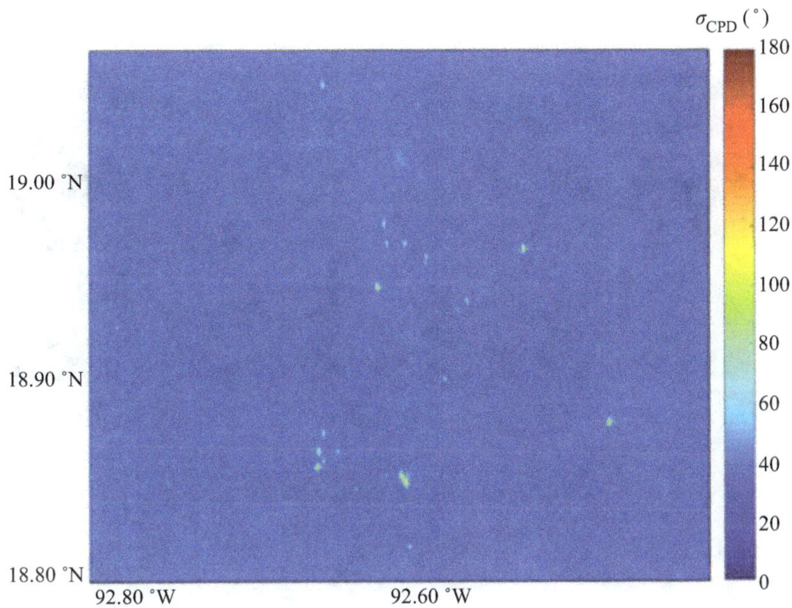

Figure 4.8 Copolarized phase difference standard deviation—σ_{CPD} (see (4.17))—image, in degrees, relevant to the full-pol SAR image shown in Figure 4.2

In addition to the eigenvalue/eigenvector TD analysis—see Figure 4.4 and 4.5—a scattering model-based TD is applied according to the Yamaguchi four component framework described in Section 4.4. Hence, the total polarimetric backscattering power is decomposed into the power of each elementary scattering mechanism, that is, surface, double bounce, volume, and helix, see (4.27). Results are shown in Figure 4.9, where (a), (b), (c), and (d) refer to P_S, P_D, P_V, and P_H, respectively. Note that dB scale is adopted. It must be noted that sea surface is characterized by a dominant surface scattering component, which is about -10 dB, while all the other components are almost negligible as witnessed by the dark blue background in

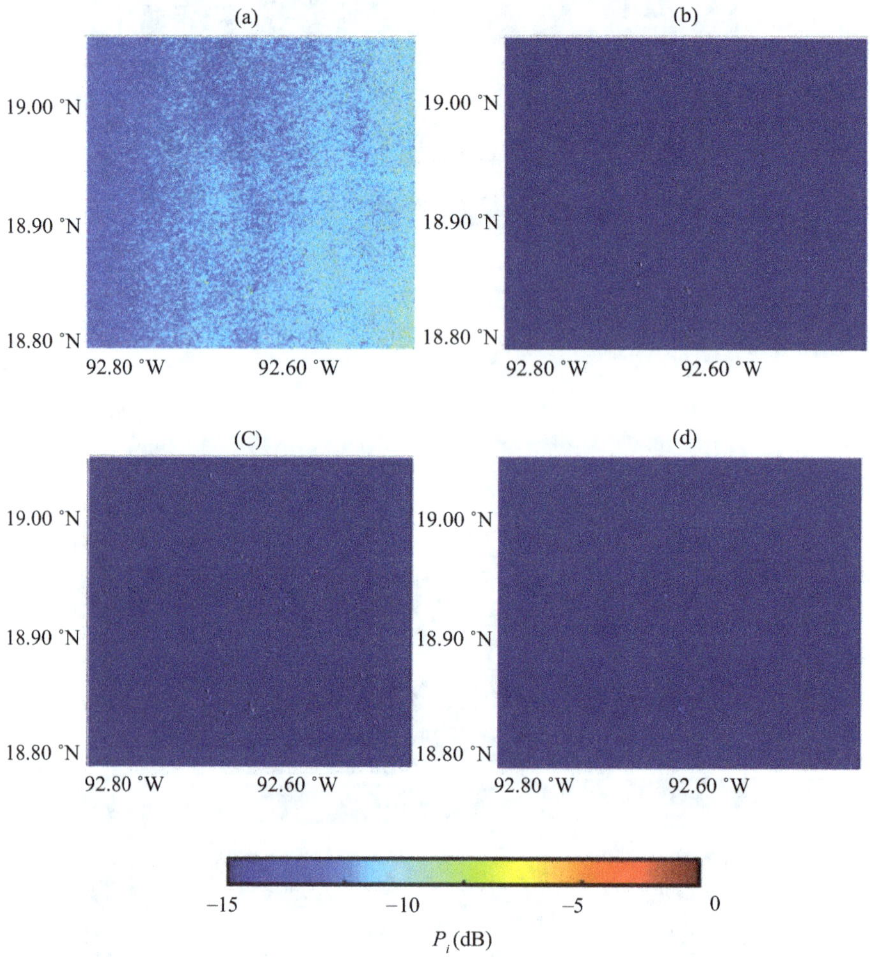

Figure 4.9 Yamaguchi 4-component target decomposition (see (4.27)) relevant to the full-pol SAR image shown in Figure 4.2. (a) P_S; (b) P_D; (c) P_V; and (d) P_H. dB scale is adopted

Figure 4.9(b)– (d) that corresponds to P_D, P_V, and P_H lower than -15 dB. Whean dealing with metallic targets at sea, instead, it is interesting to note that the amount of backscattered power associated to each elementary scattering component is strictly related to the structure of the target. Nevertheless, for most of the targets, results in Figure 4.9 show that double bounce, volume, and helix scattering components are no longer negligible and are often larger than P_S.

4.6 Conclusions

In this chapter, the fundamentals of radar polarimetry are introduced. First, polarimetric observables that deal with the description of the polarimetric scattering process are introduced. Then, a set of polarimetric features derived from polarimetric matrices and scattering decompositions obtained from model-based and eigenvalue/eigenvector-based TD approaches that aim at discussing the physical characterization of the observed targets is presented. In addition, sea surface is considered as a reference polarimetric scattering scenario: an actual FP SAR scene is processed to show the most important polarimetric properties characterizing sea surface scattering. The latter showcase is of paramount applicative importance since suggests how to exploit polarimetric information, by means of proper modelling and analysis tools, for sea targets observation purposes, including ship (see Chapter 6) and oil spill (see Chapter 9) detection.

Acknowledgments

The authors would like to thank the Earth Observation Research and Application Center (EORC) of JAXA for providing the Alos PalSAR-2 data under the research project "Multi-polarization PalSAR-2 SAR data for oil field and sea ice observation" within the framework of the 4th Research Agreement for the Advanced Land Observing Satellite-2.

List of acronyms

ALOS	Advanced land observing satellite
ASI	Italian space agency
BSA	Backscattering alignment
CAST	China academy of space technology
CONAE	Argentina national space activities commission
CP	Compact-polarimetric
CPD	Copolarized phase difference
CSA	Canadian space agency
DLR	German aerospace center
DoP	Degree of polarization

DP	Dual-polarimetric
EORC	Earth observation research and application center
ESA	European space agency
FP	Full-polarimetric
FSA	Forward scattering alignment
h	Horizontal
INTA	National institute of aerospace technique
ISRO	Indian space research organization
JAXA	Japanese aerospace exploration agency
KARI	Korea aerospace research institute
NASA	National aeronautics and space administration
NESZ	Noise equivalent sigma zero
NPH	Normalized pedestal height
NRCS	Normalized radar cross section
PalSAR	Phased array type L-band SAR
PSD	Positive semidefinite
RGB	Reed green blue
SAR	Synthetic aperture radar
SAOCOM	Microwave Observation Argentinian Satellite
SIR	Spaceborne Imaging Radar
SLC	Single-look complex
SVD	Singular value decomposition
TD	Target decomposition
v	Vertical

References

[1] Moreira A., Prats-Iraola P., Younis M., Krieger G., Hajnsek I. and Papathanassiou K.P. "A tutorial on synthetic aperture radar." IEEE Geoscience and Remote Sensing Magazine. 2013; 1(1): 6–43.

[2] Tebaldini S. and Rocca F. "Multibaseline polarimetric SAR tomography of a boreal forest at P- and L-bands." IEEE Transactions on Geoscience and Remote Sensing. 2012; 50(1): 232–246.

[3] Simard M. and Denbina M. "An assessment of temporal decorrelation compensation methods for forest canopy height estimation using airborne L-band same-day repeat-pass polarimetric SAR interferometry." IEEE Journal of Selected Topics in Applied Earth Observation and Remote Sensing. 2018; 11(1): 95–111.

[4] McNairn H. and Brisco B. "The application of C-band polarimetric SAR for agriculture: A review." *Canadian Journal of Remote Sensing*. 2004; 30(3): 525–542.

[5] Canisius F., Shang J., Liu J., *et al.*, "Tracking crop phenological development using multi-temporal polarimetric Radarsat-2 data." Remote Sensing of Environment. 2018; 210: 508–518.

[6] Bhattacharya A. and Touzi R. "Polarimetric SAR urban classification using the Touzi target scattering decomposition." *Canadian Journal of Remote Sensing*. 2011; 37(4): 323–332.

[7] Zhang L., Sun L., Zou B. and Moon W.M. "Fully polarimetric SAR image classification via sparse representation and polarimetric features." IEEE Journal of Selected Topics in Applied Earth Observation and Remote Sensing. 2015; 8(8): 3923–3932.

[8] Dabboor M., Montpetit B., Howell S. and Haas C. "Improving sea ice characterization in dry ice winter conditions using polarimetric parameters from C- and L-band SAR data." Remote Sensing. 2017; (9): 1270.

[9] Akbari V. and Brekke C. "Iceberg detection in open and ice-infested waters using C-band polarimetric synthetic aperture radar." IEEE Transactions on Geoscience and Remote Sensing. 2018; 56(1): 407–421.

[10] Migliaccio M., Nunziata F. and Buono A. "SAR polarimetry for sea oil slick observation." International Journal of Remote Sensing. 2015; 36(12): 3243–3273.

[11] Zhang T., Marino A., Xiong H. and Yu W. "A Ship detector applying principal component analysis to the polarimetric notch filter." Remote Sensing. 2018; 10: 948.

[12] Guissard A. "Phase calibration of polarimetric radars from slightly rough surfaces." IEEE Transactions on Geoscience and Remote Sensing. 1995; 32(3): 712–715.

[13] Cloude S.R. *Polarisation: Applications in remote sensing.* Oxford: Oxford University Press, 2009.

[14] Raney R.K. "Hybrid-polarity SAR architecture." IEEE Transactions on Geoscience and Remote Sensing. 2007; 45(11): 3397–3404.

[15] Cloude S.R. and Pottier E. "A review of target decomposition theorems in radar polarimetry." IEEE Transactions on Geoscience and Remote Sensing. 1996; 34(2): 498–518.

[16] Van Zyl J.J., Papas C.H. and Elachi C. "On the optimum polarisations of incoherently reflected waves." IEEE Transactions on Antennas and Propagation. 1987; 35(7): 818–825.

[17] Lee J.S. and Pottier E. *Polarimetric radar imaging: From basics to applications.* CRC Press, Boca Raton, FL, 2009.

[18] Nunziata F., Migliaccio M. and Gambardella A. "Pedestal height for sea oil slick observation." IET Radar, Sonar and Navigation. 2011; 5(2): 103–110.

[19] Buono A., Nunziata F. and Migliaccio M, "Analysis of full and compact polarimetric SAR features over the sea surface." IEEE Geoscience and Remote Sensing Letters. 2016; 13(10): 1527–1531.

[20] Schuler D.L., Lee J.S., and Hoppel K.W. "Polarimetric SAR image signatures of the ocean and gulf stream features." IEEE Transactions on Geoscience and Remote Sensing. 1993; 31(6): 1210–1221.
[21] Migliaccio M., Nunziata F. and Gambardella A. "On the co-polarised phase difference for oil spill observation." International Journal of Remote Sensing. 2009; 30(6): 1587–1602.
[22] Baum C.E. and Kritikos H.N. *Electromagnetic symmetry.* New York, NY: Taylor & Francis, 1995.
[23] Nunziata F., Migliaccio M. and Brown C.E. "Reflection symmetry for polarimetric observation of man-made metallic targets at sea." IEEE Journal of Oceanic Engineering. 2012; 37(3): 384–394.
[24] Velotto D., Migliaccio M., Nunziata F., and Lehner S. "Dual-polarized TerraSAR-X data for oil spill observation." IEEE Transactions on Geoscience and Remote Sensing. 2011; 49(12): 4751–4762.
[25] Nghiem S.V., Yueh S.H., Kwok R. and Li F.K. "Symmetry properties in polarimetric remote sensing." Radio Science. 1992; 27(5): 693–711.
[26] Yueh S.H., Kwok R. and Nghiem S.V. "Polarimetric scattering and emission properties of targets with reflection symmetry." Radio Science. 1994; 29(6): 1409–1020.
[27] Yamaguchi Y., Moriyama T., Ishido M., and Yamada H. "Four-component scattering model for polarimetric SAR image decomposition." IEEE Transactions on Geoscience and Remote Sensing. 2005; 43(8): 1699–1706.

Chapter 5
Ambiguity problems and their mitigation

Gerardo Di Martino[1], *Antonio Iodice*[1] *and Domenico Velotto*[2]

5.1 Introduction

As witnessed by the many applications described in the present book, synthetic aperture radar (SAR) provides a particularly convenient way of observing the ocean. Indeed, spaceborne SAR sensors were used from their early years to estimate physical parameters (e.g., currents and winds) and to detect the presence of ships (which usually appear as bright spots over the sea dark background) with all-weather day-and-night capabilities. In view of all these applications, it is important to obtain data with the highest possible radiometric accuracy. In this context, special attention must be paid to SAR ambiguity problems. Due to the potential appearance of displaced and attenuated replicas of strong signals ("ghosts") related to bright structures present on the coast or to ships over the sea surface, this issue is especially significant for ship-detection applications, since it may dictate a relevant increase in the number of false detected targets.

Ambiguities have been always a key aspect of SAR sensor design [1]. Two types of ambiguity may arise in an SAR system: range ambiguity and azimuth ambiguity. The former arises when the echoes of two successive pulses overlap at the receiver. To avoid this situation, a maximum limit on the pulse repetition frequency (PRF) of the system must be imposed, so that each inter-pulse interval can accommodate the reception of the complete echo of the illuminated swath. Instead, azimuth ambiguity arises when the azimuth signal is sampled with a frequency lower than the Nyquist one. To avoid this situation, a minimum limit on the PRF of the system must be imposed. The combination of these two requirements led to the introduction of the concept of the so-called "minimum antenna area" [1–4]. Actually, this constraint has been later recognized as a "soft" constraint, in the sense that it has to be strictly enforced only in the special case in which the best possible azimuth resolution and range coverage are chosen as design goals [5]. In more general design situations, resolution or coverage requirements may be relaxed to obtain less restrictive limits on the antenna area. This is particularly useful when

[1]Dipartimento di Ingegneria Elettrica e delle Tecnologie dell'Informazione, Università di Napoli Federico II, Naples, Italy
[2]SAR Signal Processing Department, German Aerospace Center (DLR), Bremen, Germany

the design of spaceborne SAR systems, which cannot accommodate very large antennas, is in order.

The system design constraints listed above are based on the definition of an ideal antenna pattern that perfectly vanishes outside the main lobe. However, actual antenna patterns present both range and azimuth sidelobes. The presence of the sidelobes is the reason for the potential appearance of ambiguity even when the mentioned design constraints are strictly followed [4,6,7]. In this situation, replicas due to antenna sidelobes are attenuated with respect to the originating target, since they are weighted by the sidelobe response, and are not perfectly focused, since they do not perfectly match with the reference signal. However, when these replicas are generated by strong targets and are located on a dark background, as is the case for sensing of the sea surface, they can reach an intensity level higher than that of the nonambiguous sea clutter [8,9]. The presence of ambiguities can be detrimental for all ocean applications, and, in particular, may be source of false alarms in ship-detection applications [10,11].

In this chapter, we focus on ambiguities due to antenna sidelobes, that is, we intentionally neglect the case of recently developed experimental low- or high-PRF modes that are based on the acquisition of images containing not only the true target but also its main-lobe range or azimuth ambiguities [12,13] (refer also to Chapter 3 for more details). These images require either an ad hoc acquisition strategy or a post-processing step to discriminate the true targets from their main-lobe high-intensity replicas. Furthermore, here we focus on azimuth ambiguities rather than on range ones. As a quite general rule, the former are characterized by azimuth displacements (in the order of kilometers for the spaceborne case) smaller than the corresponding range displacements expected for range ambiguities (in the order of tens of kilometers for the spaceborne case). For this reason, frequently range ambiguities and the target originating from them do not appear on the same image, thus hampering their identification. Moreover, range ambiguities are much more unfocused and attenuated than azimuth ones. Note that range ambiguities originating from permanent objects present on the coast can be identified by analyzing image time series looking for persistent objects on the sea surface [14]. In conclusion, this chapter is mainly devoted to sidelobe azimuth ambiguities, which represent an unwanted, potentially severe, source of radiometric distortion.

In the last decades, many techniques for azimuth ambiguity mitigation have been proposed. A first class of approaches exploits the a priori knowledge of the displacements of the ghosts with respect to the original targets, which can be evaluated as a function of the radar system parameters [6]. Accordingly, the images can be recursively searched for ambiguities through simple intensity checks [10,15]: ghosts can be then simply discarded, or the ambiguous area can be reconstructed via compressive sensing or inpainting techniques [15]. Another class of approaches is based on in-phase cancellation [7,16], where a deconvolving step is performed after pulse matched filtering, aimed at suppressing the azimuth ambiguities. In particular, [7] proposes the ideal filter approach, whereas in [16] a method for reconstructing and suppressing azimuth ambiguity based on the azimuth ambiguity factor is presented. These two classes of methods are good candidates mainly in those situations where only the ghosts due to point targets are of interest, as in the case of ocean scenes in low-resolution images, where the typical sources of ambiguities are isolated ships, whose size is comparable with image resolution. Conversely, these methods provide degraded performance in the presence of distributed targets, which typically appear in coastal areas, or also in many ocean scenes on high-resolution images, where also most ships behave as distributed targets. In these more general cases, the use of techniques based on the concept of selective filtering has been advocated [8,9,17]. These techniques use appropriate filters to select from the whole azimuth spectrum those areas which are less affected by ambiguities, so that they can be applied to ghosts due to both point targets and distributed targets, although, in general, at the cost of a resolution loss. In particular, in [8,9] and [17], different methods based on the concept of adaptive filtering have been proposed to minimize the resolution loss. Moreover, the technique of [17] has been successfully applied in acquisition modes different from the canonical stripmap one: in particular, it has been positively tested on sliding spotlight images, but it is argued that it can provide good performance also for Sentinel-1 TopSAR mode, which is of great interest for maritime applications. All the mentioned techniques require as input Single Look Complex SAR images, apart from the basic displacement-based ones. Another approach particularly well suited to remove azimuth (or range) ambiguities from detected SAR images of the ocean is based on the use of multitemporal series of SAR images. In this situation, a search for permanent targets present on the sea surface can be performed [18]: indeed, distributed permanent targets over the sea may be regarded with a high degree of confidence as ambiguities.

A special discussion must be devoted to azimuth ambiguity filtering in full-polarimetric (also known as quadrature-polarimetric, i.e., quad-pol) SAR data. These data can only be generated by SAR systems designed to transmit interleaved orthogonal polarization pulses and receive pulses simultaneously and coherently in the linear horizontal (H) and vertical (V) polarizations. In a conventional quad-pol SAR system, the transmitted orthogonal polarizations are as well linear H and V [19]. Other options for the transmitted orthogonal pulses are the left (L) and right (R) circular polarizations or (H + V) and (H – V) polarizations, leading to the so-called hybrid and $\pm\pi/4$ polarimetric SAR, respectively [20]. Each combination of transmitted and received polarization represents a measurement called complex scattering amplitude, which collected in a 2×2 matrix forms the scattering matrix **S**. With **S** completely acquired in one single flight pass of the platform, the manner of how the target transforms the incident electric field vector into the received electric field vector becomes known and the target can be polarimetrically characterized [21]. For such reason, despite the higher system complexity of quad-pol SAR, most civilian SAR missions have been designed implementing this technique, boost that is consequence of the proven benefits in numerous applications in remote sensing including ship detection.

The transmission of interleaved pulses with different polarization requires that the radar PRF must be doubled. While the simultaneous reception of the H and V polarization implies the use of two antennas (or equivalently one physical antenna electronically split in halves), these technical bounds have direct consequences in the appearance of azimuth ambiguities in quad-pol SAR data. Due to the higher side-lobes and the single complex scattering amplitude sampled with lower frequency, the probability of having an ambiguous signal exceeding the un-ambiguous one is higher, that is, there is a worsening of the azimuth ambiguity to signal ratio (AASR). Moreover, the displacement in azimuth between the "ghosts" and the real target decreases due to the processing at lower PRF, increasing the possibilities of ghosting ships docked in harbors, coastal cities, and built-up areas in the nearby calm water area. In [22], the analysis of the AASR in each polarization channel for conventional, hybrid, and $\pm\pi/4$ quad-pol SAR configuration is presented with the aim of optimizing the design of the system proposing solutions that could be implemented in future quad-pol SAR missions. However, the scope of this chapter and Section 5.4 is to review only those techniques developed to mitigate azimuth ambiguities in post-processing and analysis of conventional quad-pol SAR data of currently flying missions.

The chapter is organized as follows. In Section 5.2, we derive the basic theoretical model of azimuth ambiguities. In Section 5.3, based on the introduced model, the main approaches for the mitigation of azimuth ambiguities in single-channel SAR images are described. We classify the techniques according to the type of targets for which they were originally developed, that is, point-like or distributed targets. In Section 5.4, special attention is provided to the topic of azimuth ambiguity filtering on polarimetric SAR images. Finally, Section 5.5 is devoted to summarize the content of the chapter.

5.2 Azimuth ambiguity modeling

We here consider a SAR sensor operating in stripmap mode (see Chapter 3). Additionally, we assume that the SAR image is obtained via standard processing (i.e., matched filtering), neglecting SAR system space variance. These assumptions are made to simplify the description of azimuth ambiguities: they can be safely relaxed without substantially changing the main aspects of the following discussion. Let $\widehat{\Gamma}(\xi,\eta)$ and $H(\xi,\eta)$ be the Fourier transforms (FTs) of the image, $\widehat{\gamma}(x,r)$, and of the raw signal, $h(x,r)$, with ξ and η being the Fourier mates of azimuth x and slant range r, respectively. The expression of $H(\xi,\eta)$ has been provided in Chapter 3. However, to study azimuth ambiguity is now necessary to account for the effect of azimuth-sampling on the raw signal. Therefore, we consider $H_s(\xi,\eta)$, thta is, $H(\xi,\eta)$ sampled in azimuth at rate PRF/v, where v is the uniform sensor velocity and PRF is the pulse repetition frequency. Accordingly, it can be written as

$$H_s(\xi,\eta) = \sum_i \Gamma(\xi_i,\eta)\exp\left[j\frac{\eta^2}{4b}\right]\exp\left[-j\frac{\xi_i^2}{4a\left(1+\frac{\eta\lambda}{4\pi}\right)}\right]\mathrm{rect}\left[\frac{\eta}{bc\tau}\right]w^2\left[\frac{\xi_i-\xi_{DC}}{2aX}\right] \tag{5.1}$$

where $X = \lambda r_0/L$ is the effective azimuth footprint, r_0 being the closest range between the platform and the center of the footprint, λ is the electromagnetic wavelength, and L is the actual azimuth antenna length; $a = \frac{2\pi}{\lambda r_0}$ and $b = \frac{2\alpha}{c^2}$, where α is the chirp rate and c the speed of light; τ is the transmitted pulse duration; $\xi_i = \xi - \frac{2\pi i \mathrm{PRF}}{v}$, with i an integer number; $\xi_{DC} = 2ar_0\tan\psi$, ψ being the sensor squint angle, accounts for the Doppler centroid; $w[\cdot]$ is the two-way azimuth antenna pattern (AAP), whose effective beamwidth is $\Delta\xi = 2aX = \frac{4\pi}{L}$; and $\Gamma(\cdot)$ is the Fourier transform of the scene reflectivity.

As detailed in Chapter 3, the final SAR image is obtained by focusing the raw signal through compensation of its phase terms, that is, multiplying $H_s(\xi,\eta)$ by the conjugate of the SAR transfer function (TF) $G_0(\xi,\eta)$, whose expression is

$$G_0(\xi,\eta) = \exp\left[j\frac{\eta^2}{4b}\right]\exp\left[-j\frac{\xi^2}{4a\left(1+\frac{\eta\lambda}{(4\pi)}\right)}\right]\text{rect}\left[\frac{\eta}{bc\tau}\right]w^2\left[\frac{\xi-\xi_{DC}}{2aX}\right] \tag{5.2}$$

Therefore, we obtain the following expression for the Fourier transform $\widehat{\Gamma}(\xi,\eta)$ of the SAR image $\widehat{\gamma}(x,r)$

$$\begin{aligned}\widehat{\Gamma}(\xi,\eta) &= H_s(\xi,\eta)\text{rect}\left[\frac{\xi-\xi_{DC}}{2aX}\right]\text{rect}\left[\frac{\eta}{bc\tau}\right]\exp\left(j\frac{\xi^2}{4a\left(1+\frac{\eta\lambda}{(4\pi)}\right)}\right)\exp\left(-j\frac{\eta^2}{4b}\right)\\ &= \Gamma(\xi,\eta)w^2\left[\frac{\xi-\xi_{DC}}{2aX}\right]\text{rect}\left[\frac{\xi-\xi_{DC}}{2aX}\right]\text{rect}\left[\frac{\eta}{bc\tau}\right]\\ &\quad+\sum_{i\neq 0}\Gamma(\xi_i,\eta)\exp\left(\frac{\xi_i^2-\xi^2}{4a\left(1+\frac{\eta\lambda}{(4\pi)}\right)}\right)w^2\left[\frac{\xi_i-\xi_{DC}}{2aX}\right]\text{rect}\left[\frac{\xi-\xi_{DC}}{2aX}\right]\\ &\quad\times\text{rect}\left[\frac{\eta}{bc\tau}\right]\end{aligned} \tag{5.3}$$

Let's now analyze the terms present in (5.3). The first term accounts for the baseband focused contribution of the target spectrum, which has already been discussed in Chapter 3. The presence of ambiguities is described by the second term, which arises from the superposition of the spectral replicas of the reflectivity that, due to aliasing, fold into the processed bandwidth. Indeed, usual SAR systems are designed to satisfy the relation $\frac{2\pi\text{PRF}}{v}\geq\Delta\xi=2aX$, that is, $\text{PRF}\geq\frac{2v}{L}$, where $2v/L$ is recognized to be the Nyquist rate of the signal: in this way, the under-sampling of the signal effective azimuth bandwidth is avoided. Therefore, if $w[\xi]$ were null for $|\xi|>\Delta\xi$ (i.e., no sidelobes), we would have $w\left[\frac{\xi_i-\xi_{DC}}{2aX}\right]\text{rect}\left[\frac{\xi-\xi_{DC}}{2aX}\right]=0$ for $i\neq 0$, and no azimuth ambiguity would be present. However, actual antenna patterns can only approximately meet this condition if an antenna with a high peak-to-sidelobe ratio is used, without exactly achieving it. Therefore, azimuth ambiguity is always present even if it is negligible in most practical situations. However, in areas where the baseband focused contribution of the target spectrum (the "true" signal) is very low (e.g., the sea surface, which is here of interest), replicas due to AAP sidelobes may reach intensities higher than those of the target. In this situation, "ghost" targets appear.

To investigate the characteristics of the ambiguity, let's now examine the phase term in the summation in (5.3). Substituting the expressions of ξ_i and a reported above and assuming that $\left(1+\frac{\eta\lambda}{4\pi}\right)^{-1}\simeq\left(1-\frac{\eta\lambda}{4\pi}\right)$, we can write

$$\begin{aligned}\exp\left(\frac{\xi_i^2-\xi^2}{4a}\right) &= \exp\left(j\xi\frac{i\text{PRF}\lambda r_0}{2v}\right)\exp\left(-j\frac{(i\text{PRF})^2\lambda r_0 2\pi}{4v^2}\right)\\ &\quad\times\exp\left(-j\xi\eta\frac{i\text{PRF}\lambda^2 r_0}{8\pi v}\right)\exp\left(j\eta\frac{(i\text{PRF})^2\lambda^2 r_0}{8v^2}\right).\end{aligned} \tag{5.4}$$

The first phase term in (5.4) is responsible for the azimuth displacement of ith order azimuth ambiguities, given by

$$\Delta x_i = i\frac{\text{PRF}\lambda r_0}{2v} \tag{5.5}$$

which, if the PRF is equal to the Nyquist rate of the azimuth signal, that is, PRF $= 2v/L$, reduces to $\Delta x_i = iX$, that is, the azimuth displacement is an integer multiple of the azimuth footprint. The second exponential term is an inessential constant phase term, whereas the last one provides the range displacement of the ith order azimuth ambiguity, given by

$$\Delta r_i = \frac{\Delta x_i^2}{2r_0} \tag{5.6}$$

where the expression in (5.5) has been used. Finally, the third phase term accounts for a slight defocusing of the replicas and a residual wavenumber-dependent migration. Since this third phase term is proportional to the square of the wavelength, the defocusing of the replicas will be more severe at low frequencies. This partly justifies the fact that azimuth ambiguity is more evident on images obtained by high-frequency (i.e., X-band) sensors. However, this is also due to the increased data dynamics obtained for higher frequencies, implying that point and man-made targets will appear much brighter than sea clutter.

Summarizing, in the presence of low-reflectivity areas (such as the sea surface), ghost targets may appear. They are replicas of high-reflectivity targets displaced according to (5.5) and (5.6), weighted according to the sidelobes of the AAP, and slightly defocused. Note that even if in (5.2)–(5.6) the spectral replicas of any order are considered, replicas of order higher than one are negligible in most practical cases.

5.3 Azimuth ambiguity mitigation in single channel SAR images

The presence of azimuth ambiguity has a negative impact on SAR maritime applications: especially in ship detection, it may imply an unacceptable increase of the false-alarm rate. Several mitigation strategies have been proposed to face the problem of ghost appearance in SAR images of the sea surface. Some of them have been developed to address the problem of ambiguities due to point-like targets: these techniques are especially well-suited for ships' ghosts in low-resolution SAR images. Indeed, in this situation, ships can be mostly considered as point-like targets. In the more general case of distributed targets, which may arise for ambiguities of ships in high-resolution SAR images and of bright structures present on the land close to the coast, techniques based on the concept of selective filtering have been advocated.

5.3.1 Point-like targets

This type of techniques have been originally developed based on the point-target model of azimuth ambiguities. Therefore, the best performance is obtained in case of point targets, for example, ships on low-resolution SAR images. Nonetheless, the considered approach can be somehow extended in a wider set of actual situations, for example, in the case of high-resolution images, where most ships lose their point-like behavior. However, in these situations, the performance of these methods may be significantly degraded and/or may become less prone to automatization. Within this category, we may include techniques based on two basic strategies.

The first one makes use of (5.5) and (5.6), that is, on the knowledge of the displacement of the replicas. Most of the available ship-detection software includes this kind of approach in the discrimination step, that is, as a post-processing step following the main prescreening ship-detection step. The vessel detector software SUMO developed by the European Commission Joint Research Centre, which has been recently made available as open source software, iteratively checks if there is a brighter target at the known azimuth ambiguity displacement at each side of a detected target, and in this case, the considered detection is flagged as a potential false alarm [14,23]. This flag is jointly used with other detection quality indicators (e.g., length and width of the target) to eventually keep or discard each target. To overcome the problem that frequently ships may not be considered as point targets, note that the detection here is relevant to the brighter pixel of the ship. Note that this procedure, which in its automatized version does not take into account the shape of targets and replicas, may be prone to errors, especially in the unfortunate case of actual bright objects placed in the same position of target replicas: in this case, actual targets may be wrongly flagged as ghosts. However, this problem is significantly mitigated by the inclusion of several different quality indicators in the discrimination step.

A more advanced version of the basic displacement-based approach is proposed in [15], where a binary map is evaluated according to

$$B(x,r) = \begin{cases} 1, \ if \left|\widehat{\gamma}_{\mu}(x,r)\right|^2 > \xi_i \left|\widehat{\gamma}_{\mu}(x+\Delta x_i, r+\Delta r_i)\right|^2 \\ 0, \text{if otherwise} \end{cases} \tag{5.7}$$

where the subscript μ highlights the fact that the image has been preliminarily processed via an averaging kernel and the value of ξ_i can be selected thanks to the availability of the AAP, being related to the AASR. It is assumed that only the ambiguity due to one of the AAP sidelobes is dominant in a single image pixel, which is a reasonable assumption in most practical situations. The pixels for which the mask assumes zero value are considered as ambiguity points. Once the ambiguity areas are identified, two different approaches are selected to process and restore the ambiguous areas, according to the size in pixels of the involved area. If the region is sufficiently small a compressive imaging technique is adopted, whereas for larger clustered regions exemplar-based inpainting is used. Also for this approach, the presence of actual weak targets in strong ambiguous regions may lead to ghost detection errors, even if the introduction of the weight parameter ξ_i helps in reducing this kind of situation. Moreover, if the weak target is not completely masked by the ambiguity, the restoration step can succeed (at least, partially) in recovering the target. Note that the displacement-based approaches can be applied to detected data, which represents a relevant advantage of this class of methods.

Another basic strategy originally developed to mitigate point-like targets' ambiguity is based on in-phase cancellation and includes a deconvolution step based on an appropriately designed reference ambiguity function. This approach was proposed in [7], where according to a signal model equivalent to that of (5.3), an ideal filter was defined, allowing to obtain an unambiguous estimate of point-target backscattering. The filter pursues the goal of perfectly compensating the phase and amplitude errors introduced in standard focusing by the presence of ambiguities. The ideal filter can be expressed as

$$R_{\text{ideal}}(\xi,\eta) = \frac{\left| w^2 \left[\frac{\xi - \xi_{DC}}{2aX} \right] \right|^2}{H_s'(\xi,\eta)} \tag{5.8}$$

where $H_s'(\xi,\eta)$ is the raw signal of a point target, given by (5.1) assuming that $\Gamma(\xi_i,\eta) = 1$. Evaluation of the expression in (5.8) must be limited to the processed bandwidth. To generate R_{ideal} a very precise knowledge of the radar Doppler parameters is necessary to obtain adequate estimates of the phase terms present in (5.3): in particular, the required accuracy is significantly higher than that necessary for standard matched filter processing. Nonetheless, the necessary accuracy can be usually achieved. Even if the ideal filter derivation is performed assuming a point target, the filter can be also applied to distributed targets, resulting in a degradation of the attainable performance [7]. The technique developed in [16] is based on similar concepts: in this case, an accurate model of first-order ambiguities (i.e., those obtained setting $i = \pm 1$ in (5.3)) is introduced, leading to the definition of the azimuth ambiguity factor, which is then used to reconstruct the ambiguities starting from the original target. The technique is able to take into account the space-variant behavior of the ambiguities and has been demonstrated to be better suitable for distributed targets than the one based on the ideal-filter approach [16]. Note that both these techniques require as input an SLC SAR image.

5.3.2 Distributed targets

The main approach for the filtering of azimuth ambiguity due to distributed targets is based on the concept of selective filtering, which imposes a trade-off between ambiguity mitigation and azimuth resolution [4]. Indeed, as illustrated in Figure 5.1, AAP sidelobes are folded within the azimuth processed bandwidth, leaving an area around the folded AAP null (highlighted in red in Figure 5.1) where ambiguity energy is close to zero. A notch filter, with center frequency corresponding to these nulls, can be devised to band-pass filter the focused SLC SAR image, so that the spectral components most affected by ambiguity are discarded at the expense of a reduction in azimuth resolution (the narrower the bandwidth of the filter, the larger the resolution loss). To locate the nulls and synthesize the filter, precise knowledge of SAR system parameters and, in particular, of the AAP is required (Figure 5.1).

To minimize the loss in geometric resolution, adaptive filtering strategies may be deployed. In [8], a procedure to design the filter based on a Wiener minimum mean squared error (MMSE) solution is presented. In (5.3), the spectral replicas of any order are considered. However, replicas of order higher than one are most often negligible, so that we limit our attention to first-order replicas only, that is, those

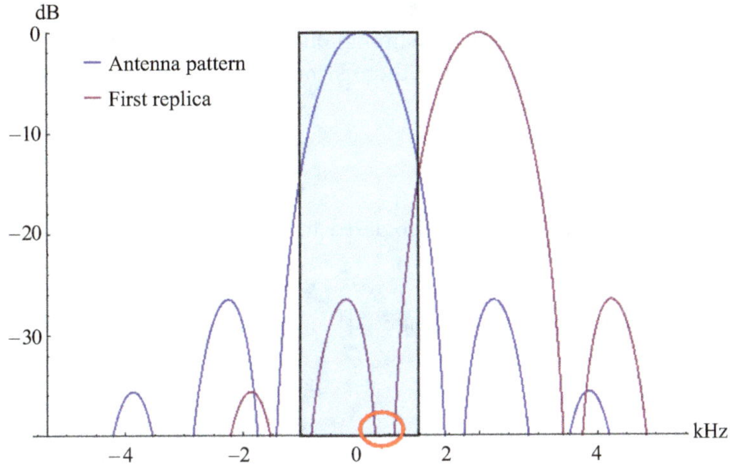

Figure 5.1 Illustration of AAP folding in the processed bandwidth. The null of the folded AAP is circled in red

due to the first folded sidelobe segments of the spectrum, obtained by setting $i = \pm 1$. Substituting (5.4) in (5.3), we get

$$\widehat{\Gamma}(\xi,\eta) = \Gamma_0(\xi,\eta)W_0[\xi,\eta] + \Gamma_+(\xi,\eta)W_+[\xi,\eta] + \Gamma_-(\xi,\eta)W_-[\xi,\eta] \tag{5.9}$$

where

$$W_0[\xi,\eta] = w^2\left[\frac{\xi-\xi_{DC}}{2aX}\right]\mathrm{rect}\left[\frac{\eta}{bc\tau}\right]\mathrm{rect}\left[\frac{\xi-\xi_{DC}}{2aX}\right] \tag{5.10}$$

$$\begin{aligned} W_\pm[\xi,\eta] = w^2 & \left[\frac{\left(\xi \mp \dfrac{2\pi\mathrm{PRF}}{v}\right)-\xi_{DC}}{2aX}\right]\mathrm{rect}\left[\frac{\eta}{bc\tau}\right]\mathrm{rect}\left[\frac{\xi-\xi_{DC}}{2aX}\right] \\ & \times \exp\left(\pm j\xi\frac{\mathrm{PRF}\lambda r_0}{2v}\right)\exp\left(-j\frac{\mathrm{PRF}^2\lambda r_0 2\pi}{4v^2}\right)\exp\left(\mp j\xi\eta\frac{\mathrm{PRF}\lambda^2 r_0}{8\pi v}\right) \\ & \times \exp\left(j\eta\frac{\mathrm{PRF}^2\lambda^2 r_0}{8v^2}\right) \end{aligned} \tag{5.11}$$

and Γ_0, Γ_+, and Γ_- are $\Gamma(\xi_i,\eta)$ for $i = 0, +1, -1$, respectively, that is, the reflectivities of the unambiguous and ambiguous targets.

The theory of Wiener filters is used to evaluate the TF of a filter which, through the use of N azimuth samples of the image, is able to filter out the ambiguities, minimizing the mean square error between the "true" unambiguous signal and the filtered one. Note that we can consider a one-dimensional problem, treating separately each fixed-range line of the SLC SAR image. For this reason, the dependence on the η variable can be omitted. Starting from the one-dimensional version of (5.9), we can define the unambiguous signal $\widehat{\Gamma}_0$ and the ambiguous data X, also affected by an additive white thermal noise T, as

$$\widehat{\Gamma}_0(\xi) = \Gamma_0(\xi)W_0[\xi] \tag{5.12}$$

$$X(\xi) = \Gamma_0(\xi)W_0[\xi] + \Gamma_+(\xi)W_+[\xi] + \Gamma_-(\xi)W_-[\xi] + T(\xi) \tag{5.13}$$

The power spectral densities associated to the unambiguous and ambiguous targets can be evaluated as follows:

$$S_0(\xi) = \mathrm{E}\left[\widehat{\Gamma}_0[\xi]\widehat{\Gamma}_0^*[\xi]\right] = \mathrm{E}\left[|\Gamma_0[\xi]|^2\right]|W_0[\xi]|^2 = \sigma_0|W_0[\xi]|^2 \tag{5.14}$$

$$S_\pm(\xi) = \mathrm{E}\left[|\Gamma_\pm[\xi]|^2\right]|W_\pm[\xi]|^2 = \sigma_\pm|W_\pm[\xi]|^2 \tag{5.15}$$

where $\mathrm{E}[\cdot]$ stands for "statistical mean," * stands for complex conjugate, and σ_0, σ_+, and σ_- are the power spectral densities of the reflectivity relevant to the unambiguous and to the two ambiguous targets, respectively, which can be considered to be independent of the azimuth wavenumber ξ in the processed bandwidth. In writing, the expressions in (5.14)–(5.15), it is assumed that the reflectivity is statistically stationary within the filter spatial extent. In this case, the expression of the Wiener filter is [8,9]:

$$\begin{aligned} H(\xi) &= \frac{\mathrm{E}\left[\widehat{\Gamma}_0(\xi)X^*(\xi)\right]}{\mathrm{E}[X(\xi)X^*(\xi)]} = \frac{S_0(\xi)}{S_0(\xi) + S_+(\xi) + S_-(\xi) + \sigma_t} \\ &= \left(1 + \frac{\sigma_+}{\sigma_0}\frac{|W_+(\xi)|^2}{|W_0(\xi)|^2} + \frac{\sigma_-}{\sigma_0}\frac{|W_-(\xi)|^2}{|W_0(\xi)|^2} + \frac{\sigma_t}{\sigma_0}\frac{1}{|W_0(\xi)|^2}\right)^{-1} \end{aligned} \tag{5.16}$$

where σ_t is the white thermal noise power density and it has been assumed that ambiguous and unambiguous data are zero-mean and uncorrelated. The ratios σ_+/σ_0 and σ_-/σ_0 represent the unknown local AASR and need to be estimated. In [8], starting from the result in (5.16), it is assumed as a worst case that these ratios are equal and very large, that is, $\sigma_+ = \sigma_- = \sigma \gg \sigma_0$. In this case, we get

$$H(\xi) = \frac{\sigma_0}{\sigma}\left(\frac{|W_+(\xi)|^2 + |W_-(\xi)|^2}{|W_0(\xi)|^2} + \frac{\sigma_t}{\sigma}\frac{1}{|W_0(\xi)|^2} + \frac{\sigma_0}{\sigma}\right)^{-1} \tag{5.17}$$

where the sum of the last two terms, even if it is very small (it can be assumed to be about −60 dB), is considered to avoid singularities in the filter TF, occurring whenever the first term approaches zero. In [8], the filter in (5.17) is applied to the SLC SAR image to obtain a coarse estimate of σ_0, which represents a necessary preliminary step for the identification of areas affected by ambiguities. In these areas, the image filtered using the filter defined in (5.17) can be used to get an estimate of σ_+/σ_0 and σ_-/σ_0. As an alternative, more advanced techniques developed for the retrieval of the AASR can be devised, for example, the spectral-based technique proposed in [24]. Once the ratios have been estimated, they can be then

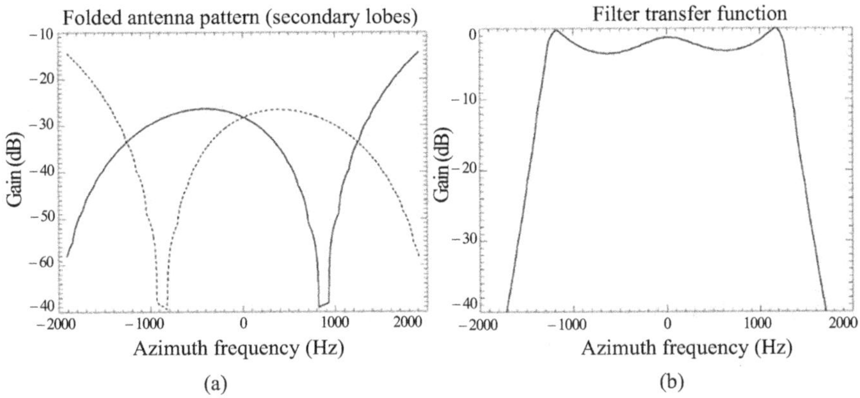

Figure 5.2 First sidelobes of the AAP folded into (a) the processed azimuth bandwidth and (b) filter TF of (5.17), in the case of TerraSAR-X data

used in (5.16) to generate the adaptive filter. The use of this filter dictates an optimal trade-off between ambiguity mitigation and resolution loss. However, the assumption of equal power for both ambiguities used to derive the filter in (5.17) may imply that this filter is not adequately selective in the frequency domain, thus providing poor filtering results in several practical situations. This occurs whenever the peak of the left (or right) folded AAP sidelobe locates almost exactly in correspondence of the null of the right (or left) one. In [9], it was verified that this may happen for many current high-resolution spaceborne SAR sensors, for example, TerraSAR-X and Cosmo/SkyMed. To illustrate this phenomenon, in Figure 5.2(a), the first sidelobes of the AAP folded into the azimuth processed bandwidth are shown, while in Figure 5.2(b), the filter TF resulting from the definition in (5.17) is presented. It can be noted that the filter TF obtained through (5.17) is almost uniform in the processed bandwidth. Whenever a situation as the one depicted in Figure 5.2 occurs, the algorithm proposed in [8] fails to identify the ambiguity areas.

To overcome the problems of the technique in [8] for high-resolution sensors in [9], a different filtering approach was proposed, based on the assumption that in each pixel only the ambiguity due to one of the folded sidelobes is present, its power being much larger than the power of the unambiguous signal. Therefore, two filters are generated in this case: either H_+ assuming $\sigma_+ \gg \sigma_0$ and $\sigma_- = 0$ or, conversely, H_- assuming that $\sigma_- \gg \sigma_0$ and $\sigma_+ = 0$:

$$H_\pm(\xi) = \frac{\sigma_0}{\sigma_\pm}\left(\frac{|W_\pm(\xi)|^2}{|W_0(\xi)|^2} + \frac{\sigma_t}{\sigma_\pm}\frac{1}{|W_0(\xi)|^2} + \frac{\sigma_0}{\sigma_\pm}\right)^{-1} \tag{5.18}$$

The filters described by the TF in (5.18) are named "asymmetric filters." Processing an SLC SAR image with both these filters, two filtered images are obtained: in each of these, the ambiguities due to one different AAP sidelobe are filtered out. In this way, it is possible to identify separately ambiguities due to each sidelobe, that is, due to interfering sources located on both sides of the considered target. In Figure 5.3, the filter TFs H_+ and H_- obtained using the same TerraSAR-

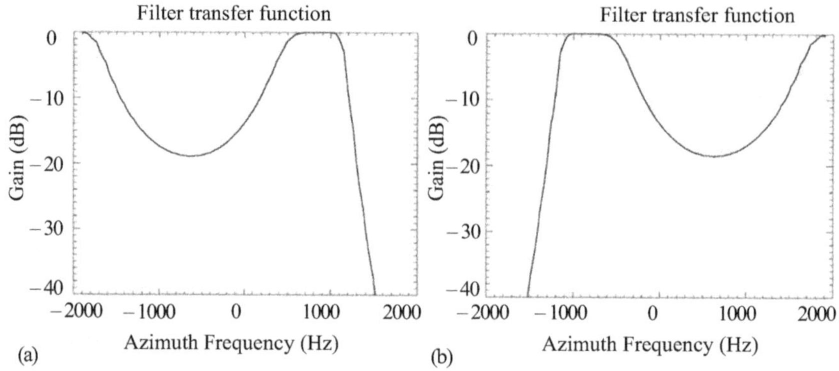

Figure 5.3 Filter TFs (a) H_+ and (b) H_-, in the case of TerraSAR-X data

X AAP considered in Figure 5.2 are shown: it can be noted that the frequency selectivity is significantly increased if compared to that of the filter reported in Figure 5.2(b). Note that, in the limiting case where a superposition in the same area of azimuth ambiguities due to both sidelobes is present, the approaches of [8] and [9] fail to properly filter out the ambiguity. According to the Asymmetric Mapping and Selective Filtering (AM&SF) approach described in [9], the two images $\widehat{\Gamma}_+$ and $\widehat{\Gamma}_-$ obtained using the filters H_+ and H_-, respectively, can be used to generate two ambiguity masks:

$$b_\pm(x,r) = \begin{cases} 1, & r_\pm(x,r) > r_t \\ 0, & r_\pm(x,r) \le r_t \end{cases} \tag{5.19}$$

where $r_\pm(x,r)$ represent properly normalized ratios between the original SLC SAR image and multilook averaged versions of $\widehat{\Gamma}_+$ and $\widehat{\Gamma}_-$, respectively, and r_t is a real number greater than one ($r_t \simeq 2$ in most practical situations). A final ambiguity-free image can be obtained by substituting in those areas of $\widehat{\Gamma}$ where $b_\pm = 1$, the appropriately normalized version of $\widehat{\Gamma}_\pm$, respectively. In this way, the azimuth ambiguities are filtered out and the geometric resolution loss is limited to the areas actually affected by ambiguities.

In Figure 5.4, azimuth ambiguity filtering results obtained with the AM&SF technique of [9] are presented. In particular, it can be noted that a large ship present in the area strongly affected by ambiguities is retained after filtering, whereas the ghosts are effectively removed. Finally, note that both the techniques of [8] and [9] have been also positively tested in the case of interferometric applications.

Both the above-discussed filters have been designed for stripmap SAR data. More recently, in [17], a technique based on similar spectral filtering concepts was developed for the more general case of acquisition modes with variant AAP weighting: in particular, the proposed technique was tested on sliding spotlight TerraSAR-X data. The applicability of this approach to Sentinel-1 TopSAR data (rather than its use for spotlight data) makes it relevant for maritime applications. Basically, in [17], a Wiener filter is generated, which takes into account the change in the AAP weighting during the acquisition time, as illustrated in Sections 3.3 and 3.5 for the sliding spotlight and TopSAR cases, respectively. This implies that, at the

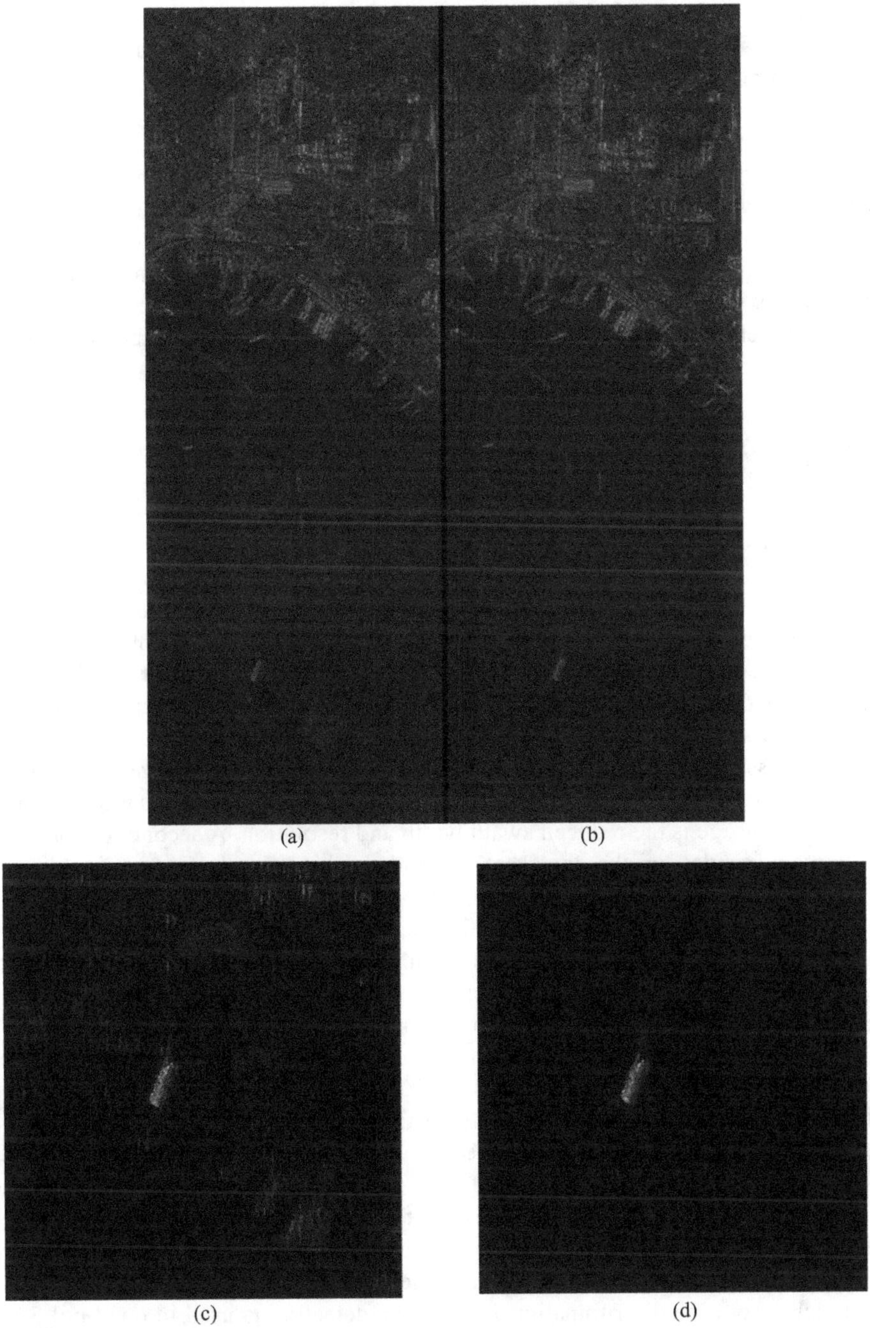

Figure 5.4 *Cosmo/SkyMed image of the Gulf of Naples (a) and AM&SF filtered version (b). Ambiguities due to the built-up areas in the upper part of the image are visible all over the sea surface in (a), while they are strongly attenuated in (b). Details of the image in the ambiguity areas: original (c) and filtered (d) image*

difference with the stripmap case, the filter will be azimuth variant. As a further step, in [17], the authors, as a way to limit resolution loss, implement an extrapolation strategy to reconstruct those parts of the image spectrum affected by ambiguities.

Another approach particularly well suited to remove land-generated azimuth (or range) ambiguities from detected SAR images of the ocean is based on the use of multitemporal series of SAR images. In this situation, a map of the ghosts can be obtained by computing the local correlation coefficient between couples of images in the time series and defining an appropriate threshold (e.g., based on maximum entropy) [18]. Indeed, highly correlated distributed permanent targets over the sea may be regarded with a high degree of confidence as ambiguities. After the identification step, in [18], an exemplar-based inpainting method is used to reconstruct the ambiguity-affected areas.

5.4 Azimuth ambiguity mitigation in polarimetric SAR images

Quad-pol SAR data have been benchmarked in the framework of the ship detection in order to quantify the relative improvement of the performance. In [25], airborne data have been processed to accomplish such task and the presented results prove that adding up information, for example, channels amplitude and phase, is beneficial in detecting the ships. Following this line of research, much effort has been devoted to developing complex ship detection algorithms that squeeze all the possible information content of quad-pol SAR data [26–28]. Although the proven effectiveness and maturity of the algorithms, the operational usage of quad-pol SAR data from space for ship monitoring is still limited. The obvious motivation of this limited usage is the reduced swath width and resolution. A second less obvious motivation resides in the fact that azimuth ambiguities become more severe implying an unacceptable increase of the false-alarm rate. In fact, the system has to operate with twice PRF to collect the four combinations of polarization composing the scattering matrix **S**, meaning that individual complex scattering amplitudes are sampled with half frequency. The worsening in the azimuth ambiguity caused by the halved sampling frequency is depicted in Figure 5.5, comparing side by side a subregion of SAR data collected almost simultaneously by the twin satellites TerraSAR-X (Figure 5.5(a)) and TanDEM-X (Figure 5.5(b)) operating with the two different sampling settings and flying over the region with a time lag of 10 s.

Several mitigation strategies have been proposed to face the problem of the azimuth ambiguity in quad-pol SAR data in order to reduce the false alarms in the ship detection application. Most of these methodologies share the common sense of generating a new image space, free of azimuth ambiguities, or where they are strongly attenuated, prior to performing the detection step. Indeed, the classical approach where the discrimination follows the detection is used in quad-pol SAR data as well. More in general, the approaches reviewed in this paragraph can be grouped into two types:

- Methods based on polarimetric analysis.
- Methods based on relation between channels.

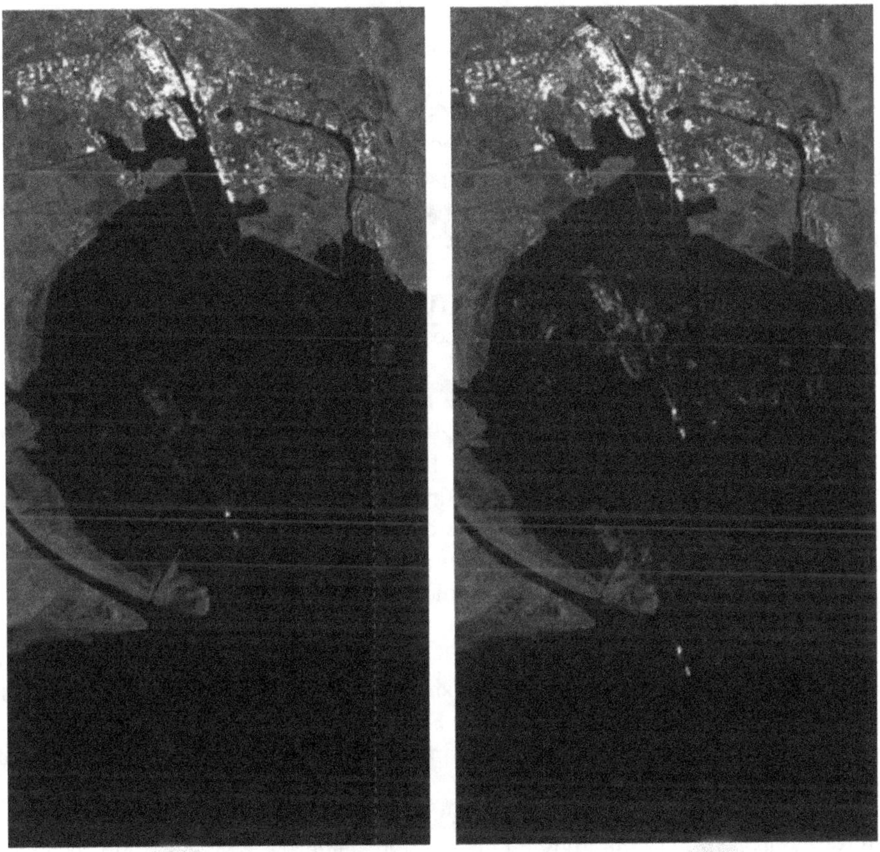

(a) (b)

Figure 5.5 Zoom-in of the single channel TerraSAR-X image of the gulf of Trieste processed with an azimuth bandwidth of 2765 Hz (a) and zoom-in of one polarization channel TanDEM-X image of the gulf of Trieste processed with an azimuth bandwidth of 1380 Hz (b). Ambiguities due to the built-up areas and ships at dock become: (1) more prominent, increasing the false detections; and (2) closer to the source targets, disrupting the image information close to the coast. © DLR (2017)

5.4.1 Method based on polarimetric analysis

The analysis of the different types of scattering mechanisms of targets and their ambiguities is proposed as discriminating strategy in [29,30]. Both proposed methods are making use of the full scattering matrix **S**. The polarimetric analysis used in [29] is based on the decomposition of the coherency matrix, while in [30], the Time-Frequency decomposition is adopted.

The base of the algorithm proposed in [29] is the distinguishable scattering mechanisms of the target and its azimuth ambiguities exploiting the eigenvalue–eigenvector decomposition method. Wang *et al.* [29] argue that azimuth ambiguities are mainly occurring when the real targets are characterized by an HH-VV phase difference of 180°, which means that in quad-pol data only very bright single or double-bounce scatterers will produce ambiguities that look like fairly bright double- or single-bounce scatterers. Ships are generally characterized by a mixture of single-bounced, double-bounced, and depolarized scattering mechanisms. Therefore, the eigenvalues corresponding to the single- or double-bounce scattering mechanisms of the ambiguities are high, but the eigenvalue corresponding to the depolarized scattering mechanisms of the ambiguities is low. Based on this rationale, Wang *et al.* [29] propose a three-steps approach for the removal of azimuth ambiguities and ship detection. The first step is to calculate the coherency matrix with optional despeckling if necessary. The second step is to obtain the eigenvalues λ_1, λ_2, and λ_3, where $\lambda_1 > \lambda_2 > \lambda_3$, decomposing the 3×3 coherency matrix. The third step is to compute the local homogeneity of the λ_3 matrix and perform a threshold-based detection. In [29], several experiments based on airborne quad-pol data are shown to demonstrate the actual capabilities of λ_3 in hampering the signal of azimuth ambiguity. In Figure 5.6, an example of the approach proposed in [29] is shown, applied on space-borne X-band quad-pol data acquired by TerraSAR-X over the San Francisco Bay, USA. Figure 5.6(a) displays the UTM map-projected data-take in Pauli RGB color coding. Figure 5.6(b) is the sub-scene highlighted by the dashed white rectangle in Figure 5.6(a) showing the HH polarization intensity, and Figure 5.6(c) is the results of the eigenvalue–eigenvector decomposition of the coherency matrix showing the smallest eigenvalues, that is, λ_3. It is easily recognizable comparing Figure 5.6(b) and Figure 5.6(c) that the signals of the azimuth ambiguity present in the sub-scene are much reduced in their intensity.

In [30], the coherent Time-Frequency analysis of quad-pol SAR data, originally introduced in [31] for studying anisotropic scattering behaviors and for characterizing urban environment, has been adopted for discriminating ships in complex backgrounds, that is, ocean area affected by ghost target and/or processing artifacts. Hu *et al.* [30] argue that ambiguities in azimuth or range directions are causing ghost echoes, which besides being located at a different spatial position and being misfocused, may result in a strong modification of the observed polarimetric features. Being the coherent Time-Frequency technique able to estimate spectral properties around specific spatial locations or equivalently spatial properties around specific spectral locations, Hu *et al.* [30] proposed such processing method of quad-pol SAR data to discriminate ships from their background by detecting and extracting coherent components from the complex random SAR responses. As a matter of fact, Hu *et al.* [30] came up with a Time-Frequency coherence indicator where real targets show high coherence over different spectral locations, whereas azimuth ambiguities do not. The Time-Frequency coherence indicator can be built using three different decompositions strategies of the 2D spectrum, namely azimuth only, range only, and both simultaneously. These decomposition methods produce an unavoidable loss of spatial or spectral resolution. In [30], all three decomposition methods are tested against effectiveness in hampering azimuth ambiguities. The experiments conducted on C-band quad-pol SAR data acquired by the space-borne RADARSAT-2 show that using both directions simultaneously is performing best, while range only could be used in alternative to save computational time.

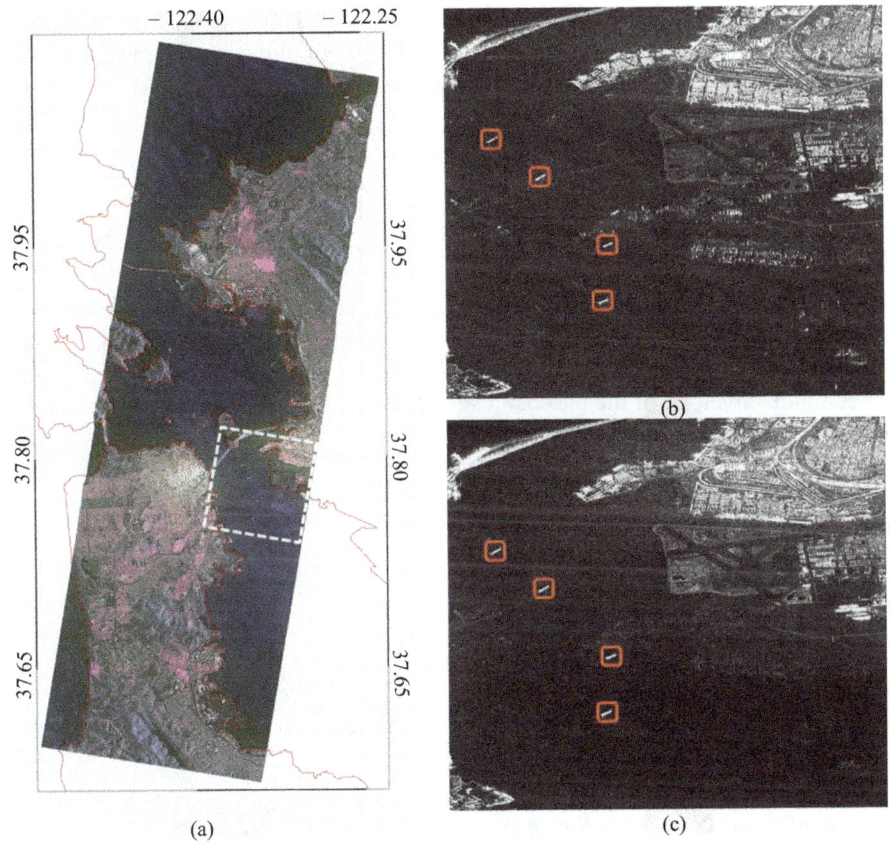

Figure 5.6 *Explanatory example of the mitigation procedure proposed in [29] applied to TerraSAR-X quad-pol data. (a) Data take of San Francisco Bay in Pauli RGB color composite, (b) sub-scene of the HH intensity channel, and (c) smallest eigenvalue derived from the target decomposition method. © DLR (2015)*

5.4.2 Methods based on relation between channels

In order to introduce the methods based on the relation between the polarimetric channels composing the SAR quad-pol data, it is worth to review the SAR point target signals for monostatic polarimetric SAR systems. In particular, the methods reviewed in this subsection exploit the relation between the cross-polarized channels, namely HV and VH. Assuming a conventional quad-pol SAR system able to transmit interleaved pulses with orthogonal polarizations and to receive both linear H and V polarization at the same antenna position (monostatic configuration) and that the first transmitted pulse is H-polarized, then HH and HV scattering amplitudes are measured at the same time, whereas the VH and VV scattering amplitudes are still measured at the same time but delayed of half the pulse repetition interval (PRI) due to the change of transmitting polarization to V. The radar moving at velocity v acquires a range line whenever it travels the distance $v \cdot \text{PRI}$, therefore the nth HV sample has azimuth position $x_n = v(n \cdot \mathrm{PRI})$, while the nth VH sample has azimuth position $x_n' = x_n + (v \cdot \mathrm{PRI}/2)$. The SAR point target signal for the HV and VH polarization at these azimuth positions are given by

$$s_{HV}(\tau', x_n) = A_w \cdot L_c \cdot S_{hv} \cdot g\left[\tau' - \frac{2}{c}s(x_n)\right] \cdot \exp\left[-i\frac{4\pi}{\lambda}s(x_n)\right] \tag{5.20}$$

$$s_{VH}\left(\tau', x_n'\right) = A_w \cdot L_c \cdot S_{vh} \cdot g\left[\tau' - \frac{2}{c}s\left(x_n'\right)\right] \cdot \exp\left[-i\frac{4\pi}{\lambda}s\left(x_n'\right)\right] \tag{5.21}$$

In (5.20)–(5.21), τ' is known as fast time (the time elapsed from each pulse transmission), A_w is the antenna pattern weighting, in elevation and azimuth direction, L_c is a factor including propagation losses and system calibration, $g(\cdot)$ is the complex envelope of the baseband signal, S_{hv} and S_{vh} are the radar cross section (RCS) coefficients, and $s(\cdot)$ is the time domain function giving the distance between the target and the radar position.

The co-registration of the polarimetric channels is obtained by a linear interpolation as following:

$$s_{\mathrm{VH}}(\tau', x_n) \cong \sum_p k_p \cdot s_{\mathrm{VH}}\left(\tau', x_n + p\frac{v\mathrm{PRI}}{2}\right) \tag{5.22}$$

where k_p is the interpolation coefficient for each odd index $p \in \{1, 3, 5, \ldots\}$. Approximating the quadratic function $s(\cdot)$ with the first two terms of its Taylor series, (5.22) becomes:

$$s_{\mathrm{VH}}(\tau', x_n) \cong A_w \cdot L_c \cdot S_{\mathrm{vh}} \cdot g\left[\tau' - \frac{2}{c}s(x_n)\right] \cdot \exp\left[-i\frac{4\pi}{\lambda}s(x_n)\right] \sum_p k_p \times \exp\left[-i\frac{2\pi}{\lambda} \cdot \mathrm{pvPRI} \cdot \frac{\delta}{\delta x_n}s(x_n)\right] \tag{5.23}$$

with $\delta/\delta x_n$ the partial derivative operator. Under the validity of the reciprocity theorem of monostatic configuration ($S_{\mathrm{hv}} = S_{\mathrm{vh}}$), that is, the scattering medium does not change the irradiated power interchanging the positions of the transmitter and receiver, comparing (5.23) with (5.20) and properly manipulating the exponential with trigonometric function, it follows that

$$s_{VH}(\tau', x_n) \cong s_{HV}(\tau', x_n) \sum_p 2k_p cos\left[\frac{2\pi}{\lambda} \cdot pvPRI \cdot \frac{\delta}{\delta x_n}s(x_n)\right] \tag{5.24}$$

For a target located zero Doppler frequency, the derivative of $s(\cdot)$ is null by definition which implies that the cosine argument in (5.24) is zero and $s_{\mathrm{VH}}(\tau', x_n) \cong s_{\mathrm{HV}}(\tau', x_n)$ as expected by the reciprocity property. Azimuth ambiguities are displaced in azimuth direction by multiple of the PRF as in (5.5). Limiting for simplicity, the case at the first order ambiguity and indicating the positive and negative location as $x_A^{\pm}$, corresponding to the positive and negative

Doppler frequency equals to $\pm$PRF, such that the phase shift between adjacent samples is $\pm 2\pi$, and the argument of the cosine function in (5.24) is

$$\frac{4\pi}{\lambda} \cdot v\text{PRI} \cdot \frac{\delta}{\delta x_A} s(x_A) = \pm 2\pi \rightarrow \frac{2\pi}{\lambda} \cdot v\text{PRI} \cdot \frac{\delta}{\delta x_A} s(x_A) = \pm \pi \tag{5.25}$$

leading to the conclusion that $s_{\text{VH}}(\tau, x_n) \cong -s_{\text{HV}}(\tau, x_n)$ for $x_n = x_A^{\pm}$. The relation between the cross-polarized scattering amplitude channels can be summarized as follows:

- HV and VH channels are approximately equal in magnitude and phase for targets' pixels.
- HV and VH channels are each other's opposite for azimuth ambiguity pixels.

This property of the cross-polarized channels has been exploited firstly in Liu and Gierull [32] for detecting moving target in quad-pol SAR data. Moreover, in [32], it is also pointed out that the phase interferogram is useful to discriminate azimuth ambiguities from real targets in maritime environment. The phase interferogram is retrieved from the two cross-polarized scattering amplitude channels of quad-pol SAR data as follows:

$$\varphi_{XPol} = \angle(\text{HV} \cdot \text{VH}^*) \tag{5.26}$$

The approach proposed in [32] is to perform the discrimination step after the detection of potential targets at sea. The likelihood ratio test (LRT) detector developed in [33] is used to analyze the quad-pol SAR data at hand and retrieve the location of potential ships in the area. Liu and Gierull [32] are pointing out in their experiments that in general, the ambiguities are also detected along with the targets, even though the decision threshold derived from the LRT is approximately optimum [33]. To identify the real ships from all detected "targets," the analysis of the phase-interferogram is adopted. Following the rationale previously exposed, the real ship is identified since φ_{XPol} of the detected pixels is near zero, while the azimuth ambiguity has φ_{XPol} near π.

The relation introduced at the beginning of this section between the cross-polarized channels of quad-pol SAR data has been exploited by Velotto *et al.* [34] to propose a new complete ship detection processing chain for high-resolution spaceborne SLC quad-pol SAR data. The proposed idea is to perform the discrimination between real targets at sea and azimuth ambiguities before the detection step. Leaving out the process employed to detect the maritime targets, the focus here is on the mitigation process being proposed in [34]. The developed mitigation approach is performed by generating a new image space free of ambiguities. This new image space is hence called HV_{free} because it exploits the cross-polarization channels property and it is free of azimuth ambiguities. In Velotto *et al.* [34] the following combination of the cross-polarized channels is proposed:

$$\text{HV}_{\text{free}} = \frac{1}{2}[(\text{HV} + \text{VH})(\text{HV} + \text{VH})^*]. \tag{5.27}$$

It is evident from (5.27) that HV_{free}, besides being free of azimuth ambiguity, has the following properties:

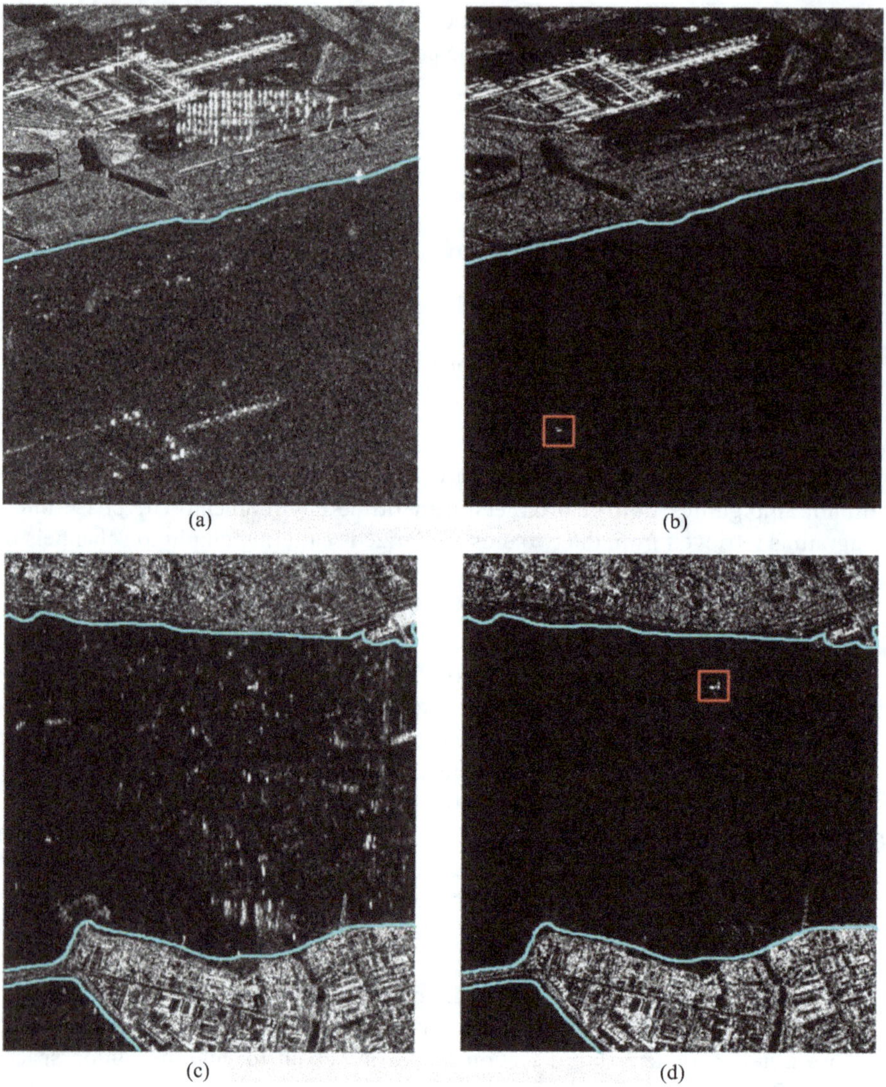

Figure 5.7 Explanatory examples of the mitigation procedure proposed in [34]. (a) and (c) are sub-scenes of TerraSAR-X quad-pol data where ghost targets due to ambiguous signal are clearly visible on the ocean surface (in cyan the coastline). (b) and (d) are the respective HV_{free} *images where the red square indicates the presence of a real target. © DLR (2015)*

- It is a pixel-based approach which doesn't require box-car filtering or reduction of the processing bandwidth, meaning that the original resolution is kept for ambiguous and unambiguous area of the SAR data.
- It is similar to a single cross-polarization channel with double rate sampling, meaning that a better signal to noise ratio (SNR) is achieved.

This mitigation approach of the azimuth ambiguity has been firstly tested in [34] by processing high-resolution quad-pol SAR data acquired by the space-borne mission TerraSAR-X. The validation conducted in [34], employing ship and maritime target ground truth, shows that the false alarms caused by targets or land azimuth ambiguities over the ocean surface can be reduced to zero.

In Figure 5.7, the azimuth ambiguity mitigation technique of [34] is illustrated. There are two TerraSAR-X sub-scenes acquired in two different locations, which show strong ambiguous signals over the ocean area. The respective HV_{free} images show that all ghosts are canceled out retaining the small and medium-size ships present in the area. In particular, it can be noted by comparing Figure 5.7(a) and Figure 5.7(b) that the area close to the airport (top of Figure 5.7(a)) is strongly affected by ambiguities hampering the visual information content, which indeed is clear in HV_{free} image (top of Figure 5.7(b)).

5.5 Summary

In this chapter, the issue of azimuth ambiguity in SAR systems has been discussed, with special attention devoted to the maritime surveillance applications. The details of the model of azimuth ambiguity on the SAR TF have been provided. The main techniques used for ambiguity filtering and mitigation have been described, analyzing in detail both the stripmap case and the polarimetric case.

Acknowledgments

TerraSAR-X data in Section 5.4 were provided by the German Aerospace Center (DLR) via AO NTI_POLI6644 (PI Domenico Velotto).

References

[1] Tomiyasu K. "Tutorial review of synthetic-aperture radar (SAR) with applications to imaging of the ocean surface." *Proc. IEEE*. 1978, vol. 66, no. 5, pp. 563–583.

[2] Curlander J. C. and McDonough R. N. *Synthetic aperture radar: Systems and signal processing*. New York: Wiley, 1991.

[3] Franceschetti G. and Lanari R. *Synthetic aperture radar processing*. New York, NY: CRC Press, 1999.

[4] Li F. K. and W. Johnson T. K. "Ambiguities in spaceborne synthetic aperture radar systems." *IEEE Trans. Aerosp. Electron. Syst.* 1983, vol. 19, no. 3, pp. 389–397.

[5] Freeman A., Johnson W. T. K., Huneycutt B., *et al.*, "The "myth" of the minimum SAR antenna area constraint." *IEEE Trans. Geosci. Remote Sens.* 2000, vol. 38, no. 1, pp. 320–324.

[6] Raney R. K. and Princz G. J. "Reconsideration of azimuth ambiguities in SAR." *IEEE Trans. Geosci. Remote Sens.* 1987, vol. 25, no. 6, pp. 783–787.

[7] Moreira A. "Suppressing the azimuth ambiguities in synthetic aperture radar images." *IEEE Trans. Geosci. Remote Sens.* 1993, vol. 31, no. 4, pp. 885–895.
[8] Monti Guarnieri A. "Adaptive removal of azimuth ambiguities in SAR images." *IEEE Trans. Geosci. Remote Sens.* 2005, vol. 43, no. 3, pp. 625–633.
[9] Di Martino G., Iodice A., Riccio D. and Ruello G. "Filtering of azimuth ambiguity in stripmap synthetic aperture radar images." *IEEE J. Sel. Topics Appl. Earth Observ. Remote Sens.* 2014, vol. 7, no. 9, pp. 3967–3978.
[10] Brusch S., Lehner S., Fritz T., Soccorsi M., Soloviev A., and van Schie B. "Ship surveillance with TerraSAR-X." *IEEE Trans. Geosci. Remote Sens.* 2011, vol. 49, no. 3, pp. 1092–1103.
[11] Avolio C., Costantini M., Di Martino G., *et al.*, "A method for the reduction of ship-detection false alarms due to SAR azimuth ambiguity." *2014 IEEE Geoscience and Remote Sensing Symposium*, IEEE, Piscataway, NJ, USA, 2014, pp. 3694–3697.
[12] Wang W. "Mitigating range ambiguities in High-PRF SAR with OFDM waveform diversity." *IEEE Geosci. Remote Sens. Lett.* 2013, vol. 10, no. 1, pp. 101–105.
[13] Iervolino P., Cohen M., Guida R., and Whittaker P. "Ship-detection in SAR imagery using low pulse repetition frequency radar." In *Proceedings of EuSAR*, VDE Verlag Gmbh, Offenbach, Germany, 2014, pp. 1–4.
[14] Santamaria C., Stasolla M., Fernandez Arguedas V., Argentieri P., Alvarez M., Greidanus H. "Sentinel-1 maritime surveillance." *European Commission Joint Research Center Technical Report*, 2015.
[15] Chen J., Iqbal M., Yang W., Wang P. and Sun B. "Mitigation of azimuth ambiguities in spaceborne stripmap SAR images using selective restoration." *IEEE Trans. Geosci. Remote Sens.* 2014, vol. 52, no. 7, pp. 4038–4045.
[16] Chen J., Wang K., Yang W. and Liu W. "Accurate reconstruction and suppression for azimuth ambiguities in spaceborne stripmap SAR images." *IEEE Geosci. Remote Sens. Lett.* 2017, vol. 14, no. 1, pp. 102–106.
[17] Wu Y., Yu Z., Xiao P. and Li C. "Suppression of azimuth ambiguities in spaceborne SAR images using spectral selection and extrapolation." *IEEE Trans. Geosci. Remote Sens.* 2018, vol. 56, no. 10, pp. 6134–6147.
[18] Leng X., Ji K., Zhou S., and Zou H. "Azimuth ambiguities removal in littoral zones based on multi-temporal SAR images." *Remote Sens.* 2017, vol. 9, no. 8, p. 866.
[19] Lee J. -S. and Pottier E. *Polarimetric radar imaging: From basics to applications.* CRC Press, Boca Raton, FL, USA, 2009.
[20] Raney R. K. "Hybrid-polarity SAR architecture." *IEEE Trans. Geosci. Remote Sens.* 2007, vol. 45, no. 11, pp. 3397–3404.
[21] Cloude S. *Polarisation: Applications in remote sensing.* Oxford: Oxford University Press, 2009.
[22] Villano M., Krieger G. and Moreira A. "New insights into ambiguities in quad-pol SAR." *IEEE Trans. Geosci. Remote Sens.* 2017, vol. 55, no. 6, pp. 3287–3308.

[23] Greidanus H., Alvarez M., Santamaria C., Thoorens F. -X., Kourti N., and Argentieri P. "The SUMO ship detector algorithm for satellite radar images." *Remote Sens.* 2017, vol. 9, no. 3, p. 246.

[24] Villano M. and Krieger G. "Spectral-based estimation of the local azimuth ambiguity-to-signal ratio in SAR images." *IEEE Trans. Geosci. Remote Sens.* 2014, vol. 52, no. 5, pp. 2304–2313.

[25] Liu C., Vachon P. W. and Geling G. W. "Improved ship detection with airborne polarimetric SAR data." *Can. J. Remote Sens.* 2005, vol. 31, no. 1, pp. 122–131.

[26] Sciotti M., Pastina D., and Lombardo P. "Polarimetric detectors of extended targets for ship detection in SAR images." In *Proc. IEEE IGARSS*, IEEE, Piscataway, NJ, USA, 2001.

[27] Touzi R., Raney R. K., and Charbonneau F. "On the use of symmetric scatterers for ship characterization." *IEEE Trans. Geosci. Remote Sens.*, 2004 vol. 42, no. 2, pp. 2039–2045.

[28] Marino A. "A Notch filter for ship detection with polarimetric SAR images." *IEEE J. Sel. Topics Appl. Earth Observ. Remote Sens.*, vol.6, no. 3, pp. 1219 1232, 2013.

[29] Wang C., Wang Y. and Liao M. "Removal of azimuth ambiguities and detection of a ship:Using polarimetric airborne C-band SAR images." *Int. J. Remote Sens.* 2012, vol. 33, no. 10, pp. 3197–3210.

[30] Hu C., Ferro-Famil L., and Kuang G. "Ship discrimination using polarimetric SAR data and coherent time-frequency analysis." *Remote Sens.* 2013, vol. 5, no. 12, pp. 6899–6920.

[31] Ferro-Famil L., Reigber A., Pottier E., and Boerner W. M. "Scene characterization using subaperture polarimetric SAR data." *IEEE Trans. Geosci. Remote Sens.* 2003 vol. 41, no. 10, pp. 2264–2276.

[32] Liu C. and Gierull C. H. "A new application for PolSAR imagery in the field of moving target indication/ship detection." *IEEE Trans. Geosci. Remote Sens.* 2007, vol. 45, no. 11, pp. 3426–3436.

[33] Liu C. and Meek A. "Likelihood ratio test polarimetric SAR ship detection application." *Defence Res. Develop. Canada*, Ottawa, ON, Canada, TM 2005-243, 2005.

[34] Velotto D., Soccorsi M., and Lehner S. "Azimuth ambiguities removal for ship detection using full polarimetric X-band SAR data." *IEEE Trans. Geosci. Remote Sens.* 2014, vol. 52, no. 1, pp. 76–88.

Part II

Applications

Chapter 6
Ship detection

Gui Gao[1], *Sheng Gao*[2], *Juan He*[3] *and Kazuo Ouchi*[4]

6.1 Introduction

The ship detection demand for maritime surveillance with high-resolution satellite sensors has increased during the past years. Starting with the launch of SEASAT in 1978, ship detection investigations using spaceborne synthetic aperture radar (SAR) have been continuously carried out, being of great environmental and economic interest for a variety of military and civilian applications including maritime safety, environmental protection, commercial fishery, and maritime traffic management. SAR is an active imaging system working day and night under all weather conditions and a powerful tool to monitor ships in the open sea and near the coast. Moreover, with the advent of new SAR spaceborne systems, such as RADARSAT-2, TerraSAR-X, COSMO-SkyMed, ALOS-PALSAR, and Sentinel-1, ocean imaging with high resolution and wide coverage can be implemented. As a result, abundant SAR data can be obtained for the ocean; hence, ship detection in spaceborne SAR images is attracting great attention worldwide.

In this chapter, the problem of ship detection from SAR images is introduced. This topic is of great significance in maritime surveillance, since it can help in the identification of noncollaborative ships. The techniques used for ship detection are mainly based on a two-step procedure: a first prescreening detection phase, followed by a discrimination phase, necessary to lower the false alarm rate due to ship-alikes (e.g., ambiguities, see Chapter 5). In the chapter, the pre-screening phase for ship detection is focused.

We start this chapter with a brief survey on detection methods of commonly used single-channel SAR images. Particularly, the constant false alarm rate (CFAR) technique is frequently used in the detection phase, where an accurate description of the sea clutter amplitude probability density function (PDF) is important for robust and reliable CFAR performance. Therefore, the recently developed sea clutter models will be presented in the following. In addition, with the evolution of radar systems, the SAR data acquisition mode has also been extended from single-channel to multichannel, such as dual polarization (DP), full polarization (FP), compact polarization (CP), and along-track

[1]Faculty of Geoscience and Environmental Engineering, Southwest Jiaotong University, Chengdu, China
[2]College of Electrical and Information Engineering, Hunan University, Changsha, Hunan, China
[3]National University of Defense Technology (NUDT), Changsha, China
[4]Space Development Department, IHI Corporation, 1 Shin-Nakahara-cho, Isogo-ku, Yokohama, Japan

interferometry (ATI). Based on this consideration, the recent detection algorithms based on multichannel SAR images will also be summarized.

6.2 Ship detection in single-channel SAR images

Essentially, ship detection in SAR images belongs to a type of anomaly detection that exploits the backscattering responses of SAR differing from the sea background. Literature about ship detection algorithms based on SAR images is abundant. A comprehensive survey of traditional studies of these techniques is given in [1]; recent advancements are partially provided in [2]. Among those algorithms, the most widely used technique is the adaptive threshold based on the constant false alarm rate (CFAR) [3–9], which is based on the fact that ships have a higher radar cross section (RCS) than the surrounding sea background due to metallic materials and corner reflection structures [10]. The pixel amplitudes/intensities of a ship are statistically larger than that of the surrounding sea clutter and can be distinguished based on a reasonable threshold. However, this assumption is not always true in practice; for example, large nonmetallic ships can have a very small radar echo [11]. Several alternative approaches were reported including those based on sublook spectral analyses [12,13], wavelet transform [14], neural network [15], segmentation [16], nonparametric models [17–19], and feature analysis [20]. Herein, several promising approaches are detailed in the following.

6.2.1 Sublook spectral analysis

The idea behind using sublook spectral analyses for detection is that the images of a ship are correlated between looks, whereas those of sea clutter are uncorrelated [2]. The advantage of this kind of method is that it does not require the backscattered echoes of a ship that are bright as long as they are correlated as illustrated in Figure 6.1. There are several different techniques in this kind of approach including sublook coherence, sublook cross-correlation, sublook entropy, and the generalized likelihood ratio (GLRT). The readers can refer to [2] and [21] for details.

The sublook coherence is defined as

$$\gamma = \frac{\left|\langle A_1 \cdot A_2^{*T} \rangle\right|}{\sqrt{\langle A_1 \cdot A_1^{*T} \rangle \langle A_2 \cdot A_2^{*T} \rangle}} \tag{6.1}$$

where $A_j (j = 1, 2)$ are the complex subimages, $*T$ means taking the complex conjugate transpose, and the angular brackets imply taking the spatial average within the moving window.

The sublook entropy is defined as follows. Denoting the vector of N subimages by $\mathbf{A} = [A_1, A_1, \cdots, A_N]$, the covariance matrix can be calculated as $[C] = \langle \mathbf{A} \cdot \mathbf{A}^{*T} \rangle$. Then, the entropy H is computed from

$$H = -\sum_{j=1}^{N} p_j \log_N p_j \tag{6.2}$$

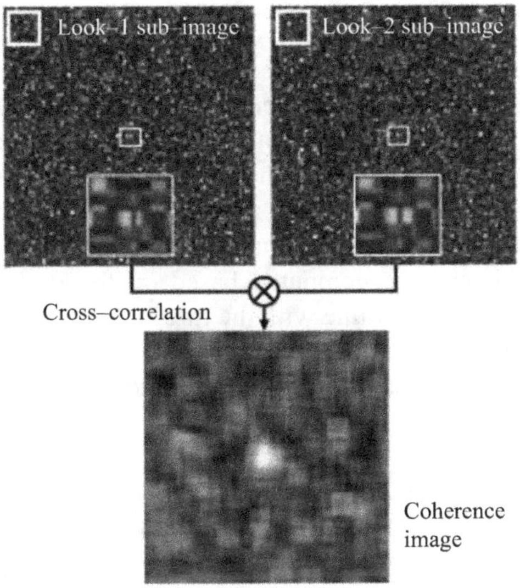

Figure 6.1 The original sublook spectral analysis for ship detection. Adapted from [12]

where p_j is the probability for jth scattering process,

$$p_j = \frac{\lambda_j}{\lambda_1 + \lambda_2 + \cdots + \lambda_N} \tag{6.3}$$

and λ_j is the jth eigenvalue obtained by the diagonalization of $[C]$.

The GLRT is a new algorithm for detection of coherent targets [2]. It is based on, as its name stands, the generalized likelihood ratio test. The GLRT is defined as

$$L_G = \frac{\left|\mathbf{a}^{*T} M^{-1} \mathbf{A}\right|^2}{\left(\mathbf{a}^{*T} M^{-1} \mathbf{a}\right)\left(\mathbf{A}^{*T} M^{-1} \mathbf{A}\right)} \tag{6.4}$$

where **a** is the vector with the location of the scatterer in the center of a resolution cell, **A** is the vector with the complex amplitudes of the pixel under test along the sublooks, and M is the theoretical normalized covariance matrix of the clutter. M depends on the number of subimages and the ratio between the total and sublook spectra. Equation (6.4) of The GLRT is derived under the assumption of Gaussian clutter. However, GLRT is robust such that it can be applied to compound-Gaussian clutter, for example, K-distribution often used to model sea clutter.

The validity of these techniques was examined using extensive SAR data over the European seas and Tokyo Bay with AIS data, indicating that on average the GLRT outperformed the other detectors for the images of small SNR as reported by Marino *et al.* [21].

6.2.2 CFAR

As shown in Figure 6.2, the whole detection process is summarized as follows. First, we select the square hollow-stencil sliding window, which is suitable for high-resolution SAR target detection. During the process of the sliding window scanning the whole image, we consider the clutter area (clutter region) surrounding the test cell in the sliding window. It is assumed that the clutter statistics obey a specific distribution. Then, the detection threshold that guarantees a prescribed false alarm probability can be computed based on the specific distribution. By comparing the current test cell value with the threshold, we obtain the binary value of the pixel, that is, if the value of the test cell exceeds the threshold, the test cell is considered to be a ship point and labeled by 1; if not, it is declared to be a clutter point and labeled by 0.

However, the choice of the sliding-window sizes in CFAR technique remains an issue. First, for the popular CFAR methods [1] using a local sliding-window to detect targets, the number of clutter pixels in the window is commonly small, to ensure an acceptable computation efficiency. Notwithstanding, this technique is extremely time-consuming. Hence, in [22], a fast algorithm of CFAR detection based on the splitting and merging characteristic of pixel in the sliding process was proposed. In addition, alternatively, in the recent literature, for example, in [23], to improve the detection speed, an operation splitting the image into blocks (250×250 clutter pixels in each block in [23]) was employed.

The SUMO (Search for Unidentified Maritime Objects) developed by JRC (Joint Research Center) uses a fast CFAR algorithm with a K-distribution noise model. Similar algorithms were adopted by CLS (Collect Localisation Satellites, France), CCRS (Canada Center for Remote Sensing), and FFI (Norwegian Defence Research Establishment). The OMW (Ocean Monitoring Workshop) developed by CCRS is essentially the same as the OceanSuite of MacDonald, Dettwiler, and Associates (MDA) and Defence Research and Development Canada (DRDC).

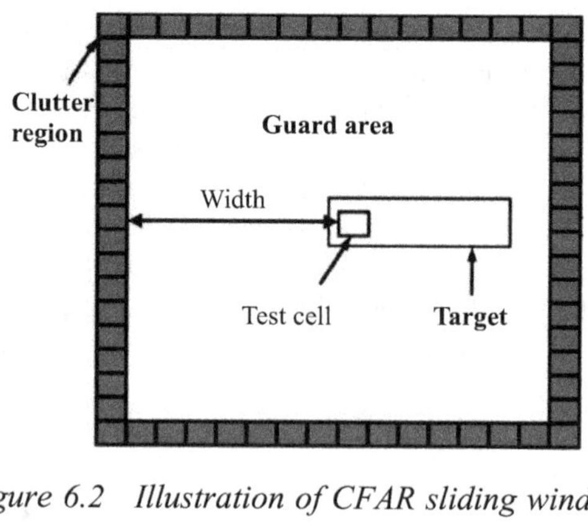

Figure 6.2 Illustration of CFAR sliding window

6.2.3 Adaptive threshold

The adaptive threshold was defined as [2]

$$z_T = \langle z \rangle + c\sigma \tag{6.5}$$

where $\langle \cdot \rangle$ stands for the ensemble average, z is the image magnitude, σ is the standard deviation of the random variable z, and c is a constant depending on SAR parameters and image areas. Commonly, the threshold in the practical detection operation is determined by a boxcar moving window. Note that this method is very simple, based on an assumption that the background is Gaussian-distributed and the targets are much brighter than the surrounding backgrounds.

The MaST (Maritime Surveillance Tool) in the United Kingdom uses a boxcar moving window similar to CFAR, where the mean and standard deviation within the signal and background windows are computed, and the threshold value is determined by comparison of these statistical quantities. Thus, the algorithm is similar to the conventional CFAR but does not assume any background noise model. Note that the information on the MaST was available only in the website, which the author could access in 2011. But, the site is not available now, maybe because of sensitive military applications.

6.3 Statistical models of sea clutter

In terms of CFAR ship detection, an accurate description of the sea clutter amplitude probability density function (PDF) is important for robust and reliable CFAR performance [24]. However, modeling sea clutter under different conditions is complex because the radar response of the sea depends on several factors such as the sea surface characteristics (e.g., sea state, wind speed, and wind direction), imaging geometry (e.g., incidence angle and aspect angle), and radar parameters (e.g., resolution, frequency, polarization, and thermal noise) [25].

6.3.1 Brief survey of state-of-the-art models

Many statistical distributions have been proposed to model the sea clutter in SAR images. Traditionally, when a fully developed speckle regime is achieved and homogeneous surfaces appear as stationary fields, the central limit theorem (CLT) can be exploited. As a result, real and imaginary parts of the received data are assumed to follow the Gaussian distribution, which leads to the Rayleigh distribution of the single-look amplitude and Nakagami distribution (i.e., square root of Gamma) for the multilook amplitude [26]. At low resolution, these distributions can provide a good fit for SAR sea clutter. However, the sea clutter histogram exhibits heavy-tailed characteristics [27] as the SAR resolution increases, which contrasts predictions based on the Rayleigh or Nakagami distributions.

To describe heavy-tailed sea clutter, several distributions, such as the Log-Normal [28], Weibull [29], and K models [27], were established. The Weibull and Log-Normal distributions are good options to model high-resolution SAR sea clutter in cases in which the Rayleigh distribution fails to describe the SAR data, although the Weibull and Log-Normal models are empirical/heuristic [30]. Another model to characterize sea clutter, more popular than the Weibull and Log-Normal models, is the K distribution based on the well-known compound model [31] in which a fast-varying speckle component is modulated by an uncorrelated and slowly varying texture component (i.e., the underlying RCS component). The speckle part is assumed to be due to the scattering of primarily capillary waves and ripples, while the texture is assumed to originate from ocean gravity waves [32]. Therefore, the K distribution is physically based and allows phenomenological interpretation while it considers a nonstationary sea surface (i.e., the corresponding RCS is not a constant). However, as pointed out in [1], the K distribution does not always fit the data of sea clutter well, especially in extremely heterogeneous sea regions such as high sea state areas with high wind and wave conditions.

Therefore, several other attempts have been made to replace the K distribution in models of the sea clutter in SAR images. They can be roughly divided into two classes. One option is that the speckle is not fully developed, which makes sense in the case of high-resolution cells because the hypothesis satisfying the CLT is no longer valid. Several distribution functions have been established for this class. For example, when there is one strong reflector in a homogeneous clutter, a Nakagami–Rice [33,34] amplitude distribution may be used. Note that the heavy-tailed Rayleigh (HTR) distribution [35] and the more recent generalized-K (GK) distribution [36] also belong to this class. The former presents the amplitude distribution by assuming that real and imaginary components follow the zero-mean symmetric alpha-stable distribution [8], which can describe impulsive data and has thicker tails than the classical Rayleigh distribution. The latter is of interest because it was recently used to describe the statistical behavior of different scattering scenarios in single channel/polarization SAR images; the K single-look distribution was considered as a special case [36–38].

Another way to adapt the K distribution is to consider a fully developed speckle hypothesis to be still valid and to add more components (e.g., thermal noise and Rayleigh residual). Several useful models, including K + noise (K+N) [39], K + Rayleigh (K+R) [40], Pareto + noise (P+N) [41], KA [42], and KK [43,44], have been developed based on the preliminary goal to modify the classical K distribution and hence capture sea-spike scattering and thermal noise. In these models, KA and KK are difficult to implement in practice because too many parameters need to be estimated [45]. Both the K and K+N models are special cases of the K+R model. Moreover, [25] and [45] found that the K+R model can provide a better fit for sea clutter than the P+N model. However, it must be noted that only the K+R and P+N models of this distribution class have been recently tested in airborne SAR images [45].

Despite considerable efforts devoted to the description of SAR image statistical properties, several critical issues remain to be discussed in the scientific community.

1. The first is related to the contradiction between the complexity of a model and its applicability in practice. The complexity of a model is commonly related to several factors such as its degree of freedom (DoF), closed-form expression, and the feasibility of parameter estimation. In general, a higher DoF of a distribution,

that is, more parameters, provides a more accurate description of real data in rationale. This also means that the effective parameter estimation would be more difficult. Therefore, in light of the applicability of a model, it is important to note that a feasible statistical model should result from a tradeoff between the complexity of the distribution and the fitting accuracy with respect to actual data. In addition, a model with a closed-form expression for the PDF is always desirable in practical applications. In several of the aforementioned models, such as HTR, K+R, and P+N, the corresponding PDFs are formulated as integral forms. Although these models fit the sea clutter in SAR images well, several disadvantages are unavoidable. On one hand, even though the parameters of these models can be accurately obtained, the computation of the corresponding PDF itself relies on numerical methods that can decrease the accuracy and are computationally expensive [43]. On the other hand, it is challenging to perform target CFAR detection using these models because an equation involving a dual-integral needs to be solved to acquire a detection threshold. This might be the reason why the HTR model was applied to some aspects of SAR image processing, as reported in [46], where the HTR model was used for SAR image filtering. However, it is not a good candidate for ship detection. Meanwhile, note that no examples of ship CFAR detection using the HTR, K+R, and P+N models have been reported in the literature. More attention was paid to determining the fitting accuracy of the K+R and P+N models for actual SAR sea clutter. Therefore, we believe that a model with a closed-form expression is preferable in practice, even though it has a slightly worse fitting ability with respect to real data than models without a closed-form expression.

2. The second is related to the modeling ability for single-look and multilook SAR images. Due to the abundance of SAR data-type products including single- and multilook, a model with the ability to model any type of product will be more useful for practical applications. Objectively, the trend extending a model from single- to multilook is rather prominent because a single-look model is generally a special case of a multilook model when the number of looks is equal to 1. One of reasons that the K distribution has been widely used in several operational ship detection systems, such as the Ocean Monitoring Workstation (OMW) [47] developed by the Canada Center for Remote Sensing (CCRS), SUMO [11], and AEGIR [2] developed by the Norwegian Defense Research Establishment (FFI), is that it can deal with both single- and multilook satellite SAR products. This trend was also observed in several recent models. For example, the K+N, P+N, and KK models were extended from single- to multilook in [39,41], and [44], respectively. Therefore, the foregoing analysis implies that a model that can describe single- and multilook SAR data is always expected.
3. The third is related to the distinction between physically based and empirical models. Some of the aforementioned models rely on the derivation based on the physical mechanism; others are rather empirical. Generally, physical guidance always helps with challenging tasks regarding statistical models of SAR sea clutter. However, on one hand, it is currently not possible to predict the PDF of sea clutter under different conditions using physical and electromagnetic models of

sea clutter [11] and, on the other hand, many so-called physically based models rely on hypotheses about empirical scattering mechanisms. For example, it is difficult to explain why an inverse gamma distribution is suitable for the description of the texture component in the P+N model and why the symmetric alpha-stable distribution can be used to model real and imaginary components in the HTR model. This means that the distinction between physically based and empirical models is less rigorous. It might be preferable to call most physically based models semi-empirical distributions. In a word, any distribution is only a mathematical approximation of the scattering behavior of sea clutter. Therefore, a pure mathematical model lacking physical justification can be accepted as long as it can provide a good fit of the real data. An example of a successful empirical model for land clutter is the Fisher distribution [34]; it has been demonstrated that this distribution better describes a wide range of terrains from natural areas to man-made objects on land than many physically based models [34].

6.3.2 Several known models

6.3.2.1 K Distribution

The K distribution is a popular and widely accepted model that appropriately fits observed sea clutter statistics with spiky properties [27]. The resulting amplitude PDF is given by [48,49]

$$p(x) = \frac{4\lambda L}{\Gamma(L)\Gamma(\alpha)}(\lambda L x)^{L+\alpha-1}K_{\alpha-L}\left(2\sqrt{\lambda L x}\right), \alpha, \lambda, L, x > 0 \tag{6.6}$$

where λ is a scale parameter, α is an order parameter, L is the number of looks, $\Gamma(\cdot)$ is the gamma function, and $K_{\alpha-L}(\cdot)$ is a modified Bessel function of the second kind with order $\alpha - L$.

It is worth noting that the K distribution has three parameters, as characterized by (6.6). Commonly, the number of looks L, that is, the nominal number, for a large SAR image is known a priori [44,50], although it can also be estimated by the equivalent number of looks (ENL) [51]. However, the assumption that L is a fixed parameter in a large SAR image is not suitable. On one hand, multilook processing of SAR images is commonly implemented by spatial averaging adjacent pixels/samples. Since adjacent pixels are statistically dependent, the number of looks is usually smaller than its nominal value (average number of samples) [52]. On the other hand, L should be estimated locally, that is, it should be adapted for different local regions within the same SAR image. The reasons for that have been sufficiently discussed in [52]. To summarize, the number of looks should be regarded as a distribution parameter to be estimated, which not only improves the adaptability of the model to describe sea clutter [30] but also enhances the fitting accuracy of the model with respect to actual data. Moreover, a recent trend in estimating the parameters of the K distribution is to use the MoLC [48], which is better suitable than traditional methods such as the method of moments (MoM). Parameter estimators of the K distribution based on the MoLC can be found in [49] and [53].

6.3.2.2 GK distribution

The three-parameter GK distribution has been proposed to characterize the statistics of full-resolution single-look complex (SLC) SAR marine scenes. It is an extension of the well-known single-look K distribution used for cases of weak scattering regimes in which a nonuniform distribution of the phase is considered. The amplitude PDF of the GK distribution is given by [36]

$$p(x) = \frac{4\alpha}{\Gamma(\alpha)\eta^{\alpha+1}}\left(\frac{\alpha}{1+\left(\frac{v^2}{4\alpha}\right)}\right)^{\frac{\alpha-1}{2}} x^{\alpha} I_0\left(\frac{v}{\eta}x\right) K_{\alpha-1}\left\{\frac{2x}{\eta}\left[\left(1+\frac{v^2}{4\alpha}\right)\alpha\right]^{\frac{1}{2}}\right\},$$
$$\alpha, \eta, v, x > 0 \tag{6.7}$$

where α, η, and v are the shape, slope, and departure from the uniform distribution of the phase parameters [36], respectively, and $I_0(\cdot)$ is the first kind zero-order modified Bessel function. It is worth noting that the GK model can only be used to characterize single-look SAR images because this model does not consider the case of multilook data. The parameter estimation method for the GK model was reported in [36].

6.3.2.3 K + R distribution

The K+R distribution was recently developed to capture both the thermal noise and extra Rayleigh component [40]. It also is a compound model and uses the following amplitude PDF expression [45]:

$$p(x) = \frac{2b^{v}x}{\Gamma(v)}\int_0^{\infty}\frac{u^{v-1}}{u+p}\exp\left(-\frac{x^2}{u+p} - bu\right)du, v, b, x > 0 \tag{6.8}$$

where v is the shape parameter and b is the scale parameter, and $p = p_n + p_r$ indicates the Rayleigh-plus-thermal noise term in which p_n is the thermal noise component and p_r is the residual Rayleigh component. As shown in (6.8), the K+R distribution has also three parameters because $p_n + p_r$ can be treated as a single component in the distribution. Moreover, up to now, only the single-look K+R distribution characterized by (6.8) has been derived in literature. Unfortunately, the K+R distribution has no closed-form expression and numerical integration is required to obtain the solution of the integral. Parameter estimators for the K+R distribution based on the MoM are given in [45].

6.3.2.4 $\mathcal{G}_{AO}$ Distribution

From a purely mathematical point of view, irrespective of the physical properties of radar clutter backscatter, the PDF of the $\mathcal{G}_{AO}$ distribution with three parameters for modeling SAR amplitude images of sea clutter is given by [54]

$$p(x) = \frac{2^{\mu}\Gamma\left(\mu+\frac{\lambda}{2}\right)}{\beta^{-\lambda}\Gamma(\mu)\Gamma\left(\frac{\lambda}{2}\right)}\frac{x^{\mu-1}\left(\sqrt{\beta^2+x^2}+x\right)^{1-\mu-\lambda}}{\sqrt{\beta^2+x^2}}, \mu, \beta, \lambda, x > 0 \tag{6.9}$$

where μ is a scale parameter, and β and λ are both shape parameters.

As derived in [54], the estimates of the underlying parameters in the $\mathscr{G}_{\mathrm{AO}}$ model can be obtained by using the first three order sample log-cumulants of the $\mathscr{G}_{\mathrm{AO}}$:

$$\ln\left(\frac{\widehat{\beta}}{2}\right) - \frac{1}{2}\Psi\left(\frac{\widehat{\lambda}}{2}\right) + \Psi(\widehat{\mu}) - \frac{1}{2}\Psi\left(\frac{\widehat{\lambda}}{2} + \widehat{\mu}\right) = \widehat{c}_1 \tag{6.10}$$

$$\frac{1}{4}\Psi\left(1, \frac{\widehat{\lambda}}{2}\right) + \Psi(1, \widehat{\mu}) - \frac{1}{4}\Psi\left(1, \frac{\widehat{\lambda}}{2} + \widehat{\mu}\right) = \widehat{c}_2 \tag{6.11}$$

$$-\frac{1}{8}\Psi\left(2, \frac{\widehat{\lambda}}{2}\right) + \Psi(2, \widehat{\mu}) - \frac{1}{8}\Psi\left(2, \frac{\widehat{\lambda}}{2} + \widehat{\mu}\right) = \widehat{c}_3 \tag{6.12}$$

where $\widehat{\mu}$, $\widehat{\beta}$, and $\widehat{\lambda}$ denote the estimates of the underlying parameters μ, β, and λ, respectively.

6.3.2.5 Generalized gamma distribution model

The GΓD was first proposed by [55], and it has been widely used in many fields, including speech, signal, and image processing [56–61]. A recent implementation of the GΓD with three parameters was given in [62], and its PDF can be expressed by

$$p(x) = \frac{|\nu|\kappa^{\kappa}}{\sigma\Gamma(\kappa)}\left(\frac{x}{\sigma}\right)^{\kappa\nu-1}\exp\left\{-\kappa\left(\frac{x}{\sigma}\right)^{\nu}\right\}, \quad \sigma, |\nu|, \kappa, x > 0 \tag{6.13}$$

where σ, ν, and κ refer to the scale, power, and shape parameters, respectively; $\Gamma(\cdot)$ is the gamma function. Although this distribution is associated with an empirical model, it has been demonstrated [62] to be relatively versatile because it is the general case of several well-known distributions that are used for modeling SAR images, such as the Rayleigh ($\nu = 2, \kappa = 1$), exponential ($\nu = 1, \kappa = 1$), square root of gamma ($\nu = 2$), gamma ($\nu = 1$), Weibull ($\kappa = 1$), LN ($\kappa \to \infty$), square root of inverse gamma ($\nu = -2$), and inverse gamma ($\nu = -1$) distributions [63–65].

6.4 Ship detection in multichannel SAR images

6.4.1 Brief survey on detection methods of conventional multipolarization

The dual- or quad-polarimetric SAR systems have also been validated to provide a more complete description of the illuminated target scattering behavior than could be provided by traditional single-channel SAR systems with HH-, HV-, or VV-polarization alone. Herein, HV denotes horizontal transmitting and vertical receiving polarization, and HH and VV are similarly defined. Therefore, ship detection using dual- or quad-polarimetric SAR data might be more reliable at appropriate radar incidence angles than using single-polarimetric SAR data.

Many investigations have been carried out on the detectability of ships using airborne or spaceborne polarimetric SAR data [60]. The results, which are now well understood, seem to be widely accepted, and can be summarized as follows:

1. For an individual channel: In terms of HV-, HH-, or VV-polarization radar backscatter from sea surface, the cross-polarized channels, that is, HV or VH, exhibit a higher detection performance at small incidence angles in contrast with co-polarized channels. On the contrary, for large incidence angles, HH tends to be easier in detecting ships than other channels, and VV is more suitable for sea surface characterization.
2. For additional polarization information: Single or dual-polarization products are available with larger swath widths than that of full-polarization products. From the viewpoint of detection performance, a fully polarimetric system is superior to both single- and multipolarization configurations for ship detection in some specific detectors, for example, the statistical decision method of the polarimetric scattering matrix [66–68], and the methods based on coherent target decomposition (CTD) [69,70]. Moreover, dual-polarimetric (dual-pol) SAR systems under some specific polarization combinations seem to provide better detection performance than single-channel systems [67–72]. To summarize, all these investigations indicate that the introduction of additional polarization information can improve detection performance if appropriate detectors are defined.
3. For assessment of detectability: The ship-sea contrast (also known as the signal-clutter-ratio, SCR) is a key index to assess the detectability using SAR data. In general, to realize the detection, an effective detector must be constructed which transforms the SAR target information from the data domain into the detector domain. Ship-sea contrast improvement is a central task of devising a useful detector [1,69], in which case, a good discrimination of ships against a sea-clutter background is expected to be obtained in the image produced by the detector.

Currently, the launch of new advanced SAR sensors allows the use of dual and quad polarizations for ship detection. Famous contemporary satellite SAR systems (e.g., TerraSAR-X, COSMO-SkyMed, ALOS-PALSAR, and RADARSAT-2) and many airborne SAR platforms (e.g., Convair-580 SAR and NASA/JPL AIRSAR) support, at least, the dual-polarimetric (dual-pol) acquisition mode. According to the abundant data obtained from these sensors, a wide variety of methods have been proposed for detecting ships in polarimetric SAR images [73]. These detection methods were often developed by circumfusing two basic ideas: one involves incorporating three channels of polarimetric SAR data into a single decision criterion, and the other uses a multichannel detection scheme that jointly exploits the information of the polarimetric data. For the former, popular detectors including the polarimetric whitening filter, SPAN detector, and minimum mean square error strategy were developed [74,75] and have demonstrated the ability to easily discriminate targets from clutter. The latter includes several unique techniques, such as the polarimetric Gaussian generalized likelihood ratio test detector [68], polarization entropy [76], polarimetric cross entropy [77], polarimetric coherence [78], co-polarization phase difference [67], coherent target decomposition [79], and the symmetric scattering characterization method [69]. The usefulness of these methods has been confirmed by SAR images obtained using various airborne and spaceborne platforms, such as Convair-580, AIRSAR, SIR-C/X, ENVISAT ASAR, and RADARSAT-2.

Recently, studies of ship detection using polarimetric SAR data were extended to several representative methods [37,80–83]. Although most of the developed methods focus on the utilization of quad polarimetric SAR data, it is also true that ship detection generally needs large swath widths to satisfy the requirement of wide-ocean-area monitoring. Thus, SAR image products with dual polarization, a relatively low resolution, and wider swaths are currently preferred [48]. Both amplitude- and single-look complex (SLC)-type products are common for all current commercial spaceborne SAR satellites involving a dual-polarization working mode. Many researchers have examined both types of products.

6.4.2 Several recent methods of conventional multipolarization

6.4.2.1 Polarimetric SAR data description

The fundamental quantities measured by a polarimetric SAR are the four elements of the complex scattering-matrix, which, in complete form, can be expressed by

$$\mathbf{S} = \begin{bmatrix} S_{hh} & S_{hv} \\ S_{vh} & S_{vv} \end{bmatrix} \tag{6.14}$$

where S_{pq} is the scattering element with p transmit and q receive polarizations; p and q denote either linear horizontal (h) or vertical (v) polarization. In a reciprocal medium, the cross-polar elements of the scattering are identical, that is, $S_{hv} = S_{vh}$.

To reduce the influence of speckle, SAR data are often multilook averaged. As the polarimetric information can also be represented by a covariance matrix, the n-look sample covariance matrix for dual-pol cases is defined as

$$\mathrm{C} = \frac{1}{n}\sum_{k=1}^{n}\begin{bmatrix} |S_1(k)|^2 & S_1(k)S_2(k)^* \\ S_1(k)^*S_2(k) & |S_2(k)|^2 \end{bmatrix} \tag{6.15}$$

where the superscript * stands for complex conjugate, n is the number of looks, and $S_1(k)$ or $S_2(k)$ is the kth single-look pixel in the averaging window. The symbol $|\cdot|$ represents the modulus of the complex signal. The diagonal elements $\frac{1}{n}\sum_{k=1}^{n}|S_i(k)|^2, i = 1,2$ indicate the multilook intensity of the ith polarimetric channel. For the sake of convenience, we use S_1 to indicate the co-polar scattering element (S_{hh} or S_{vv}) and S_2 to represent the cross-polar scattering element (S_{hv} or S_{vh}).

6.4.2.2 Reflection symmetry metric

The RSM r is defined as the modulus of the correlation between the co- and cross-polarized scattering amplitudes (i.e., the modulus of the off-diagonal elements in (6.15)):

$$r = \left|\langle S_1 S_2^* \rangle\right| = \left|\frac{1}{n}\sum_{k=1}^{n}\left[S_1(k)S_2(k)^*\right]\right| \tag{6.16}$$

where $\langle\cdot\rangle$ stands for the ensemble average, that is, the average of n samples. Reflection symmetry is accounted for since most natural targets, that is, the sea surface, satisfy this property. Consequently, the values of r are expected to be close to zero. On the contrary, for ships, which may consist of oriented planes and dihedrals, the reflection symmetry is violated; thus, the corresponding r values are significantly larger than those of the pure sea surface.

6.4.2.3 PMA detector

A PMA detector is defined as the product of two multilook amplitudes (i.e., the square root of the multilook intensity, see (6.15)):

$$\xi = \sqrt{\left(\frac{1}{n}\sum_{k=1}^{n}\left|S_1(k)\right|^2\right)} \cdot \sqrt{\left(\frac{1}{n}\sum_{k=1}^{n}\left|S_2(k)\right|^2\right)} \tag{6.17}$$

Comparing (6.16) and (6.17), it can be seen that the utilization of the phase information is absent in the PMA, compared to the RSM, since multilook processing is employed for the data.

6.4.2.4 Notch filter

The covariance matrix can be estimated as

$$\mathbf{C} = \langle \mathbf{k}\mathbf{k}^{*T} \rangle \tag{6.18}$$

where the superscript * is the complex conjugate and "T" represents the transpose. The $\langle\cdot\rangle$ operator denotes statistical averaging (in SAR polarimetry, averaging is usually done spatially) and $\mathbf{k}$ is the scattering vector and formed of measured values, that is, scattering elements or their linear combinations. For example, in the case of a reciprocal medium and monostatic sensor, $\mathbf{k} = [k_1, k_2, k_3]^T = \left[S_{hh}, \sqrt{2}S_{hv}, S_{vv}\right]^T$ is in the lexicographic basis. Other definitions of $\mathbf{k}$ are possible because the vector can be linearly transformed to emphasize physical interpretations of elements (i.e., Pauli basis), or a low-dimension version ($d = 2$) for dual-polarization SAR.

Furthermore, vectorization of the covariance matrix is accomplished by eliminating elements with the same information in (6.18) to completely characterize partial target results in a so-called feature partial scattering vector, which can be expressed by

$$\mathbf{t} = [t_1, t_2, t_3, t_4, t_5, t_6]^T \tag{6.19}$$

In the original version of the $\mathbf{t}$ definition, introduced in [81], the real part of each element in $\mathbf{t}$ was given using an intensity form, for instance, $t_1 = \langle |k_1|^2 \rangle$, where $|\cdot|$ is the modulus of the complex signal. In this chapter, we adopt an amplitude version of $\mathbf{t}$:

$$\mathbf{t} = \left[\sqrt{\langle |k_1|^2 \rangle}, \sqrt{\langle |k_2|^2 \rangle}, \sqrt{\langle |k_3|^2 \rangle}, \sqrt{|\langle k_1^* k_2 \rangle|} \exp\left(j\phi_{\langle k_1^* k_2 \rangle} \right), \right.$$
$$\left. \sqrt{|\langle k_1^* k_3 \rangle|} \exp\left(j\phi_{\langle k_1^* k_3 \rangle} \right), \sqrt{|\langle k_2^* k_3 \rangle|} \exp\left(j\phi_{\langle k_2^* k_3 \rangle} \right) \right]^T \tag{6.20}$$

where ϕ denotes the complex signal phase. In contrast to the original version of $\mathbf{t}$ constructed in [84], the amplitude version defined in (6.20) might be preferable for the following three reasons. First, it uses equivalent information contained in the second-order statistics and has no information loss for observed targets and thus can sufficiently describe a partial target. In this sense, the expression in (6.20) is another form of a partial vector exploiting second-order statistics. Second, the notch filter based on (6.20) can share the same derivation procedure with the notch filter of the original version of $\mathbf{t}$. This can be clearly seen from [81] and [84] because the notch filter derivation is only based on $t_i, i = 1, 2, \cdots, 6$ and has the advantage of not being related to any specific assumption regarding the polarimetric signature of t_i. More explicitly, the definition in (6.20) does not influence the final expression of the notch filter. Third, the notch filter is a physical detector essentially reflecting residual power. Taking vector $\mathbf{t}$ into account, the modulus square of $\mathbf{t}$ or equivalently $\mathbf{t}^{*T}\mathbf{t}$ is equal to the sum of the power of each element in the vector. In general, the intensity of a pixel in a SAR complex image is indicated as the pixel power in the field of radar signal processing [1]. Therefore, the definition of the amplitude version for $\mathbf{t}$, as shown in (6.20), indicates that $\mathbf{t}^{*T}\mathbf{t}$ can be regarded as the synthesis of the power/intensity of each element; we note that the synthesis is not rigorously identical to the power of the covariance matrix that is in theory proportional to its Frobenius norm. This is very similar to the SPAN technique [74,75] and possibly has a better physical justification for power than the original version. To summarize, in comparison with the original version of $\mathbf{t}$, the definition in (6.20) has the following properties: it provides descriptions of partial targets without loss of polarimetric information, does not change the final expression of the notch filter, and has a better physical justification for the partial vector power. Furthermore, the normalized version $\widehat{\mathbf{t}}$ with the unit modulus value for the vector $\mathbf{t}$ is $\widehat{\mathbf{t}} = \mathbf{t}/\|\mathbf{t}\|$, where $\|\cdot\|$ is the modulus of a vector.

After a series of mathematical manipulations based on subspace analysis, the final expression of the notch filter is [24]

$$\gamma = \frac{1}{\sqrt{1 + \mathrm{RedR} \frac{1}{\mathbf{t}^{*T}\mathbf{t} - \left|\mathbf{t}^{*T}\widehat{\mathbf{t}}_{sea}\right|^2}}} \tag{6.21}$$

where $\mathbf{t}$ is the partial vector extracted from the observed scene, $\widehat{\mathbf{t}}_{sea}$ is the normalized version of $\mathbf{t}_{sea}$ (i.e., the partial vector of sea clutter), and RedRis the Reduction Ratio parameter. Clearly, the total power of the observed vector $\mathbf{t}$ can be written using $P_{tot} = \mathbf{t}^{*T}\mathbf{t}$. Moreover, $|\mathbf{t}^{*T}\widehat{\mathbf{t}}_{sea}|$ represents the geometric length of the projection of $\mathbf{t}$ in the sea clutter subspace. As a result, the sea clutter power is $P_{sea} = |\mathbf{t}^{*T}\widehat{\mathbf{t}}_{sea}|^2$. Thus, the power of the target of interest (e.g., a ship) is equal to that of the total power minus the power of the sea clutter, that is, $P_T = P_{tot} - P_{sea} = \mathbf{t}^{*T}\mathbf{t} - |\mathbf{t}^{*T}\widehat{\mathbf{t}}_{sea}|^2$; its value will be subject to the scattering of PolSAR data when the observed vector belongs to the signature of the target of interest.

In summary, the idea behind the notch filter is to build an algorithm that is able to identify any partial target differing from the background clutter. In the case of ship detection using the notch filter, the differences in polarimetric scattering properties in $\mathbb{C}^6$ between ships and sea clutter are translated into the responses of the notch filter in one-dimensional space. Therefore, proper decision criteria based on a single threshold can be employed to distinguish targets of interest and sea clutter in the given data.

6.4.3 Brief survey on detection methods of compact polarization

Recently, it has been proposed that compact polarimetric (CP) SAR [85] is a good option for SAR applications. Although CP is a specific dual-polarization configuration, it can provide more information than traditional linear dual-polarimetric SAR while maintaining the benefits (wide swath coverage and less energy budget) that FP SAR systems cannot provide [85,86]. Currently, there are three CP configurations: the $\pi/4$ [87], dual circular polarization (DCP) [88], and hybrid modes [89,90]. Particularly, the $\pi/4$ mode cannot ensure a rotationally robust backscattered field with respect to the shape of the observed scene and the DCP mode has certain engineering disadvantages. However, the hybrid-polarimetric (HP) mode can retain all advantages of the DCP mode by transmitting circular polarization but receiving two linear polarizations, although it is based on simpler hardware implementation [90].

Nowadays, CP SAR has been greatly promoted, from concept to actual satellite platforms. Several spaceborne SAR systems with CP capability have been in orbit. For example, the C-band Indian Space Research Organization Radar Imaging Satellite (Risat-1) launched in 2012 provides the HP mode [91]. The Japanese Aerospace Exploration Agency ALOS-2 allows both the $\pi/4$ and hybrid modes operated in the L band. The Canadian Radarsat Constellation Mission (RCM) SAR will be launched in 2018 and will be equipped with a polarimetric SAR operating in HP mode [92]. Therefore, CP SAR data exploitation including ship detection is attracting increasing attention.

For ship observations on the open sea, several approaches using CP SAR data have been proposed. Initially, these detection methods were developed by combining two basic ideas. The first employed existing FP detection algorithms to pseudo-quad-polarimetric (PQ) data reconstructed by CP data in which a crucial step of the detection performance is to obtain a 4×4 PQ covariance matrix from a 2×2 CP covariance matrix. Several pioneering reconstruction algorithms were established such as the Souyris [87], Nord [89], and Yin [93] algorithms. However, it was demonstrated that the assumptions underlying these reconstruction algorithms are not valid for ocean scenes [94,95]. Therefore, a simple reconstruction strategy based on an empirical model was used in [94] and [96], which is suitable for sea background analysis. Moreover, by linking the likelihood ratio test with the Neyman–Pearson criterion [97], the ship detection performance based on reconstructed PQ data was also assessed [94,96].

Similar to other practical applications of CP SAR, such as classification [85], the idea of reconstructing PQ SAR data with the aforementioned methods has been strongly criticized by Raney [98]. The reasons can be explained as follows. On one hand, reconstruction is a process that does not add any valid information from CP to PQ data. Thus, reconstruction is not helpful for the discrimination of targets from background clutter, even if it is sufficiently accurate. More importantly, because reconstruction algorithms are generally restricted by specific assumptions that might be far from the truth, reconstruction itself can introduce uncertainty and affects the rebuilding accuracy. This will lead to unconvincing detection results. Accordingly, Raney suggested [98] that the processing should start with a 2×2 CP covariance matrix without any assumption and data expansion.

The second idea is devoted to the direct utilization of CP SAR data, which in fact follows (or agrees with) the suggestions of Raney, although most of the methods in this category were published before Raney's criticism [98]. Most analyses of this category focus more on the extraction of child parameters from the Stokes vector to exploit the wave polarimetric difference between ships and sea clutter because the four real parameters of the Stokes vector capture all information inherent to the two orthogonal channels [86]. The child parameters can include the degree of polarization (DoP) m, circular polarization ratio μ_C, relative phase δ between two receiving channels, complex correlation coefficient μ_{xy}, and other parameters [90,95]. Several investigations based on the parameters derived from the Stokes vector have been reported in the literature. For example, the effectiveness of the DoP in detecting ships was explored for the first time in [99] and [100]. This work was further extended in [101] by defining an excursion of the DoP to provide a significant ship–sea contrast increase. The relative phase δ was used for the discrimination between wind turbines and ocean wakes in [102], although this study's aim was not ship detection. However, a more recent report [103] provided a detailed investigation of the detection capability of δ together with m, μ_C, $|\mu_{xy}|$, and entropy H_w. Note that the entropy can also be derived from the Stokes vector of CP SAR data [104]. The strategies in [105] based on $m - \delta$ and $m - \chi$ decompositions seem to improve the performance of ship detection compared with the former individual Stokes parameter. However, this conclusion is exclusively based on visual inspection of composite images, has not been followed up in later work by other researchers, but deserves further investigation. Moreover, it was suggested in [95] that the $m - \delta$ and $m - \chi$ decompositions should be used to exploit the physics of ocean and ship backscattering for the discrimination of ships from ocean false alarms in discrimination rather than detection stages.

Essentially, the fundamental principle of the foregoing methods based on the Stokes parameters is to extract effective features characterizing the response difference between ships and background. It is not surprising that all issues of pattern recognition involve a core step, feature extraction. There are several recent examples of the extraction of effective features to detect ships in CP SAR data. Three parameters/features estimated from the extended Bragg (X-Bragg) model were reported to perform well in ship detection in [106]. A vision saliency feature was used to discriminate ships from the sea background in [107]. Two features, the entropy H_w and scattering angle α, were extracted by $H-\alpha$ decomposition in [108]. Preliminary discussions indicated the potential of the two features.

In summary, the aforementioned studies indicated the feasibility of feature extraction for ship detection using CP SAR data. However, it was also suggested, for example, in [106] and [103], that the detection capability is very limited when using any single feature/parameter for popular methods based on the Stokes vector. One of the main reasons is that the former Stokes child parameters (m, μ_C, δ, μ_{xy}, and H_w) can only exploit wave polarimetric characterizations. Moreover, the information of the features might be redundant (e.g., m and μ_{xy} is strongly related). It is well known that the radar response of an object can contain not only wave polarimetric information but also scattering energy/power information. The former is related to the object's structures such as the single-bounce, double-bounce, and volumetric reflections, and can be described with many parameters (m, μ_C, δ, μ_{xy}, and H_w) extracted from the Stokes vector. The latter is also related to the material of the object. In terms of ships, the scattering power is an important physical quantity for the distinction between ships and sea clutter. Therefore, the fusion of wave polarimetric characterizations and the backscattering power might be a feasible way to improve the detectability of ships.

Based on this consideration, a novel decomposition strategy (Power–Entropy decomposition) was developed. Based on the strategy, the high-entropy scattering amplitude (HESA) component can be obtained and used as a potential detector. It was further mathematically proved that the detector is an effective physical quantity indicating the distinction of ships and sea background from the perspective of SCR improvement. The GΓD was then found suitable for the characterization of the HESA statistics of sea clutter with a wide range of homogeneity. Based on the same benchmark of GΓD and CFAR operations, experiments performed on HP measurements emulated from L-band ALOS-PALSAR and C-band RADARSAT-2 FP data validated the soundness and superiority of the proposed CFAR method based on the HESA detector. Because the HESA detector is a combination of two orthogonal features (wave polarimetry and total scattering power) in the feature space, HESA is a feasible detector for the improvement of the detection performance in both theoretical analysis and experimental validation.

6.4.4 Brief survey on detection methods of along-track interferometry

Noticeably, SAR in along-track interferometry (ATI) mode has been widely investigated because of its potential in measuring ocean currents and waves in the last several decades. Commonly, ATI-SAR employs two identical antennas that are physically separated in platform flight path direction by a short distance (baseline). The phase difference between the complex data of two channels for the same point in a scene is proportional to the Doppler shift of the backscattered signal and thus to the line-of-sight target velocity, that is, the radial velocity. In this sense, ATI-SAR systems do not only provide the radar cross section (RCS), similar to a traditional SAR sensor, but also velocity information of the target. Therefore, ATI-SAR is more advantageous for the purpose of ship observation on the ocean surface than the conventional SAR; it allows the enhanced detection of ships by utilizing both the RCS and phase difference (i.e., ATI phase) information because ships on the ocean are rarely stationary in practice.

In recent years, more and more air- and spaceborne platforms were equipped with ATI-SAR such as RADARSAT-2 (RS2), TerraSAR-X (TSX) or TanDEM-X (TDX), AIRSAR, PAMIR, and Pi-SAR2. It is desirable to develop effective and adaptive techniques for ship detection in ATI-SAR. Along with several missions [109–111], the techniques have been extensively investigated [112–116]. Popular processing approaches can be divided into two main categories [117]. One is executed in the complex image domain after SAR compression; the other is operated in the range-compression raw data domain. The former includes conventional techniques such as clutter cancellation using a displaced phase-center antenna (DPCA) [116], phase difference detection with the classical ATI [118], and the fusion of DPCA and ATI [109,110]. When more than two receiving channels are available, the extended DPCA (EDPCA) [119] was developed to provide a substantial improvement in detection performance [111]. It should be noted that these techniques were originally designed for two-channel radar systems (i.e., identical antenna patterns). The latter include space-time adaptive processing (STAP) methods [120–122], which are optimal (at least in principle) and can be applied for an arbitrary number of receiver channels.

However, current and near-future satellite missions, such as TSX, TDX, RS2, and Spanish SEOSAR/PAZ [123], have a limit of two physical receivers collocated in along-track direction due to cost and technological restrictions [116]. Moreover, operator-friendly image-type products are sometimes preferred [124]. Accordingly, the detection techniques in the image domain for the dual receiver antenna mode are currently of particular interest.

Although the aforementioned methods were initially used for target (e.g., cars) detection in land, the results were also rather inspiring when applied to maritime situations. In this situation, the ATI and DPCA are good candidates due to their relative implementation simplicity and widespread usage. However, essentially, the DPCA is a clutter cancellation technique utilizing phase and magnitude difference between the two spatially co-registered channels. After channel balance, stationary objects/terrains, appearing at the same spatial position, can be completely canceled with an ideal, noise-free radar. On the contrary, the echoes by moving targets are preserved. This implies that the principle of DPCA operation is very suitable for the detection of land moving vehicles because the internal motion of land scenes can be negligible and hence the DPCA clutter cancellation is sufficiently effective. However, unlike land scenes, the sea surface is not stationary at all due to the impact of wind, waves, and currents, which can principally lead to severe clutter residues in the resulting DPCA image and makes the detection of ships very challenging, especially in high sea state conditions.

In terms of ATI technique, interferometric magnitude and phase exploit different types of information relating to targets and background clutter. In the case of GMTI, whose objective is to detect dim, discrete targets against a stationary background, the SCR in magnitude domain might be very low because of a multitude of complicated terrain contents, particularly in the urban regions that abound in man-made buildings. These render ground moving target detection unsuccessful when only the magnitude information is considered. On the contrary, the ATI shows great promise for GMTI in that the interferometric phases of stationary clutter are expected to be zero ideally, but the ones of moving targets are not zero after interferometric processing of the two channels. This property enables discrimination between interferometric phase domains of moving targets and stationary background if only the targets have significant absolute-speed. This explains why most of the previous researchers used ATI to solve the GMTI problems. Of course, using a combination of magnitude and phase can indeed reduce the number of false alarms in a simple background (for instance, the highways in rural regions). However, no such situation exists on the oceanic surface. On the one hand, the ships generally exhibit wide ship-sea contrast (i.e., high SCR value), except under some specific circumstances, such as the following: extremely bad sea environment, some small-size ships in low incidence angle, and spiky sea clutter in HH polarization (where sea clutter can be sometimes much worse than ground clutter). This allows using magnitude information for ship detection, although the false alarm ratio (FAR) is generally high, if one does not consider estimating velocity. On the other hand, unlike ground background, sea surface is not stationary at all. The entire ocean surface commonly acts as a single and large target moving at a nearly uniform velocity in a specific region. Consequently, the sea surface motion contributes a nonzero interferometric phase leading to a shift of phase center in the histogram of interferometric phase. Therefore, conventional ATI technique fails to meet the expected level of performance in GMTI; besides, it will produce severe false alarms when an improper phase threshold is selected. Yet, ATI technique is quite attractive for ship detection, even in situations with strong sea clutter, where pure magnitude-based methods can fail.

Therefore, in [125] and [126], the authors aimed at combining magnitude and phase information expecting to achieve improvement in the accuracy of detecting moving ships. This led to proposing an interesting SIMP metric for the detection of moving ships in ATI-SAR images. The results of the experiment validated the effectiveness of the SIMP metric. The problem of moving ship detection in SAR with ATI mode, assembled on the practical airborne and spaceborne platforms (e.g., AirSAR, TerraSAR-X, and RadarSAT-2), can be resolved by using the proposed SIMP metric because the underlying principle is the same. Because airborne and spaceborne geometries and parameters are different, the sensitivity and velocity estimation accuracy of those systems are also significantly different.

References

[1] Crisp D. J. *The state-of-the-art in ship detection in synthetic aperture radar imagery*. DSTO. Dept. Defence, Australian Government, Canberra, Australia. Public Release Document DSTO-RR-0272. 2004.

[2] Ouchi K. 'Current status on vessel detection and classification by synthetic aperture radar for maritime security and safety'. *Symposium on Remote Sensing for Environmental Sciences*. Gamagori, Aichi, Japan, September 2016. pp. 5–12.

[3] Brusch S., Lehner S., Fritz T., Soccorsi M., Soloviev A., and Schie, B. V. 'Ship surveillance with TerraSAR-X'. *IEEE Trans. Geosci. Remote Sens.* 2011, vol. 49(3), pp. 1092–1102.

[4] Eldhuset K. 'An automatic ship and ship wake detection system for spaceborne SAR images in coastal regions'. *IEEE Trans. Geosci. Remote Sens.* 1996, vol. 34(4), pp. 1010–1019.

[5] Wackerman C. C., Friedman K. S., Pichel W. G., Clemente-Colon P., and Li X. 'Automatic detection of ships in RADARSAT-1 SAR imagery'. *Can. J. Remote Sen.* 2011, vol. 27(5), pp. 568–577.

[6] Vachon P. W., Thomas S. J., and Cranton J. 'Validation of ship detection by the RADARSAT synthetic aperture radar and the ocean monitoring workstation'. *Can. J. Remote Sen.* 1999, vol. 27(5), pp. 568–577.

[7] Wang C., Liao M., and Li X. 'Ship detection in SAR image based on the alpha-stable distribution'. *Sensors.* 2008, 4948–4960.

[8] Liao M., Wang C., Wang Y., and Jiang L. 'Using SAR images to detect ships from sea clutter'. *IEEE Geosci. Remote Sens. Lett.* 2008, vol. 5(2), pp. 194–198.

[9] Cui Y. 'On semiparametric clutter estimation for ship detection in synthetic aperture radar images'. *IEEE Trans. Geosci. Remote Sens.* 2013, vol. 51(5), pp. 3170–3180.

[10] Margarit G., Mallorqui J., Guasch J. F., and Lopez-Martinez C. 'Phenomenological vessel scattering study based on simulated inverse SAR imagery'. *IEEE Trans. Geosci. Remote Sens.* 2009, vol. 47(4), pp. 1212–1222.

[11] Greidanus H., Alvarez M., Santamaria C., Thoorens F., Kourti N., and Argentieri P. 'The SUMO ship detector algorithm for satellite radar images'. *Remote Sens.* 2017, vol. 9(246), pp. 1–27.

[12] Ouchi K., Tamaki S., Yaguchi H., and Iehara M. 'Ship detection based on coherence images derived from cross correlation of multilook SAR images'. *IEEE Geosci. Remote Sens. Lett.* 2004, vol. 1(3), pp. 184–187.

[13] Wang Y., and Liu H. 'A hierarchical ship detection scheme for high-resolution SAR images'. *IEEE Trans. Geosci. Remote Sens.* 2012, vol. 50(10), pp. 4173–4184.

[14] Tello M., López-Martínez C., and Mallorquí J. J. 'A novel algorithm for ship detection in SAR imagery based on the wavelet transform'. *IEEE Geosci. Remote Sens. Lett.* 2005, vol. 2(2), pp. 201–205.

[15] Leung H., Dubash N., and Xie N. 'Detection of small objects in clutter using a GA-RBF neural network'. *IEEE Trans. Aerosp. Electr. Syst.* 2002, vol. 38(1), pp. 98–118.

[16] Lombardo P., and Sciotti M. 'Segmentation-based technique for ship detection in SAR images'. *IEE Proceedings on Radar, Sonar and Navigation.* 2001; 148 (3):147–159.

[17] Gao G. 'A Parzen-window-kernel-based CFAR algorithm for ship detection in SAR images'. *IEEE Geosci. Remote Sens. Lett.* 2011, vol. 8(3), pp. 557–561.

[18] Cui Y. 'On semiparametric clutter estimation for ship detection in synthetic aperture radar images'. *IEEE Trans. Geosci. Remote Sens.* 2013, vol. 51(5), pp. 3170–3180.

[19] Pelich R., Longepe N., Mercier G., Hajduch G., and Garello R. 'AIS-based evaluation of target detectors and SAR sensors characteristics for maritime surveillance'. *IEEE J. Sel. Topics Appl. Earth Observ. Remote Sens.* 2015, vol. 8(8), pp. 3892–3901.

[20] Wang C., Jiang S., Zhang H., Wu F., and Zhang B. 'Ship detection for high-resolution SAR images based on feature analysis'. *IEEE Geosci. Remote Sens. Lett.* 2014, vol. 11(1), pp. 119–123.

[21] Marino A., Sanjuan-Ferrer M. J., Hajnsek I., and Ouchi K. 'Ship detection with spectral analysis of synthetic aperture radar: A comparison of new and well-known algorithm'. *Remote Sens.* 2015, vol. 7, pp. 5416–5439.

[22] Gao G., Liu L., Zhao L., Shi G., and Kuang G. 'An adaptive and fast CFAR algorithm based on automatic censoring for target detection in high-resolution SAR images'. *IEEE Trans. Geosci. Remote Sens.* 2009, vol. 47(6), pp. 1685–1697.

[23] Martín-de-Nicolás J., Jarabo-Amores P., del-Rey-Maestre N., Gomez-del-Hoyo P., and Bárcena-Humanes, J. L. 'Robustness of a generalized gamma CFAR ship detector applied to TerraSAR-X and Sentinel-1 images'. In *Proc. IEEE Int. Conf. on Computer as a Tool (EUROCON)*; 2015, pp. 1–6.

[24] Gao G., 'Statistical Modeling of SAR Images: A Survey'. *Sensors.* 2010; 10:775–795.

[25] Fiche A., Angelliaume S., Rosenberg L., and Khenchaf A. 'Statistical analysis of low grazing angle high resolution X-band SAR sea clutter'. *IEEE Int. Radar Conf.*; 2014, pp. 1–6.

[26] Oliver C. J., and Quegan S. *Understanding synthetic aperture radar images.* Norwood, MA: Artech House, 1998.

[27] Jakeman E., and Pusey P. N. 'A model for non-Rayleigh sea echo'. *IEEE Trans. Antennas Propagat.* 1976, vol. 24, pp. 806–814.

[28] Goldstein G. 'False-alarm regulation in log-normal and Weibull clutter'. *IEEE Trans. Aerosp. Electron. Syst.* 1973, vol. AES-9(1), pp. 84–92.

[29] Trunk G. V., and George S. F., 'Detection of targets in non-Gaussian sea clutter'. *IEEE Trans. Aerosp. Electron. Syst.* 1970, vol. AES-6(5), pp. 620–628.

[30] Messina M., Greco M., and Pinelli G. 'Statistical characterization of high resolution SAR images under different sea state conditions'. *Presented at the IEEE Gold Remote Sens. Conf.*; 2010, pp. 1–5.
[31] Ward K. D. 'Compound representation of high resolution sea clutter'. *Electron. Lett.* 1981, vol. 17(16), pp. 561–563.
[32] Farshchian M., and Posner F. L. 'The Pareto distribution for low grazing angle and high resolution X-band sea clutter'. In *Proc. IEEE Radar Conf.*; 2010, pp. 789–793.
[33] Dana R., and Knepp D. 'The impact of strong scintillation on space based radar design II: Noncoherent detection'. *IEEE Trans. Aerosp. Electron. Syst.* 1986, vol. AES-22, pp. 34–36.
[34] Tison C., Nicolas J. M., Tupin F., and Maitre H. 'A new statistical model for Markovian classification of urban areas in high-resolution SAR images'. *IEEE Trans. Geosci. Remote Sens.* 2004, vol. 42(10), pp. 2046–2057.
[35] Kuruoglu E. E., and Zerubia J. 'Modeling SAR images with a generalization of the Rayleigh distribution'. *IEEE Trans. Image Process.* 2004, vol. 13(4), pp. 527–533.
[36] Migliaccio M., Ferrara G., Gambardella A., Nunziata F., and Sorrentino A. 'A physically consistent speckle model for marine SLC SAR images'. *IEEE J. Ocean. Eng.* 2007, vol. 32(4), pp. 839–847.
[37] Ferrara G., Migliaccio M., Nunziata F., and Sorrentino A. 'GK-based observation of metallic targets at sea in full-resolution SAR data: A multi-polarization study'. *IEEE J. Ocean. Eng.* 2011, vol. 36(2), pp. 195–204.
[38] Velotto D., Soccorsi M., and Lehner S., 'Azimuth ambiguities removal for ship detection using full polarimetric X-band SAR data'. *IEEE Trans. Geosci. Remote Sens.* 2014, vol. 52(1), pp. 76–88.
[39] Sahed M., Mezache A., and Laroussi T. 'A novel [zlogz]-based closed form approach to parameter estimation of K-distributed clutter plus noise for radar detection'. *IEEE Trans. Aerosp. Electron. Syst.* 2015, vol. 51(1), pp. 492–505.
[40] Rosenberg L., Watts S., and Bocquet S. 'Application of the K+Rayleigh distribution to high grazing angle sea-clutter'. *Presented at the IEEE Int. Radar Conf.*; 2014, pp. 1–6.
[41] Rosenberg L. and Bocquet S., 'Application of the Pareto plus noise distribution to medium grazing angle sea-clutter'. *IEEE J. Sel. Topics Appl. Earth Observ. Remote Sens.* 2015, vol. 8(1), pp. 255–261.
[42] Middleton D. 'New physical-statistical methods and models for clutter and reverberation: The KA distribution and related probability structures'. *IEEE J. Ocean. Eng.* 1999, vol. 24(3), pp. 261–284.
[43] Dong Y. *Distribution of X-band high resolution and high grazing angle sea clutter*. Def. Sci. Technol. Org., Canberra, ACT, Australia, Res. Rep. DSTO-RR-0316, 2006.
[44] Rosenberg L., Crisp D. J., and Stacy N. J. 'Analysis of the KK-distribution with medium grazing angle sea-clutter'. *IET Radar, Sonar Navigat.* 2010, vol. 4(2), pp. 209–222.

[45] Fiche A., Angelliaume S., Rosenberg L., and Khenchaf A. 'Analysis of X-band SAR sea-clutter distributions at different grazing angles'. *IEEE Trans. Geosci. Remote Sens*. 2015, vol. 53(8), pp. 4650–4660.

[46] Achim A., Kuruoglu E. E., and Zerubia J. 'SAR image filtering based on the heavy-tailed Rayleigh model'. *IEEE Trans. Image Process*. 2006, vol. 15(9), pp. 2686–2693.

[47] Vachon P. W., Thomas S. J., and Cranton J. 'Validation of ship detection by the RADARSAT synthetic aperture radar and the ocean monitoring workstation'. *Can. J. Remote Sen*. 1999, vol. 27(5), pp. 568–577.

[48] Brekke C., and Anfinsen S. N. 'Ship detection in ice-infested waters based on dual-polarization SAR imagery'. *IEEE Geosci. Remote Sens. Lett.* 2011, vol. 8(3), pp. 391–395.

[49] Nicolas J. M. 'Introduction to second kind statistic: Application of log-moments and log-cumulants to SAR image law analysis'. *Trait. Signal.* 2002, vol. 19(3), pp. 139–167.

[50] Crisp D. J, Rosenberg L., Stacy N. J., and Dong Y. 'Modelling X-band sea clutter with the K-distribution: Shape parameter variation'. *IEEE Int. Radar Conf.*; 2009, pp. 1–6.

[51] Paes R. L., Lorenzzetti J. A., and Gherardi D. F. M., 'Ship detection using TerraSAR-X images in the Campos Basin (Brazil) '. *IEEE Geosci. Remote Sens. Lett.* 2010, vol. 7(3), pp. 545–548.

[52] Gierull C. H., and Sikaneta I. C. 'Estimating the effective number of looks in interferometric SAR data'. *IEEE Trans. Geosci. Remote Sens*. 2002, vol. 40 (8), pp. 1733–1742.

[53] Krylov V. A., Moser G., Serpico S. B., and Zerubia J. 'On the method of logarithmic cumulants for parametric probability density function estimation'. *IEEE Trans. Image Process*. 2013, vol. 22(10), pp. 3791–3806.

[54] Gao G., Gao S., Ouyang K., He J., and Li G. 'Scheme for characterizing clutter statistics in SAR amplitude images by combining two parametric models'. *IEEE Trans. Geosci. Remote Sens*. 2018, vol. 56(10), pp. 5636–5646.

[55] Stacy W. 'A generalization of the gamma distribution'. *Ann. Math. Statist.* 1962, vol. 33(3), pp. 1187–1192.

[56] Choy S. K., and Tong C. S. 'Statistical wavelet subband characterization based on generalized gamma density and its application in texture retrieval'. *IEEE Trans. Image Process*. 2010, vol. 19(2), pp. 281–289.

[57] Cui S., and Datcu M. 'Statistical wavelet subband modeling for multi-temporal SAR change detection'. *IEEE J. Sel. Topics Appl. Earth Observ. Remote Sens*. 2012, vol. 5(4), pp. 1095–1109.

[58] Hirose H. 'Maximum likelihood parameter estimation by model augmentation with applications to the extended four-parameter generalized gamma distribution'. *Math. Comput. Simul.* 2000, vol. 54, pp. 81–97.

[59] Shin J. W., Chang J. H., and Kim N. S. 'Statistical modeling of speech signals based on generalized gamma distribution'. *IEEE Signal Process. Lett.* 2005, vol. 12(3), pp. 258–261.

[60] Chang J. H., Shin J. W., Kim N. S., and Mitra S. K. 'Image probability distribution based on generalized gamma function'. *IEEE Signal Process. Lett.* 2005, vol. 12(4), pp. 325–328.

[61] Eltoft T., 'Modeling the amplitude statistics of ultrasonic images'. *IEEE Trans. Med. Imag.* 2006, vol. 25(2), pp. 229–240.

[62] Li H. C., Hong W., Wu Y. R., and Fan P. Z., 'On the empirical-statistical modeling of SAR images with generalized gamma distribution'. *IEEE J. Sel. Top. Signal Process.* 2011, vol. 5(3), pp. 386–397.

[63] Qin X., Zhou S., Zou H., and Gao G., 'Statistical modeling of sea clutter in high-resolution SAR images using generalized gamma distribution'. In *Proc. IEEE Comput. Vision in Remote Sen. (CVRS);* Xiamen, China, 2012, pp. 306–310.

[64] Martín-de-Nicolás J., Jarabo-Amores M. P., Mata-Moya D., del-Rey-Maestre N., and Bárcena-Humanes J. L. 'Statistical analysis of SAR sea clutter for classification purposes'. *Remote Sens.* 2014, vol. 6(10), pp. 9379–9411.

[65] Gao G., Ouyang K., Luo Y., Liang S., and Zhou S. 'Scheme of parameter estimation for generalized gamma distribution and its application to ship detection in SAR images'. *IEEE Trans. Geosci. Remote Sens.* 2017, vol. 55 (3), pp. 1812–1832.

[66] Gao G., Shi G., Li G., and Cheng J. 'Performance comparison between reflection symmetry metric and product of multilook amplitudes for ship detection in dual-polarization SAR images'. *IEEE J. Sel. Topics Appl. Earth Observ. Remote Sens.* 2017, vol. 10(11), pp. 5026–5038.

[67] Liu C., Vachon P. W., and Geling G. W., 'Improved ship detection with airborne polarimetric SAR data'. *Canadian Journal of Remote Sensing.* 2005; 31:122–131.

[68] Liu C., and Meek A. *Likelihood ratio test polarimetric SAR ship detection application*. Defense R&D Canada, Ottawa, Technical Memorandum, TM 2005-243, 2005.

[69] Touzi R., Charbonneau F., Hawkins R. K., and Vachon P. W. 'Ship detection and characterization using polarimetric SAR'. *Canada Journal of Remote Sensing.* 2004; 30:552–559.

[70] Touzi R., and Charbonneau F. 'Characterization of target symmetric scattering using polarimetric SARs'. *IEEE Trans.Geosci.Remote Sens.* 2002, vol. 40(11), pp. 2507–2516.

[71] Velotto D., Nunzaita F., Migliaccio M., and Lehner S. 'Dual-polarimetricTerraSAR-X SAR data for target at sea observation'. *IEEE Geosci. Remote Sens. Lett.* 2013. vol. 10(5), pp. 1114–1118.

[72] Nunziata F., and Migliaccio M. 'On the COSMO-SkyMedPingPong mode to observe metallic targets at sea'. *IEEE J. Ocean. Eng.* 2013, vol. 38(1), pp. 71–79.

[73] Arnesen T., and Olsen R. B. *Literature review on vessel detection*. Forsvarets Forskningsinstitutt, Kjeller, Norway, FFI-Rapport-2004/02619, 2004.

[74] Novak L. M., and Burl M. C. 'Optimal speckle reduction in polarimetric SAR imagery'. *IEEE Trans. Aerosp. Electron. Syst.* 1990, vol. 26(2), pp. 293–305.

[75] Chaney R. D., Bud M. C., and Novak L. M. 'On the performance of polarimetric target detection algorithms'. *IEEE Aerosp. Electron. Syst.Magazine.* 1990, vol. 5(11), pp. 10–15.
[76] Sciotti M., Pastina D., and Lombardo P. 'Polarimetric detectors of extended targets for ship detection in SAR images'. *IGARSS'01.* Sydney, Australia, 2001, pp. 3132–3134.
[77] Chen J., Chen Y., and Yang J., 'Ship detection using polarization cross-entropy'. *IEEE Geosci. Remote Sens. Lett.* 2009, vol. 6(4), pp. 723–727.
[78] Marino A., Walker N., and Woodhouse I. 'Ship detection with RADARSAT-2 quad-pol SAR data using a notch filter based on perturbation analysis'. *IGARSS'10*, 2010, pp.3704–3707.
[79] Cloude R., and Pottier E. 'A review of target decomposition theory in radar polarimetry'. *IEEE Trans. Geosci. Remote Sens.* 1996, vol. 34(2), pp. 498–518.
[80] Gao G., Luo Y., Ouyang K., and Zhou S. 'Statistical modeling of PMA detector for ship detection in high-resolution dual-polarization SAR images'. *IEEE Trans.Geosci. Remote Sens.* 2016, vol. 54(7), pp. 4302–4313.
[81] Marino A. 'A notch filter for ship detection with polarimetric SAR data'. *IEEE J. Sel. Topics Appl. Earth Observ. Remote Sens.* 2013, vol. 6(3), pp. 1219–1232.
[82] Marino A., and Hajnsek I. 'Statistical tests for a ship detector based on the polarimetric notch filter'. *IEEE Trans. Geosci. Remote Sens.* 2015, vol. 53(8), pp. 4578–4595.
[83] Marino A., Sugimoto M., Ouchi K., and Hajnsek I. 'Validating a notch filter for detection of targets at sea with ALOS-PALSAR data: Tokyo Bay'. *IEEE J. Sel. Topics Appl. Earth Observ. Remote Sens.* 2014, vol. 7(12), pp. 4907–4918.
[84] Marino A., Cloude S. R., and Woodhouse I. H. 'Detecting depolarized targets using a new geometrical perturbation filter'. *IEEE Trans. Geosci. Remote Sens.* 2012, vol. 50(10), pp. 3787–3799.
[85] Charbonneau F. J., Brisco B., Raney R. K., et al. 'Compact polarimetry overview and applications assessment'. *Can. J. Remote Sen.* 2010, vol. 36(s2), pp. S298–S315.
[86] Raney R. K. 'Dual-polarized SAR and Stokes parameters'. *IEEE. Geosci. Remote Sens. Lett.* 2006, vol. 3(3), pp. 317–319.
[87] Souyris J. C., Imbo P., Fjortoft R., Mingot S., and Lee J. S. 'Compact polarimetry based on symmetry properties of geophysical media: The $\pi/4$ mode'. *IEEE Trans. Geosci. Remote Sens.* 2005, vol. 43(3), pp. 634–646.
[88] Ghods S., Shojaedini S. V., and Maghsoudi Y. 'A modified H-α classification method for DCP compact polarimetric mode by reconstructing quad H and α parameters from dual ones'. *IEEE J. Sel. Topics Appl. Earth Observ. Remote Sens.* 2016, vol. 9(6), pp. 2233–2241.
[89] Nord M. E., Ainsworth T. L., Lee J. S., and Stacy N. J. S. 'Comparison of compact polarimetric synthetic aperture radar modes'. *IEEE Trans. Geosci. Remote Sens.* 2009, vol. 47(1), pp. 174–188.

[90] Raney R. K. 'Hybrid-polarity SAR architecture'. *IEEE Trans. Geosci. Remote Sens*. 2007, vol. 45(11), pp. 3397–3404.

[91] Raney R. K., Cahill J. T. S., Patterson G. W., and Bussey D. B. J. 'The *m-chi* decomposition of hybrid dual-polarimetric radar data with application to lunar craters'. *J. Geophys. Res.* 2012, vol. 117(E12), Art. No. E00H21.

[92] Dabboor M., and Geldsetzer T. 'Towards sea ice classification using simulated RADARSAT constellation mission compact polarimetric SAR imagery'. *Remote Sens. Environ.* 2014, vol. 140, pp. 189–195.

[93] Yin J., Yang J., and Zhang X. 'On the ship detection performance with compact polarimetry'. In *Proc. IEEE Radar Conf.*; 2011, pp. 675–680.

[94] Collins M. J., Denbina M., and Atteia G. E. 'On the reconstruction of quad-pol SAR data from compact polarimetry data for ocean target detection'. *IEEE Trans. Geosci. Remote Sens*. 2013, vol. 51(1), pp. 591–600.

[95] Atteia G. E. *On the Use of Hybrid Compact Polarimetric SAR for Ship Detection*. Ph.D. dissertation, Univ. Calgary, Alberta, 2014.

[96] Atteia G. E., and Collins M. J. 'On the use of compact polarimetry SAR for ship detection'. *ISPRS J. Photogramm. Remote Sens*. 2013, vol. 80, pp. 1–9.

[97] Liu C., Vachon P. W., English R. A., and Sandirasegaram N. *Ship detection using RADARSAT-2 fine quad mode and simulated compact polarimetry data*. Defence R&D Canada, Ottawa, ON, Canada, 2009, Technical Memo.

[98] Raney R. K., 'Comparing compact and quadrature polarimetric SAR performance'. *IEEE. Geosci. Remote Sens. Lett.* 2016, vol. 13(6), pp. 861–864.

[99] Shirvany R., Chabert M., and Tourneret J. Y. 'Ship and oil-spill detection using the degree of polarization in linear and hybrid/compact dual-pol SAR'. *IEEE J. Sel. Topics Appl. Earth Observ. Remote Sens*. 2012, vol. 5(3), pp. 885–892.

[100] Shirvany R., Chabert M., and Tourneret J. Y., 'Comparison of ship detection performance based on the degree of polarization in hybrid/compact and linear dual-pol SAR imagery'. *Presented at the IGARSS Conf.*; Vancouver, BC, Canada, 2011, pp. 3550–3553.

[101] Touzi R., and Vachon P. W. 'RCM polarimetric SAR for enhanced ship detection and classification'. *Can. J. Remote Sen*. 2015, vol. 41, pp. 473–484.

[102] Li H., Perrie W., He Y., Lehner S., and Brusch S. 'Target detection on the ocean with the relative phase of compact polarimetry SAR'. *IEEE Trans. Geosci. Remote Sens*. 2013, vol. 51(6), pp. 3299–3305.

[103] Paes R. L., Nunziata F., and Migliaccio M. 'On the capability of hybrid-polarity features to observe metallic targets at sea'. *IEEE J. Ocean. Eng.* 2016, vol. 41(2), pp. 346–361.

[104] Cloude S. R., Goodenough D. G., and Chen H. 'Compact decomposition theory'. *IEEE. Geosci. Remote Sens. Lett.* 2012, vol. 9(1), pp. 28–32.

[105] Yin J., and Yang J. 'Ship detection by using the *m-chi* and *m-delta* decompositions'. *Presented at the IGARSS Conf.*; Quebec City, QC, Canada, 2014, pp. 2738–2741.

[106] Yin J., Yang J., Zhou Z. S., and Song J. 'The extended Bragg scattering model-based method for ship and oil-spill observation using compact

polarimetric SAR'. *IEEE J. Sel. Topics Appl. Earth Observ. Remote Sens.* 2015, vol. 8(8), pp. 3760–3772.
[107] Xu L., Zhang H., Wang C., Zhang B., and Tian S. 'Compact polarimetric SAR ship detection with m-δ decomposition using visual attention model'. *Remote Sens.* 2013, vol. 8(751), pp. 1–16.
[108] Xie L., Zhang H., Wang C., Wu F., Zhang B., and Tang Y. 'Maritime application using H-α decomposition in compact and dual-pol SAR'. *Presented at the APSAR Conf.*; 2013, pp. 563–566.
[109] Lukowski T. I. 'Detection of vessels in littoral waters in two-channel XWEAR SAR imagery'. In *Proc. IEEE Radar Conf.*; 2009, pp. 1–5.
[110] Suchandt S., Runge H., and Steinbrecher U. 'Ship detection and measurement using the TerraSAR-X dual-receive antenna mode'. In *Proc. IGARSS*; 2010, pp. 2860–2863.
[111] Kojima S., 'Evaluation of the ship detection by dual polarimetric along-track interferometry'. In *Proc. APSAR*; 2013, pp. 589–592.
[112] Schulz K., Soergel U., and Thoennessen U. 'Segmentation of moving objects in SAR-MTI data'. In *Proc. SPIE*; 2001, vol. 4382, pp. 174–181.
[113] Arnaud A. 'Ship detection by SAR interferometry'. In *Proc. IGARSS*; 1999, vol. 5, pp. 2616–2618.
[114] Baumgartner S. V., and Krieger G. 'Ship detection and motion parameter estimation with TanDEM-X in large along-track baseline configuration'. In *Proc. SEASAR;* Tromso, Norway, 2012, pp. 1–4.
[115] Cerutti-Maori D., Sikaneta I., and Gierull C. H. 'Ship detection with spaceborne multi-channel SAR/GMTI radars," In *Proc. EUSAR*; 2012, pp. 400–403.
[116] Makhoul E., Broquetas A., Rodon J. R., Zhan Y., and Ceba F., 'A performance evaluation of SAR-GMTI missions for maritime applications'. *IEEE Trans. Geosci. Remote Sens.* 2015, vol. 53(5), pp. 2496–2509.
[117] Cerutti-Maori D., and Sikaneta I. 'A generalization of DPCA processing for multichannel SAR/GMTI radars'. *IEEE Trans. Geosci. Remote Sens.* 2013, vol. 51(1), pp. 560–572.
[118] Zhang Y. *Along track interferometry synthetic aperture radar (ATI-SAR) techniques for ground moving target detection.* Air Force Research Laboratory/SNRT, Rome New York, Public Release Document AFRL-SN-RS-TR-2005-410, 2006.
[119] Cerutti-Maori D., Sikaneta I., and Gierull C. H. 'Optimum SAR/GMTI processing and its application to the radar satellite RADARSAT-2 for traffic monitoring'. *IEEE Trans. Geosci. Remote Sens.* 2012, vol. 50(10), pp. 3868–3881.
[120] Gracheva V., and Ender J. 'Multichannel analysis and suppression of sea clutter for airborne microwave radar systems'. *IEEE Trans. Geosci. Remote Sens.* 2016, vol. 54(4), pp. 2385–2399.
[121] Ender J. H. G. 'Space-time processing for multichannel synthetic aperture radar'. *Electron. Commun. Eng. J.* 1999, vol. 11(1), pp. 29–38.

[122] Ender J. H. G., 'Detection and estimation of moving target signals by multichannel SAR'. *Int. J. Electron. Commun.* 1996, vol. 50(2), pp. 150–156.

[123] Gonzalez A. S., Labriola M., Soteras J. C., and Palma J. S. 'PAZ instrument design and performance'. In *Proc. APSAR*; 2011, pp. 1–4.

[124] Gierull C. H. 'Analysis of the multimode feedhorn concept for multichannel SAR-GMTI'. *IEEE Trans. Geosci. Remote Sens.* 2011, vol. 49(10), pp. 3611–3621.

[125] Gao G., Wang X., and Lai T. 'Detection of moving ships based on a combination of magnitude and phase in along-track interferometric SAR-Part I: SIMP metric and its performance'. *IEEE Trans. Geosci. Remote Sens.* 2015, vol. 53(7), pp. 3565–3581.

[126] Gao G., Wang X., and Lai T. 'Detection of moving ships based on a combination of magnitude and phase in along-track interferometric SAR-Part II: Statistical modeling and CFAR detection'. *IEEE Trans. Geosci. Remote Sens.* 2015, vol. 53(7), pp. 3582–3599.

Chapter 7

Monitoring of intertidal areas and coastal habitats

Martin Gade[1]

Intertidal zones are coastal areas that fall dry once during each tidal cycle. This chapter demonstrates how high-resolution synthetic aperture radar (SAR) imagery of coastal zones can be used to study morphodynamical changes in these zones, since exposed intertidal flats show up as dark or bright patches on SAR imagery. In addition, submerged sandbanks can cause SAR image signatures through variations of the surface current field. Multifrequency SAR data can be used for the derivation of surface roughness parameters, and the polarimetric decomposition of dual-copolarization SAR data helps in identifying coastal habitats. And finally, we demonstrate how high-resolution SAR imagery can be used for archaeological surveys in areas that are difficult to access.

7.1 Introduction

Intertidal zones at the interface between the land and the open sea are particularly vulnerable to anthropogenic impacts and to global change [1]. They can be found in several coastal regions worldwide, including the European North Sea and Atlantic coasts, the Korean and Chinese coasts, and the North-American Atlantic coast. Intertidal zones fall dry once during each tidal cycle, and they usually consist of fine sandy or muddy sediments, or a mixture of both, but may also contain rocky areas. These sensitive ecosystems may be partly vegetated and show a high biodiversity, thereby making them a key site for millions of migratory birds [2]. The Wadden Sea on the North Sea coast of Continental Europe is a UNESCO World Heritage, and national and international directives require a continuous monitoring [3–5]. However, because of the limited accessibility, which allows in situ monitoring and assessment only during short periods around low tide, satellite remote sensing techniques have been the method of choice for the monitoring and assessment of intertidal zones since the early 1980s [6]. Traditional methods used for the monitoring and classification of coastal environments are based on optical and infrared (i.e., multispectral) sensors.

[1]Institut für Meereskunde, Universität Hamburg, Hamburg, Germany

More recently, the use of SAR data as complementing input to existing monitoring and classification schemes was suggested [6,7].

As an example, Figure 7.1 shows a composite of two Sentinel-1A SAR-C images (VV-polarization, Interferometric Wide swath mode) of the German and Danish North Sea coast acquired on 21 November 2018 at 17:09 UTC. Exposed intertidal flats can be seen as dark and bright patches all along the coast, reaching into the open sea or connecting the islands with the mainland. Note that these exposed intertidal flats, particularly their edges, appear much brighter in the lower part of the SAR image composite, that is, in the southern part of the area. Because of the propagation of the tidal wave from southwest toward north, low tide in the southwestern part of the area had already been before the image acquisition, while it was shortly after image acquisition in the northern part. Consequently, the sediments on the exposed flats in the southwest are dryer (i.e., the soil moisture is lower) than in the north. Given that the sediment composition is about the same in most parts of the imaged area, this example demonstrates the difficulties that can arise when SAR imagery of coastal wetlands and intertidal flats is analyzed and when those analyses are based upon the absolute radar backscatter value.

Only exposed sandbanks, mudflats, or rocks can be seen on SAR imagery directly, whereas submerged topography may cause SAR signatures through surface current variations [8]. If a clear identification of the waterline (i.e., the border between the sea and the exposed flats) is possible, SAR data can be used for the detection of shorelines [9] and for topographic mapping [10]. This identification, however, may be difficult, when the radar return from the wind-roughened sea is similar to that of the exposed flats. In order to better understand the radar backscattering from exposed intertidal flats, therefore, SAR signatures of different bare sediments were investigated. It was shown that the surface roughness and thus, the strength of the radar return, decreases with the amount of mud and increases with the median grain size [11]. Additional features influencing the surface roughness are biological activity and puddles of remnant water, whose surface roughness may dominate over that of the bare sediment [12].

SAR signatures retrieved by different sensors were related to surface roughness measurements and model simulations [13–15]. Here, the Integral Equation Model (IEM) [16,17] was shown to provide the best estimates of the radar backscattering from exposed intertidal flats and it was already used for a crude sediment classification [18]. However, the application of analytical models such as the IEM to multiple SAR images may be hampered, when those SAR data were acquired from different platforms and at different times: since most intertidal zones show a very flat topography, only slight differences in the local water level may result in different backscatter patterns and thus, may make any sediment classification more difficult.

In addition to the analyses of various SAR images acquired at single or multiple frequencies, it was already demonstrated that multipolarization SAR data can be used to distinguish vegetated areas from bare soil [19], to classify sand- and mudflats [20] and to detect bivalve beds [14,21,22]. More recently, the influence of environmental conditions and imaging geometry on dominant radar backscattering mechanisms on intertidal flats was investigated using dual-copolarization SAR data [23].

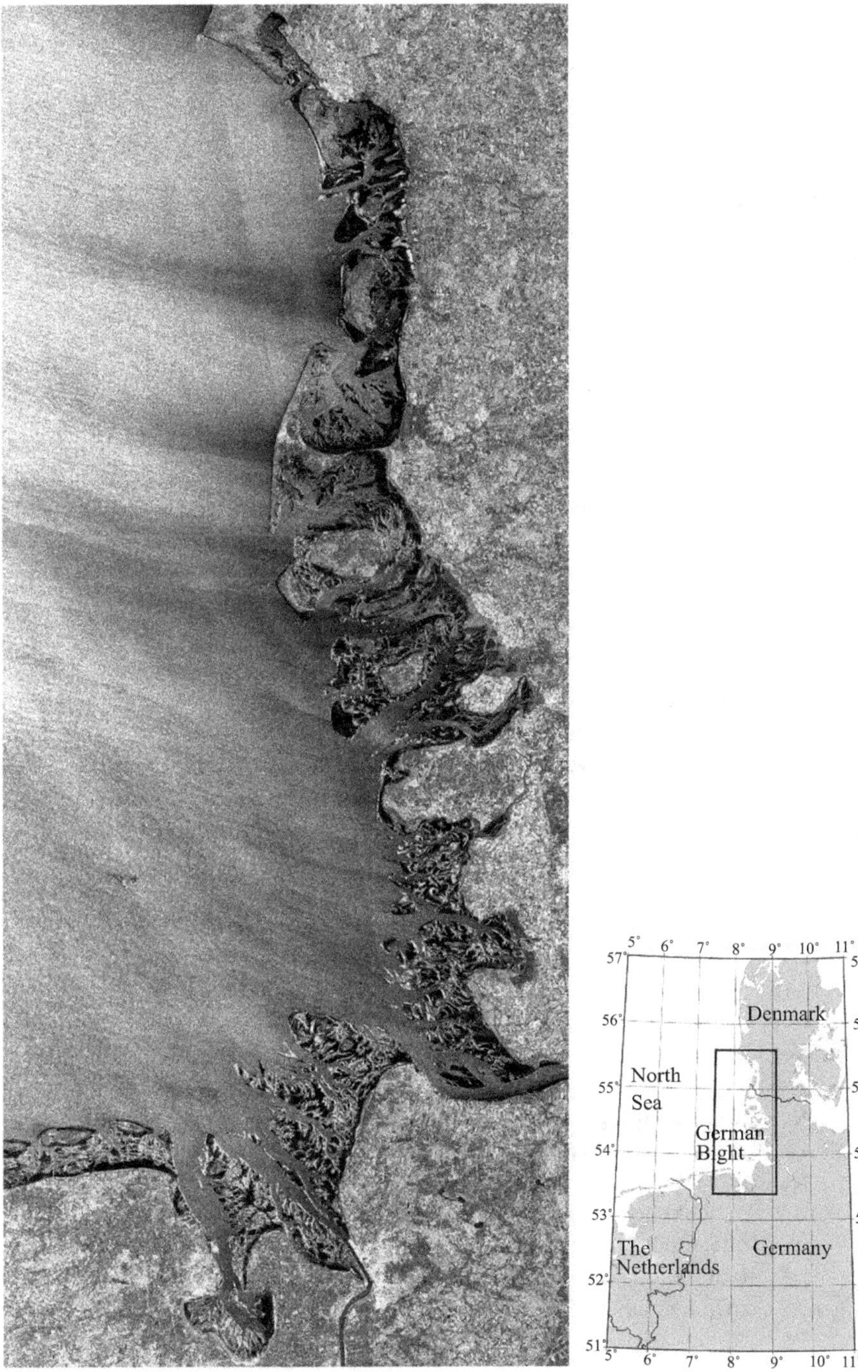

Figure 7.1 Composite of two Sentinel-1A SAR-C images of the German Bight of the North Sea (see map) acquired on 21 November 2018. Exposed intertidal flats can be seen all along the coast. © ESA 2018

In this chapter, we give a broad overview of the way, in which coastal sandbanks, submerged or exposed, bare or vegetated, show up on SAR imagery, and how routine SAR monitoring can aid in the identification of morphological changes and in the detection of habitats. Section 7.2 provides examples of early SAR images of sea bottom topography, adjacent to exposed sand flats; Section 7.3 shows how temporal changes and manmade objects can be seen on SAR imagery. Sections 7.4 and 7.5 demonstrate how roughness parameters can be derived and how bivalve beds can be detected using multifrequency and multipolarization SAR imagery, respectively. High-resolution SAR imagery is then used in Section 7.6 to demonstrate how SAR data can complement archaeological surveys. Finally, a brief summary is given Section 7.7.

7.2 Signatures of sea bottom topography

The American Seasat was the first satellite mission that was particularly devoted to remote sensing of the World's oceans using, among others, an L-band SAR operating at horizontal (HH) polarization. Although the satellite functioned for only 105 days, from June through October 1978, it delivered a wealth of SAR data that provided new views at the oceans and at coastal waters in particular.

Among the many features that had not been seen on and studied using SAR imagery before are those shown in Figure 7.2. The SAR image was acquired on 19 August 1978 and shows signatures of elongated underwater sandbars in the English Channel (see the map). The strong tidal currents in that area are modulated by the underwater topography, which in turn causes a modulation of the sea surface roughness sensed by the SAR [24]. Basically, if the environmental conditions (wind speed and direction, tidal current field) at the time of SAR image acquisition are known, and if some depth measurements are available to tune the system, such SAR data can be inverted into bathymetric maps [25].

Two more examples of Seasat SAR images with surface features associated with underwater sandbanks are shown in Figure 7.3. The same area, the Dutch coast and the West Frisian Islands north of the IJsselmeer (Lake IJssel), was imaged at different tidal phases and thus different water levels and current conditions. Note the different current patterns in the lower left image parts, north off the Affsluitdijk (the major dike that closes the IJsselmeer from the North Sea).

Also note the several dark patches, between the islands and the mainland, that are due to exposed sandflats: the different sea levels during both image acquisitions manifest in different areas that are exposed. Obviously, those sandflats cause a low radar backscatter at L-band and therefore, they can be easily discriminated from flooded areas when the sea surface is roughened by the wind. Such SAR data can be used to generate topographic maps, if precise information about the local water levels is available (e.g., from tide gauges and high-resolution numerical models), along with a sufficient amount of SAR images acquired at different tidal phases [10].

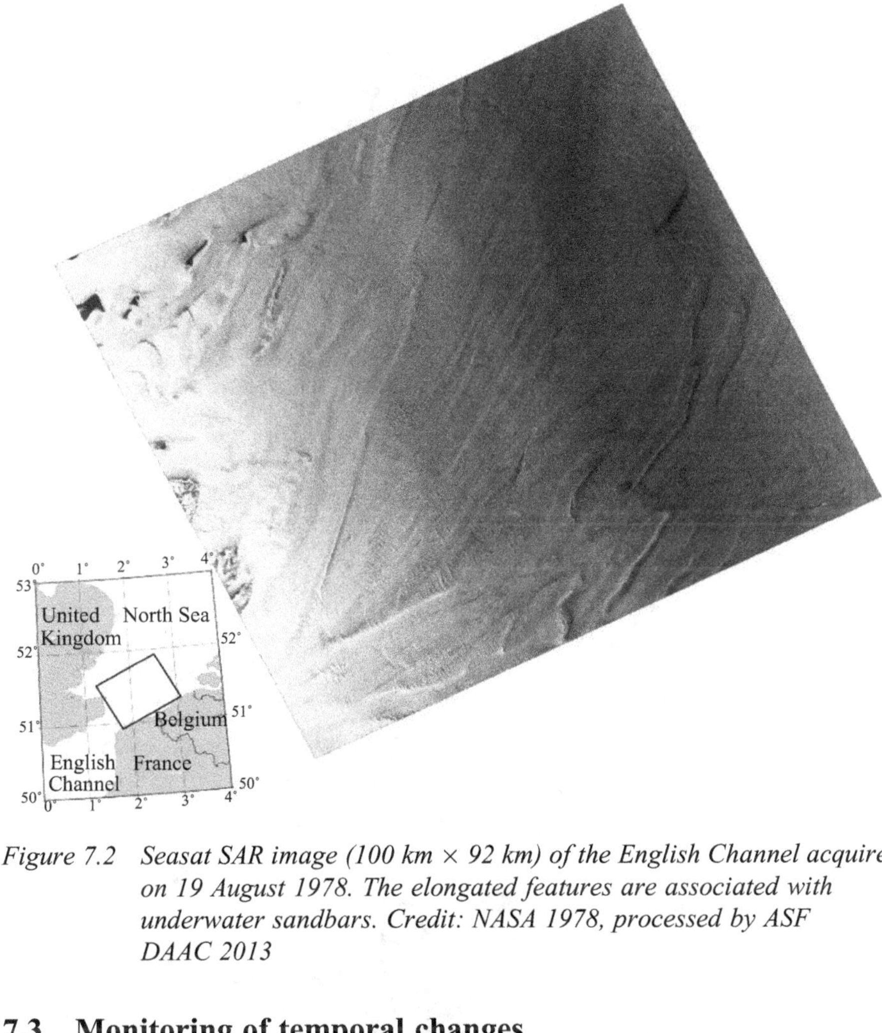

Figure 7.2 Seasat SAR image (100 km × 92 km) of the English Channel acquired on 19 August 1978. The elongated features are associated with underwater sandbars. Credit: NASA 1978, processed by ASF DAAC 2013

7.3 Monitoring of temporal changes

Care has to be taken when multiple SAR images of the same spot in an intertidal zone are analyzed and compared: not only the water level may be (slightly) different, causing different fractions of the flats to be exposed, but also the soil moisture may vary, because of either different water level or different wind speed or a different history of both [23]. The soil moisture, however, is one of the parameters influencing the radar backscattering, as can be seen in Figure 7.1.

From the early 1990s on, various European (ERS-1/2, EnviSAT, Sentinel-1A/B), Japanese (JERS-1, ALOS-1/2), Canadian (Radarsat 1/2), and German (TerraSAR-X, TanDEM-X) satellite missions have provided a wealth of SAR data that can be used to investigate the radar backscattering from the same (or similar) targets at different frequencies and polarizations, but also to monitor temporal changes. In order to

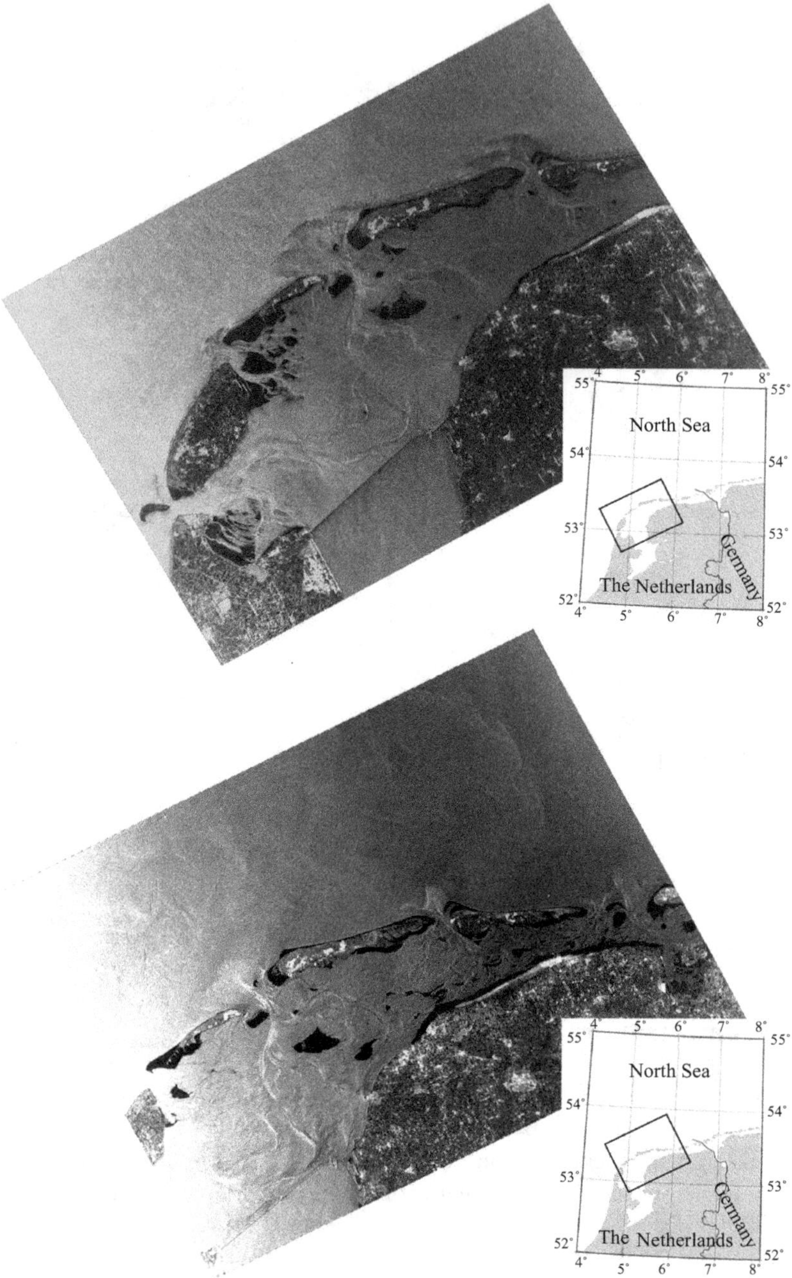

Figure 7.3 *Two Seasat SAR images (upper: 100 km × 77 km, 28 August 1978; lower: 100 km × 73 km, 9 October 1978) of the Dutch coast and the West Frisian Islands. Elongated and curved bright and (slightly) dark features are due to underwater sandbanks, and dark patches are exposed intertidal flats. Credit: NASA 1978, processed by ASF DAAC 2013*

demonstrate this, Figure 7.4 shows two C-band SAR images of the same site on the Chinese coast, off Rudong County, Jiangsu Province, north of the Yangtse estuary. The SAR images were acquired by the ERS-1 SAR on 10 June 1993 (upper panel) and by the Sentinel-1A SAR-C on 19 December 2018 (lower panel), that is, more than 25 years later.

Exposed intertidal flats along the coast can be seen on both SAR images in Figure 7.4, along with an extended intertidal area in the image's upper (northern) part. This latter extended system of sandbanks and mudflats is sometimes referred to as "Xinchuan Gang Shoals," It is obvious that the wind speed was higher during the first image acquisition, resulting in a stronger radar backscatter from the sea (higher image brightness). This stronger wind may also be responsible for the distinct textures on the open sandbanks, that is, the strong variation between dark and bright areas on the exposed flats, likely corresponding to higher and lower sediment moisture, respectively [23].

The intertidal flats along the coast, in the lower part of both SAR images, underwent only smaller changes during those 25 years between the image acquisitions, and only minor differences can be seen at some places. Major changes, however, can be found in areas of strong coastal developments, which are marked by two letters "B" in the small panel on the middle right of Figure 7.4. Here, extensive land reclamation has resulted in changes in the coastline and therefore, of the topography in the intertidal range. The letter "G" in the panel's center marks the Rudong offshore LNG (liquefied natural gas) terminal that was built in the late 2000s (operational since 2011). The terminal is clearly visible as a bright construction in the center of the Sentinel-1A SAR-C image (lower panel).

Letters in the small panel in the middle right of Figure 7.4 also indicate other areas, where temporal changes can be identified. The major tidal channel marked by letter "C" has changed its position and width, and extensive aquacultures started in recent years. Figure 7.5 shows (25 km × 25 km) details of two ERS-1 SAR images (upper row) and an EnviSat ASAR and a Sentinel-1A SAR-C image (lower row) that demonstrate these changes in greater detail.

It is obvious that the morphological changes in that area between 1993 and 2000 (upper two panels) were smaller than thereafter. The major tidal channel (marked "C" in Figure 7.4) appears in the upper left corner of the EnviSat ASAR image acquired in 2009, and it had extended toward south by the time of the Sentinel-1A SAR-C acquisition in 2017. This development occurred along with a general shift of the intertidal flats toward south-east, resulting in more exposed flats in the depicted area in 2017. Note that the developing, but still submerged, sandbanks caused surface roughness variations (as described earlier) in the EnviSat ASAR image acquired in 2009 (lower left). Similar brightness variations can also be seen in the first ERS-1 SAR image (upper left), but are less pronounced, because of the higher wind speed at that time.

Also note that the exposed flats appear flatter on the Sentinel-1A SAR-C image (lower right), with less variable radar backscattering (image brightness). Strip-shaped patterns can be identified on most of them, indicating extensive aquaculture (mainly seaweed; letters "A" in Figure 7.4). The growing area of those strip-shaped patterns demonstrates the ongoing intensification of aquaculture in the study area.

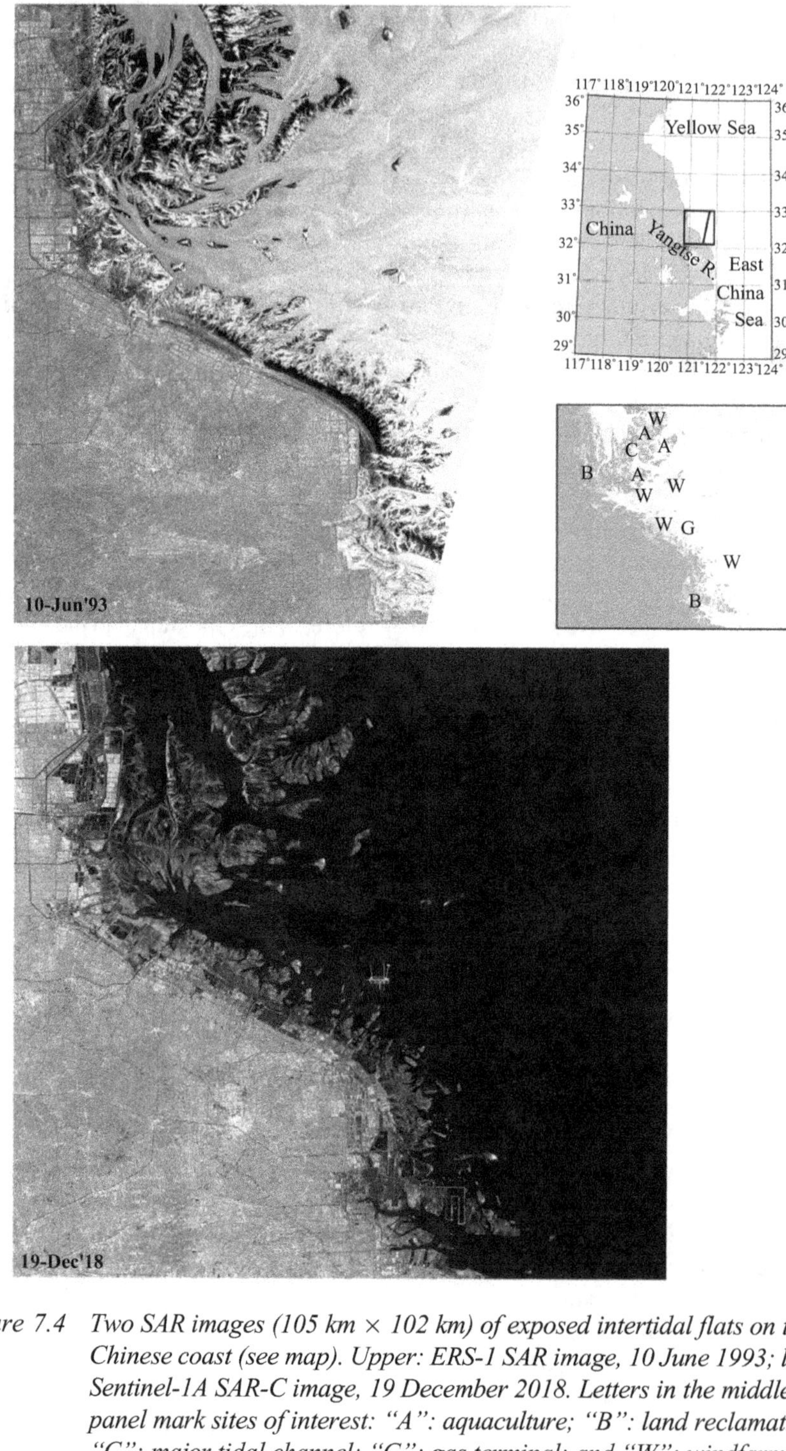

Figure 7.4 Two SAR images (105 km × 102 km) of exposed intertidal flats on the Chinese coast (see map). Upper: ERS-1 SAR image, 10 June 1993; lower: Sentinel-1A SAR-C image, 19 December 2018. Letters in the middle right panel mark sites of interest: "A": aquaculture; "B": land reclamation; "C": major tidal channel; "G": gas terminal; and "W": windfarms. SAR images. © ESA 1993, 2018

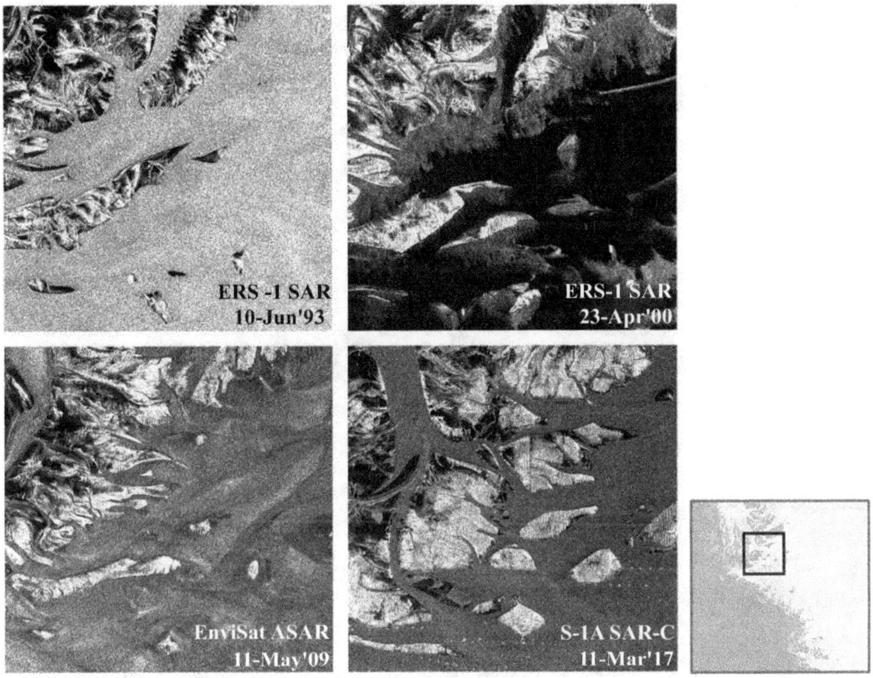

Figure 7.5 Morphological changes and aquaculture on extended intertidal flats off Rudong County. Details (25 km × 25 km) of two ERS-1 SAR (upper row), an EnviSat ASAR and a Sentinel-1A SAR-C (lower row) images acquired between 1993 and 2017 show that major changes occurred in recent years. The lower right panel depicts the location inside the SAR images shown in Figure 7.4. © ESA 1993, 2000, 2009, 2017

In recent years, a number of offshore windfarms, as marked by letters "W" in Figure 7.4, were built in the coastal area off Rudong County. The first one, the Jiangsu Rudong intertidal offshore wind farm, was built from 2011, and it was followed by more offshore windfarms in that area, which, by late 2018 (i.e., by the time of the acquisition of the second SAR image in Figure 7.4) comprised more than 400 wind turbines in total.

Offshore wind parks are known to have an impact on the (downstream) surface wind and current fields, as they are causing long wakes [26], but also turbulence both in the air [27] and in the water [28]. Figure 7.6 shows the details of three Sentinel-1A SAR-C images that demonstrate these effects.

The upper left panel (image acquired on 21 April 2016) shows that the monopiles in the water enhance the turbulence locally. This effect appears to be more pronounced at the western part of the wind park, where the water depth is shallowest (well below 10 m; note the exposed flats just west of the wind park). The upper right panel (17 December 2016) demonstrates that the wind turbines can cause elongated wakes in the atmosphere. When

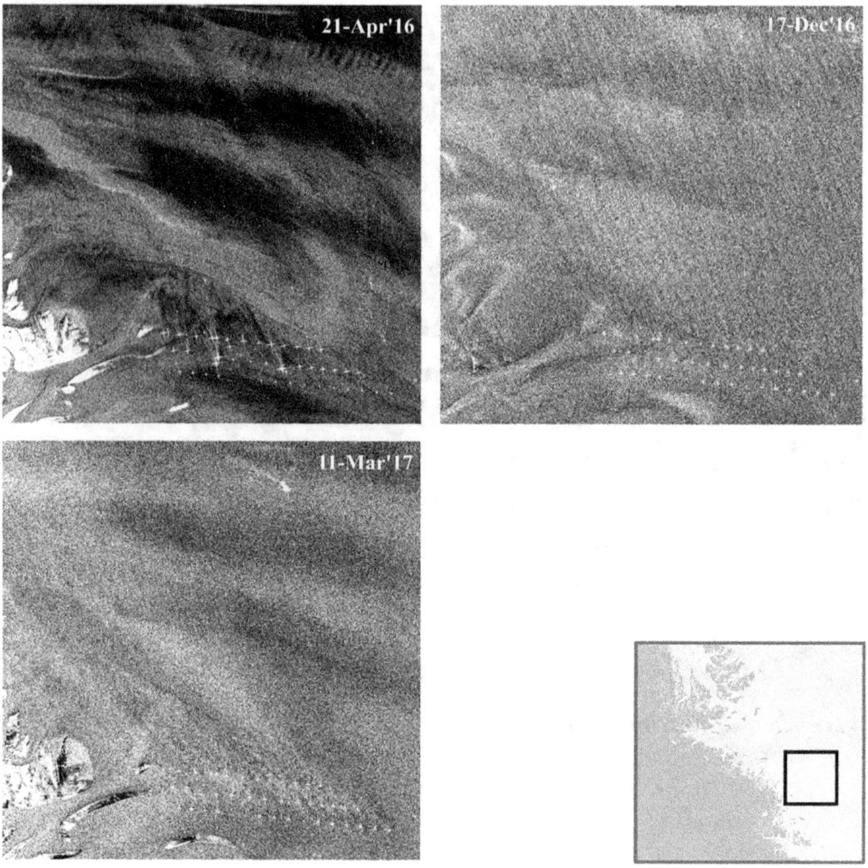

Figure 7.6 *Offshore windfarm "CGN Rudong" off the Chinese coast in the Southern Yellow Sea (bright dots on the bottom of each panel). The (25 km × 25 km) details of three Sentinel-1A SAR-C images acquired in 2016 and 2017 show the influence of the wind turbines on the local wind and current field. The lower right panel depicts the location inside the SAR images shown in Figure 7.4. © ESA 2016, 2017*

the wind direction is aligned along the mesh structure of the wind park, this effect may be amplified, resulting in wind wakes that exceed 20 km in length [27]. The lower left panel (11 March 2017) shows shorter wakes, but also slightly increased image brightness behind the wind turbines indicating some downstream atmospheric turbulence.

7.4 Derivation of roughness parameters

During the two Spaceborne Imaging Radar - C / X-band SAR (SIR-C/X-SAR) missions in 1994, simultaneous SAR acquisitions at L-, C-, and X-band were made from aboard the space shuttle "Endeavor." The two shuttle missions are still the only spaceborne missions in history that provided simultaneous multifrequency SAR data. Figure 7.7 shows three VV-polarization SIR-C/X-SAR images of the German Wadden Sea. On the L-band SAR image (upper panel), the exposed intertidal flats show up as dark patches (lower radar backscattering than the surrounding sea surface), similar to those imaged by

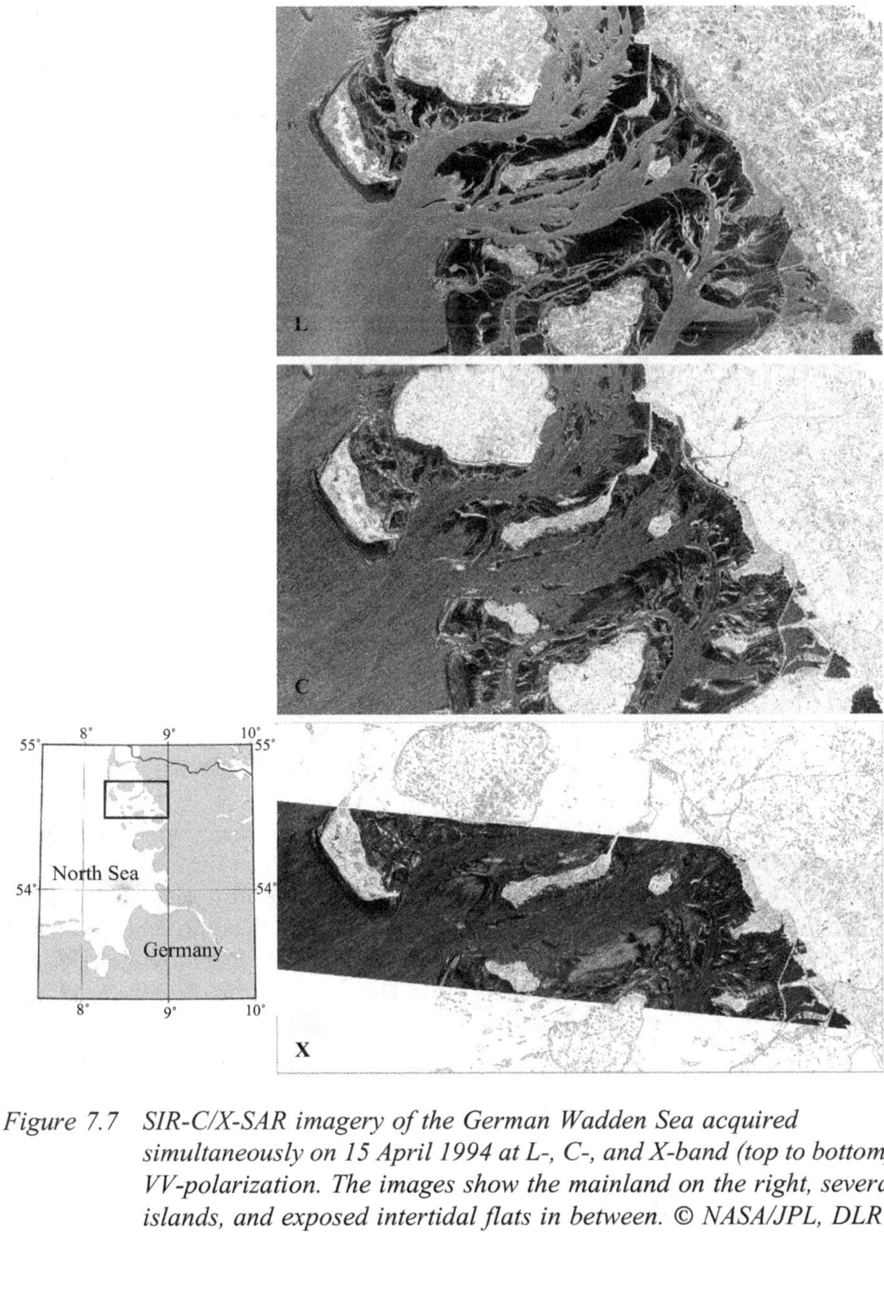

Figure 7.7 SIR-C/X-SAR imagery of the German Wadden Sea acquired simultaneously on 15 April 1994 at L-, C-, and X-band (top to bottom), VV-polarization. The images show the mainland on the right, several islands, and exposed intertidal flats in between. © NASA/JPL, DLR

Seasat (Figure 7.3). In contrast, the radar signatures at C- and X-band are more variable, similar to those acquired by the C-band SARs aboard ERS-1, EnviSat, and Sentinel-1A (Figures 7.3 and 7.4).

Assuming that the moisture of the exposed sediments is high enough (at least 40%) and surface scattering is the dominant scattering mechanism, one can use pairs of such images for an inversion of the IEM to derive the surface roughness parameters, autocorrelation length, and root mean square (rms) height [18]. The copolarization backscatter coefficient, σ^0_{pp}, is then given as [16,17]:

$$\sigma^0_{\mathrm{pp}} = \frac{k^2}{4\pi}\exp\left(-2k^2h^2\cos^2\theta\right)\sum_{n=1}^{\infty}\left|I^n_{\mathrm{pp}}\right|^2 \cdot \frac{W^{(n)}(2k\sin\theta, 0)}{n!}$$

where

$$I^n_{\mathrm{pp}} = (2kh\cos\theta)^n f_{pp}\exp\left(-k^2h^2\cos^2\theta\right) + (kh\cos\theta)^n F_{\mathrm{pp}}, \tag{7.1}$$

k is the radar wavenumber, h is the rms height, θ is the incidence angle, $W^{(n)}$ is the Fourier transform of the nth power of the surface autocorrelation function, and f_{pp} and F_{pp} are field coefficients for polarization pp [17]. The IEM predicts that the radar backscattering at C- and X-band shows similar dependencies on those roughness parameters (in accordance with SAR observations, also see Figure 7.7), so that best results are obtained when L-band data are used together with either C- or X-band data [18]. Moreover, the autocorrelation length has proven to be a suitable proxy for sediment classes, that is, it can be used for a crude sediment classification [18].

However, whereas SIR-C/X-SAR provided SAR images acquired under the same environmental and geometrical conditions (i.e., the same wind speed and direction, water level and the same incidence and look angles), SAR images acquired from different platforms (satellites) typically have the disadvantage that they were acquired under different conditions. Nevertheless, attempts were made to derive surface roughness parameters through an IEM inversion using pairs of SAR data from different platforms, and the results show evidence that this approach also yields realistic estimates [7].

Figure 7.8 shows maps (20.1 km × 20.6 km) of the correlation length (upper panel) and the rms height (lower panel), as derived through an IEM inversion using ERS-2 SAR (C-band, VV) and ALOS-1 PALSAR (L-band, HH) data acquired around low tide on 15 May 2008 and 23 May 2008, respectively. A land-and-water mask was applied, which was generated using the corresponding cross-polarization PALSAR data for waterline detection [15]. The variation of the rms heights is generally small, the correlation lengths are largest in sandy areas.

These results demonstrate that estimates of surface roughness parameters are possible, even when SAR data acquired at different times and from different platforms are used. For an assessment of the feasibility of the approach, however, in situ measurements of those roughness parameters are required, and care has to be taken when sandy sediments are dried out under windy conditions, thereby changing their dielectric properties.

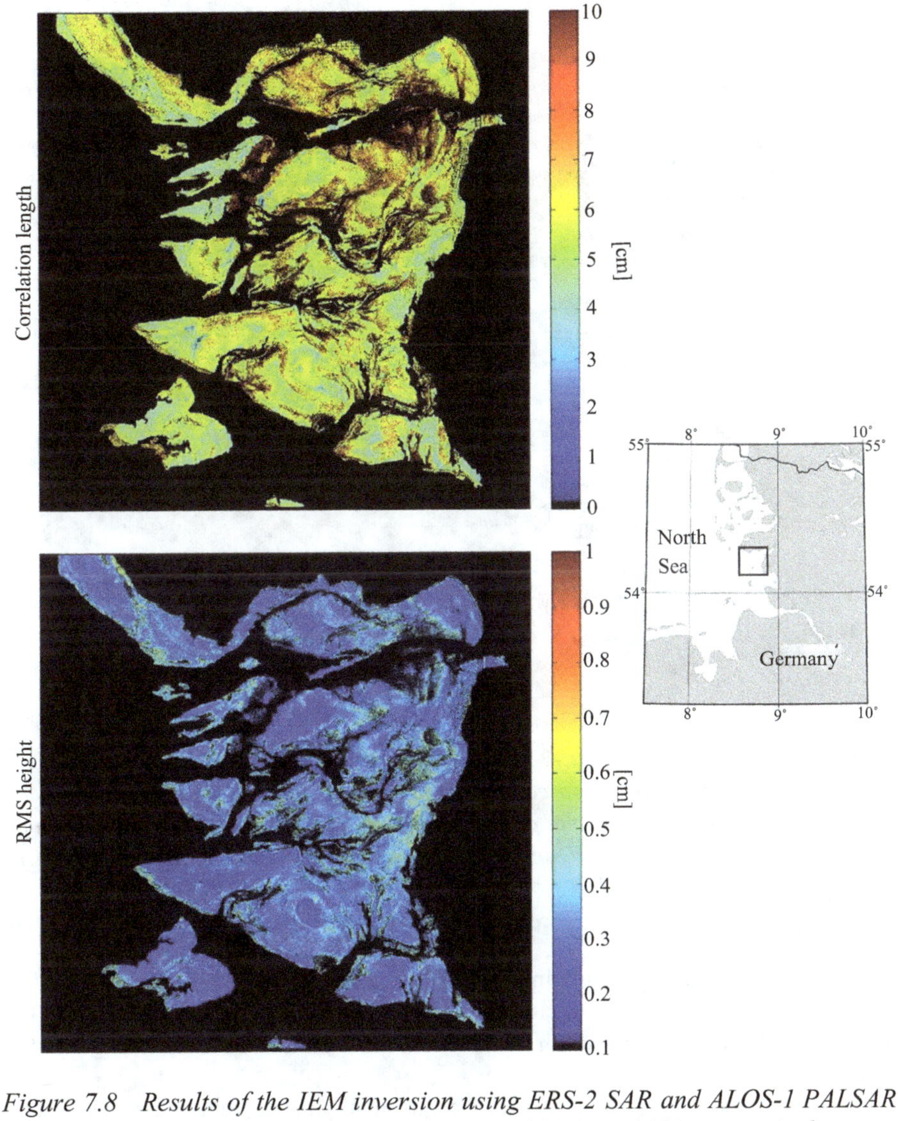

Figure 7.8 Results of the IEM inversion using ERS-2 SAR and ALOS-1 PALSAR data acquired on 15 May 2008 and 23 May 2008, respectively, shortly before or after low tide. Upper panel: autocorrelation length; lower panel: rms height; and dimensions: 20.1 km × 20.6 km

7.5 Detection of habitats

In addition to SAR data acquired at different frequencies, single-frequency polarimetric SAR data also can be used to infer roughness parameters on exposed intertidal flats [29]. Different targets on intertidal flats can be identified through polarimetric decomposition

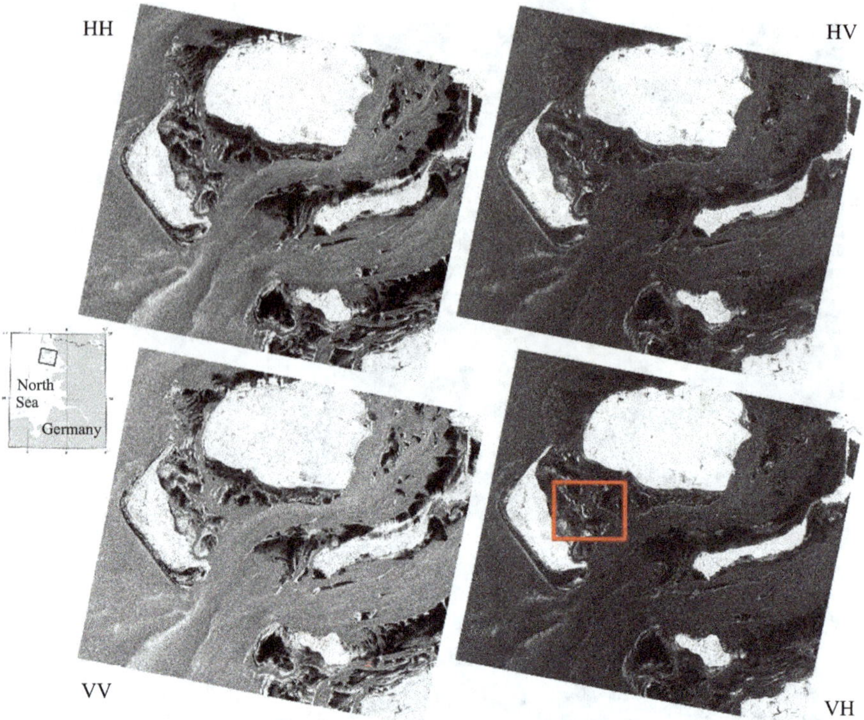

Figure 7.9 Quadpol Radarsat-2 SAR images of exposed intertidal flats on the German North Sea coast acquired on 24 December 2015. The red rectangle marks the area shown in Figure 7.10. Radarsat-2 data and products. © MacDonald, Dettwiler and Associates Ltd. (2015) – All rights reserved

[30], and in particular, fully polarimetric SAR data can be used to detect bivalve beds [14,22]. Figure 7.9 shows polarimetric Radarsat-2 SAR data of an area on the German North Sea coast, in which bivalve beds are encountered (red rectangle in the lower right panel).

The four panels in Figure 7.9, again, show that bathymetric features can manifest on SAR imagery. The data were acquired shortly after low tide and the tidal current entering the area caused dark and bright SAR image features in the panel's lower left part. Exposed flats between the islands mostly show up as dark patches with sharp boundaries, but can also cause bright sharp patches, where the exposed sandy sediment is dryer. Note that such features had already been discussed for the Sentinel-1A SAR-C data of the Chinese coast (Figures 7.4 and 7.5) and appear to be symptomatic for C-band SAR imagery of exposed sandflats. Some of the bright features, however, are due to bivalve

Figure 7.10 Photograph of extended oyster beds on the German North Sea coast (formed by the invasive Pacific oyster, Crassostrea gigas). In between the oysters are puddles of remnant water. Photo Credit: M. Gade

(oyster) beds, particularly in an area marked by the red rectangle in the lower right panel of Figure 7.9. Because of the complex radar backscattering in those bivalve beds, they often cause a strong signal in cross-polarization data (right column of Figure 7.9). A photograph of extended oyster beds on the German North Sea coast (formed by the invasive Pacific oyster, *Crassostrea gigas*) is shown in Figure 7.10. Note that water puddles remain between the oysters. If large enough, those puddles, which are also found on the exposed sandflats, can influence the radar signal.

Naturally, full polarimetric decompositions can only be applied to quad-pol SAR data such as those shown in Figure 7.9. However, many SAR sensors currently in orbit routinely operate in dual-copolarization (HH+VV) or dual-polarization (either HH+HV or VV+VH) mode only. In order to overcome this constraint and to take benefit of a larger data basis, the use of dual-copolarization SAR data for the routine monitoring of intertidal flats, especially for the classification of sediments and detection of bivalve beds, was recently suggested [20,23].

A general approach for the use of polarimetric SAR data is based on the decomposition of the Kennaugh matrix $[K]$, which contains ten independent elements K_i [31]. In the case of dual-copolarization SAR data only four Kennaugh elements contain information:

$$\begin{aligned}
K_0 &= \frac{1}{2}\left\{|S_{\mathrm{HH}}|^2 + |S_{\mathrm{VV}}|^2\right\} \\
K_3 &= -Re\left\{S_{\mathrm{HH}}S_{\mathrm{VV}}^*\right\} \\
K_4 &= \frac{1}{2}\left\{|S_{\mathrm{HH}}|^2 - |S_{\mathrm{VV}}|^2\right\} \\
K_7 &= Im\left\{S_{\mathrm{HH}}S_{\mathrm{VV}}^*\right\}
\end{aligned} \tag{7.2}$$

where $|S_{jj}|^2$ is the SAR image intensity at copolarization jj, the asterisk (*) denotes the complex conjugate, and the factor ½ appears for symmetry reasons. K_0 is (half of) the total image intensity, K_3 reflects the difference between even- and odd-bounce scattering, K_4 the relation between horizontal and vertical dipoles, and K_7 the phase difference between the odd- and even-bounce scattering events [31]. It is convenient to use the normalized Kennaugh elements, defined as

$$\begin{aligned}
k_0 &= \frac{I-1}{I+1} \\
k_i &= \frac{K_i}{K_0} \qquad i = 3;4;7
\end{aligned} \tag{7.3}$$

where I is the total intensity. The normalized Kennaugh elements range from -1 to $+1$, thereby making the results more comparable.

The four normalized Kennaugh elements of dual-copolarization SAR data can be used to gain further insight into the radar backscattering from exposed intertidal flats of different sediment composition, but care has to be taken, since the normalized radar cross section (NRCS) of the same intertidal flat may vary, depending on environmental conditions and imaging geometry [23]. The use of the Kennaugh element framework for the detection of bivalve beds, however, appears to be a robust approach.

The area marked by the red rectangle in Figure 7.9 was imaged in 2015 and 2016 by the Radarsat-2 SAR (24 December 2015, quad-pol), by ALOS-2 PALSAR (29 February 2016, quad-pol), and by TerraSAR-X (20 June 2016, dual-copolarization). From those data, the four normalized Kennaugh elements defined in (7.2) and (7.3) could be derived on a pixel-by-pixel basis. A simple mathematical operation then allowed the detection of bivalve beds: in the maps of the normalized fourth Kennaugh element, k_4 (previously introduced as polarization coefficient, PC [32]), the mean and standard deviation of a moving 11 pixels × 11 pixels window was calculated. The absolute value of their product P defined as [33]

$$P = |\mu_{PC}| \cdot \sigma_{PC} \tag{7.4}$$

was then attributed to the window's center coordinate so that new maps as shown in Figure 7.11 were generated. The normalized fourth Kennaugh element (or polarization coefficient) had proven to be an indicator for bivalve beds when derived from TerraSAR-X data [32]. Figure 7.11 demonstrates that also data from SAR

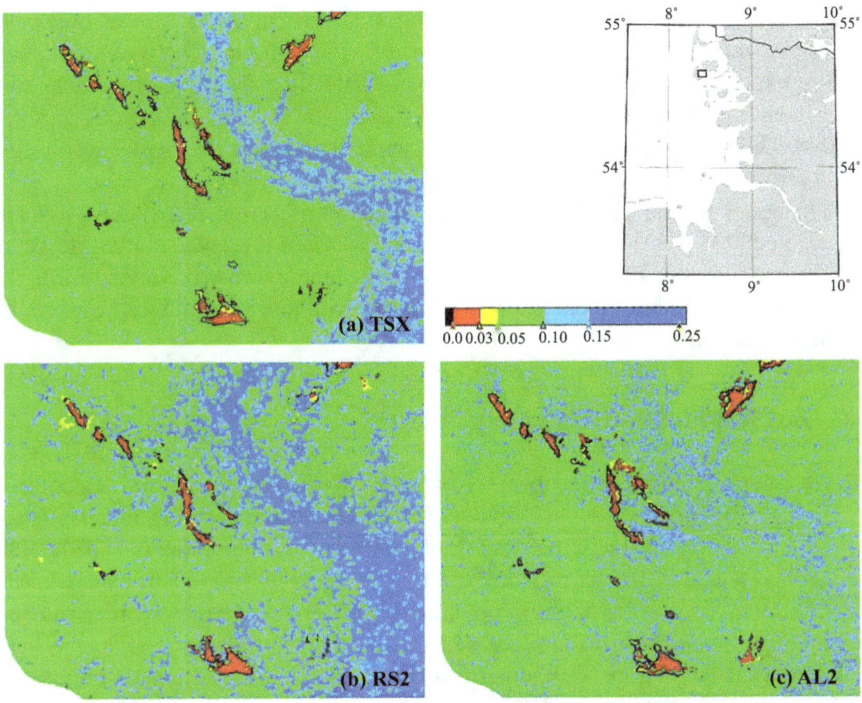

Figure 7.11 Detection of bivalve beds (in red) using the normalized fourth Kennaugh element (see text) derived from different SAR data: (a) TerraSAR-X, 20 June 2016; (b) Radarsat-2, 24 December 2015; and (c) ALOS-2, 29 February 2016. The thin black lines mark the borders of bivalve beds found during field campaigns. Redrawn after [33]

sensors working at other radar bands (C-band: Radarsat-2 SAR and L-band: ALOS-2 PALSAR) can be used for this approach [33].

In Figure 7.11, an arbitrary color map was chosen so that the values of the product P smaller than 0.03 appear in red. It is obvious that those values were obtained only in areas, where bivalve (oyster) beds were found during previous field campaigns, see the black lines inserted into each panel. Due to their rough surface and the heterogeneous orientation of the oysters sticking out of the (sandy) sediment (Figure 7.10), the radar backscattering from oyster beds is similarly strong at both copolarizations, HH and VV, and therefore, k_4 [see (7.2)] is very small. Since this effect is stable inside the oyster beds, the standard deviation of the moving window is small also, thereby resulting in minimal values of the product P. The detection accuracy of this approach was found to be 87.7%, 85.6%, and 84.7% for X-band, C-band, and L-band data, respectively [33].

Also note that large values of the product P ($P \geq 0.10$, marked by blue colors in Figure 7.11) were found inside the tidal creeks. Here, the radar backscattering from the wind-roughened water surface is stronger at vertical (VV) than at horizontal (HH) polarization, resulting in larger values of k_4. The borders between the creeks and the exposed flats are sharper for X-band data, that is, for shorter radar (and Bragg) wavelengths, whereas they are diffused for L-band data (longer wavelengths). This can be explained by the short wind fetch inside the creeks, resulting in only a small-scale roughness.

The use of Kennaugh elements for the detection of sea-grass meadows was also investigated [23], but the results are not as clear as those presented above. However, there is evidence that the decomposition of the Kennaugh matrix yields results that can be used as additional input to existing classification systems, thereby improving the final results obtained.

7.6 Archaeological surveys

The new generation of spaceborne X-band SAR sensors is capable of acquiring SAR images with pixel sizes well below 0.1 m^2 that can be used to identify small, narrow features on exposed intertidal flats. Such features comprise (small) habitats such as bivalve beds, the narrow arms of tidal creeks, but also spatial variations in surface roughness that are of anthropogenic origin. On the German North Sea coast, the latter can be due to historic settlements and remains of former drainage systems and of peat cutting, dating back to periods before major storm surges hit the coast in the 14th and 17th centuries [34].

Figure 7.12 demonstrates that the northern part of the German coast has undergone severe changes in 1362 and 1634, when those major storm surges, called "Grote Mandrenke" (great drowning of men), caused that vast areas were lost to the sea [35]. Marked in green is the coastline before the first "Grote Mandrenke," when most parts of the North Frisian coast were marshland, cultivated, and still connected to the mainland. Red lines denote the coastline after the first, and before the second, "Grote Mandrenke," when most parts of that marshland had disappeared and when only a number of small islands were left over. Finally, marked in blue is the coastline at the beginning of the nineteenth century, with several (diked and undiked) islands in that area. The horizontal arrow marks an area, which was cultivated and used as farmland, with small villages and farmhouses, until the second "Grote Mandrenke" hit the coast in 1634.

What was left from the former settlements, farmhouses and farmland was buried under sediments and appeared again at the surface only when those sediments were driven away. This strong erosion under the action of tidal currents, wind and waves, however, also causes that those remains may be completely lost, shortly after they appeared again, which is why frequent archaeological surveys are needed. Here, high-resolution satellite SAR imagery has been proposed to complement archaeological surveys in intertidal regions [36]. Figure 7.13 is an example of such high-resolution SAR images of that area, acquired on 13 October 2019 and showing remains of the historic land use as linear features all over the (2 km $\times$ 2 km) image.

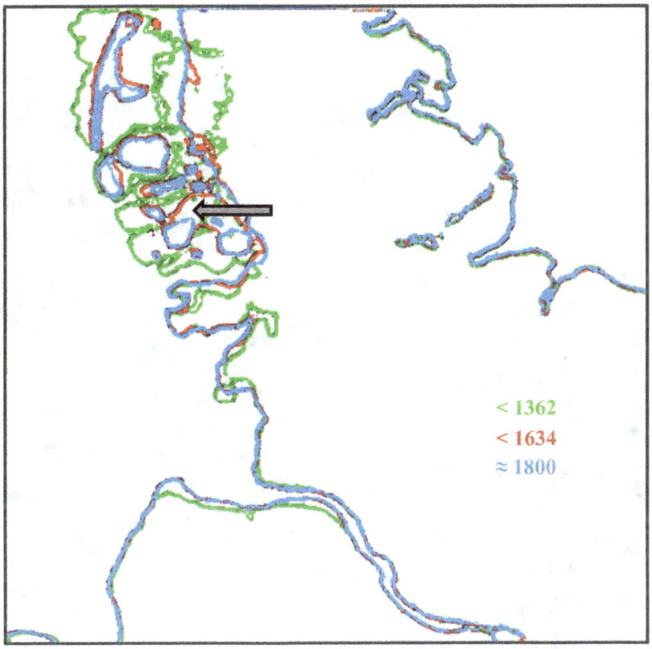

Figure 7.12 *Coastline changes on the German North Sea coast since the fourteenth century. The green line denotes the coastline before the major storm surge in 1362, the red line before the second major storm surge in 1634, and the blue line at the beginning of the nineteenth century. The horizontal arrow marks the area of interest. Adapted from [35,36]*

Because of the high spatial resolution, with 28 cm pixel spacing, these features are well resolved on the SAR image.

The letters "A" in Figure 7.13 mark areas with parallel bright and dark lines that are remnants of the former drainage system, that is, they are due to former didges and lanes. Those remnants are marked by denser sediment causing higher surface roughness [36], which in turn causes enhanced radar backscattering. In addition, remnant water along those lines (mainly in their wind shadows) may cause reduced radar backscattering, thereby enhancing the contrast of the linear features. Letters "B" in Figure 7.13 marks signatures of remains of former settlements in that area. Foundations of farmhouses and stables, as well as remnants of wells and cisterns, can be identified through bright features [36].

The dramatic changes and strong morphodynamics can be inferred from Figure 7.14, which shows the same spot as Figure 7.13 but imaged less than 5 years before. Letters "A" and "B" mark the same features at the same locations. It is evident that the farmhouse remnants east of the tidal creek (marked by the lower right letter "B") have disappeared, mainly because the creek has moved eastward.

Figure 7.13 Excerpt (2 km × 2 km) of a TanDEM-X SAR image (X-band, VV-polarization) acquired on 13 October 2019 in the German Wadden Sea, north of the island of Pellworm (see map). The high-resolution Staring Spotlight mode (with 28 cm pixel spacing) allows identifying remains of historical land use as linear features in most parts of the image. © DLR

Instead, west of the creek "new" features have appeared, after the top sediment layer was driven away. The same holds for the many remnants of the drainage system (letters "A"), whose signatures are better visible in the recent SAR image (Figure 7.13), but which are also subject to strong erosion.

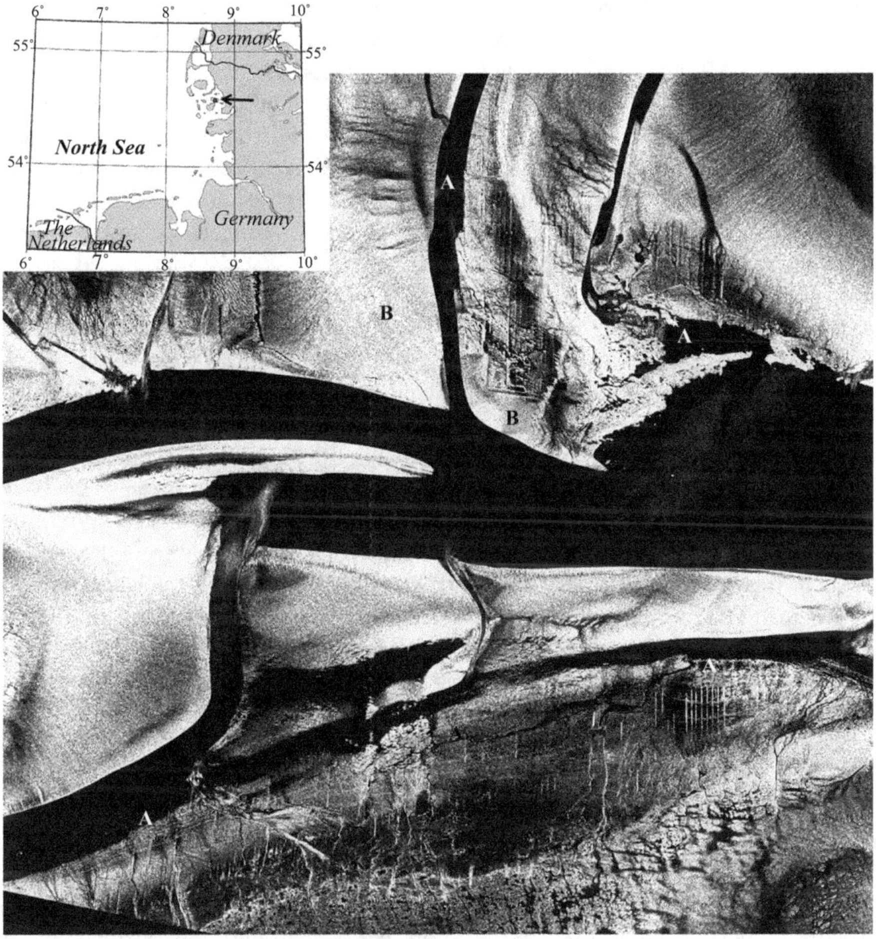

Figure 7.14 TanDEM-X SAR image of the same spot as in Figure 7.13, but acquired on 20 January 2015. © DLR

7.7 Summary

SAR data can be used to monitor coastal ecosystems, exposed intertidal flats, and coastal habitats. We have demonstrated that exposed intertidal flats, consisting of sandy, muddy, and mixed sediment, show up on SAR imagery as bright or dark patches, depending on the sediment type, water level (soil moisture), wind speed (if strong enough), and radar parameters (mainly the radar band). Submerged sandbars can cause distinct SAR signatures as well, due to a modulation of the sea surface current and therefore, of the sea surface roughness.

SAR imagery acquired over shorter periods can be used to infer topography maps, while SAR data from longer periods provide information on morphodynamical changes in the intertidal zone. Further, the influence of manmade objects such as offshore wind farms on the local wind and current field can be observed and demonstrated using SAR data. Here we note that those objects may cause additional, though smaller, morphological changes that will be observed using future SAR imagery.

The use of SAR data acquired at different radar bands for an inversion of analytical backscattering models such as the IEM appears to be a powerful approach to infer roughness parameters, which in turn can help in classifying surface sediment types. Here, the underlying idea is that the surface roughness parameters, namely the rms height and the autocorrelation length, depend on the grain size distribution and thus, on the sediment type. However, we also note that the radar backscattering from the same sediment (type) is increasing with decreasing moisture. The reason may be a change of those roughness parameters (e.g., when there is no more remnant water in the troughs of the sand ripples), but also of the dielectric properties that depend, among others, on the soil moisture. Further developments of the presented approach will depend on the availability of in situ measurements for the tuning of the inversion scheme.

Multipolarization SAR data can be used to infer information on radar backscattering mechanisms and to detect areas, in which the mechanisms strongly differ from those in their vicinity. Bivalve (oyster and mussel) beds are good examples for such areas, if their borders are sharp enough, because their surface is much rougher than that in their surroundings. Here, the polarimetric decomposition based on the Kennaugh matrix yielded reliable detection results that are in very good agreement with assessments made during field campaigns. There is evidence that the same polarimetric decomposition can also be used to detect sea-grass meadows, but only if those meadows are exposed during low tide and if the grass changes (increases) the surface roughness.

SAR data of highest resolution (below 1 m^2 pixel size) can be used to detect small-scale features on exposed intertidal flats. Among those features are remnants of historical land use, including didges of drainage systems and foundations of (former) farmhouses and stables. Since in situ observations of such features are difficult to perform, because of the short periods around low water, satellite observations, and SAR-based surveillance in particular, are complementing archaeological surveys in intertidal zones that are difficult to access.

The broad overview given in this chapter has demonstrated that SAR data can be a valuable addition to existing monitoring systems that are purely based on optical data. Because of the necessary condition to use only remote sensing data acquired around low tide, and because the use of optical data requires cloud-free, daylight conditions, SAR data may often be advantageous over optical and infrared (IR) data, but only if the very radar backscattering mechanisms are known. This is still subject to ongoing research.

References

[1] CWSS – Common Wadden Sea Secretariat. Nomination of the Dutch-German Wadden Sea as world heritage site. Volume 1. Besemann: Wittmund; 2008, 163 pp.

[2] Koh C.H., and Khim J.S. "The Korean tidal flat of the Yellow Sea: Physical setting, ecosystem and management." Ocean & Coastal Management. 2014; 102: 398–414.

[3] European Commission. "Council Directive 92/43/EEC of 21 May 1992 on the conservation of natural habitats and of wild fauna and flora." Official Journal of European Communities. 1992; L 206: 1–66.

[4] European Commission. "Directive 2000/60/EC of the European Parliamant and of the Council of 23 October 2000 establishing a framework for Community action in the field of water policy." Official Journal of European Communities. 2000; L 327: 1–72.

[5] European Commission. "Directive 2008/56/EC of the European Parliamant and of the Council of 17 June 2008 establishing a framework for community action in the field of marine environmental policy (Marine Strategy Framework Directive)." Official Journal of European Communities. 2008; L 164: 19–40.

[6] Müller G., Stelzer K., Smollich S., *et al.* "Remotely sensing the German Wadden Sea: A new approach to address national and international environmental legislation." Environmental Monitoring and Assessment. 2016; 188 (10): 1–17.

[7] Gade M., Melchionna S., Stelzer K., and Kohlus J. "Multi-frequency SAR data help improving the monitoring of intertidal flats on the German North Sea coast." Estuarine, Coastal and Shelf Science. 2014; 140: 32–42.

[8] Fu L.L., and Holt B. "Seasat views oceans and sea ice with synthetic-aperture radar." JPL Publication, NASA Jet Propulsion Laboratory, California Institute of Technology Pasadena, CA, USA. 81–120;1982. 204 pp.

[9] Mason D. C. and Davenport I.J. "Accurate and efficient determination of the shoreline in ERS-1 SAR images." IEEE Transactions on Geoscience and Remote Sensing. 1996; 34(5): 1243-1253.

[10] Heygster G., Dannenberg J., and Notholt J. "Topographic mapping of the German tidal flats analyzing SAR images with the waterline method." IEEE Transactions on Geoscience and Remote Sensing 2010; 48(3): 1019–1030.

[11] van der Wal D., and Herman P.M.J. "Regression-based synergy of optical, shortwave infrared and microwave remote sensing for monitoring the grain-size of intertidal sediments." Remote Sensing of Environment. 2007; 111: 89–106.

[12] Kim D., Moon W.M., Kim G., Park S.E., and Lee H. "Submarine groundwater discharge in tidal flats revealed by space-borne synthetic aperture radar." Remote Sensing of Environment. 2011; 115: 793–800.

[13] van der Wal D., Herman P.M.J., and Ysebaert T. "Space-borne synthetic aperture radar of intertidal flat surfaces as a basis for predicting benthic macrofauna distribution." EARSeL eProceed 2004; 3(1): 69–80.

[14] Choe B.H., Kim D., Hwang J.H., Oh Y., and Moon W.M. "Detection of oyster habitat in tidal flats using multi-frequency polarimetric SAR data." Estuarine, Coastal and Shelf Science. 2012; 97: 28–37.

[15] Deroin J.P. "Combining ALOS and ERS-2 SAR data for the characterization of intertidal flats: Case study from the Baie des Veys, Normandy, France." International Journal of Applied Earth Observation and Geoinformation. 2012; 18: 183–194.

[16] Fung A.K., Li Z., and Chen K.S. "Backscattering from a randomly rough dielectic surface." IEEE Transactions on Geoscience and Remote Sensing. 1992; 30: 356–369.

[17] Fung A.K., and Chen K.S. "An update on the IEM surface backscattering model." IEEE Geoscience and Remote Sensing Letters. 2004; 1: 75–77.

[18] Gade M., Alpers W., Melsheimer C., and Tanck G. "Classification of sediments on exposed tidal flats in the German Bight using multi-frequency radar data." Remote Sensing of Environment. 2008; 112: 1603–1613.

[19] Lee Y.K., Park J.W., Choi J.K., Oh Y., and Won J.S. "Potential uses of TerraSAR-X for mapping herbaceous halophytes over salt marsh and tidal flats." Estuarine, Coastal and Shelf Science. 2012; 115: 366-376.

[20] Wang W., Yang X., Li X., *et al.* "A fully polarimetric SAR imagery classification scheme for mud and sand flats in intertidal zones." IEEE Transactions on Geoscience and Remote Sensing. 2017; 55: 1734–1742.

[21] Lee S.K., Hong S.H., Kim S.W., Yamaguchi Y., and Won J.S. "Polarimetric features of oyster farm observed by AIRSAR and JERS-1." IEEE Transactions on Geoscience and Remote Sensing. 2006; 44: 2728–2735.

[22] Cheng T.Y., Yamaguchi Y., Chen K.S., Lee J.S., and Cui Y. "Sandbank and oyster farm monitoring with multi-temporal polarimetric SAR data using four-component scattering power decomposition." IEICE Transactions on Communications. 2013; 96: 2573–2579.

[23] Gade M., Wang W., and Kemme L. "On the imaging of exposed intertidal flats by single- and dual-co-polarization synthetic aperture radar." Remote Sensing of Environment. 2018; 205: 315–328.

[24] Alpers W., and Hennings I. "A theory of the imaging mechanism of underwater bottom topography by real and synthetic aperture radar." Journal of Geophysical Research. 1984; 89: 10529–10546.

[25] Wensink H., and Campbell G. Bathymetric map production using the ERS SAR. Backscatter. 1997; 8(1): 17–22.

[26] Hasager C.B., Vincent P., Badger J., *et al.* "Using satellite SAR to characterize the wind flow around offshore wind farms" Energies. 2015; 8(6): 5413–5439.

[27] Djath B., Schulz-Stellenfleth J., and Cañadillas B. "Impact of atmospheric stability on X-band and C-band synthetic aperture radar imagery of offshore windpark wakes." Journal of Renewable and Sustainable Energy. 2018; 10: 043301.

[28] Vanhellemont Q., and Ruddick K. "Turbid wakes associated with offshore wind turbines observed with Landsat 8." Remote Sensing of Environment. 2014; 145:105–115.

[29] Park S.E., Moon W.M., and Kim D.J. "Estimation of surface roughness parameter in intertidal mudflat using airborne polarimetric SAR data." IEEE Transactions on Geoscience and Remote Sensing. 2009; 47(4): 1022–1031.

[30] Geng X.M., Li X.M., Velotto D, and Chen K.S. "Study of the polarimetric characteristics of mud flats in an intertidal zone using C- and X-band spaceborne SAR data." Remote Sensing of Environment. 2016; 176: 56–68.

[31] Schmitt A., Wendleder A., and Hinz S. "The Kennaugh element framework for multi-scale, multi-polarized, multi-temporal and multi-frequency SAR image preparation." ISPRS Journal of Photogrammetry and Remote Sensing. 2015; 102: 122–139.

[32] Gade M., and Melchionna S. "Joint use of multiple synthetic aperture radar imagery for the detection of bivalve beds and morphological changes on intertidal flats." Estuarine, Coastal and Shelf Science. 2016; 171: 1–10.

[33] Wang W., Gade M., and Yang X. "New polarimetric SAR indicators for bivalve beds on exposed intertidal flats." Remote Sensing. 2017; 9: 1047.

[34] Bantelmann A. "Die Landschaftsentwicklung an der schleswig-holsteinischen Westküste, dargestellt am Beispiel Nordfriesland." Die Küste 1966; 14: 1–182.

[35] Behre K.-H. Landschaftsgeschichte Norddeutschlands: Umwelt und Siedlung von der Steinzeit bis zur Gegenwart, Wachholtz, Neumünster, Germany, 2009; 308 pp.

[36] Gade M., Kohlus J., and Kost C. "SAR imaging of archaeological sites on intertidal flats in the German Wadden Sea." Geosciences. 2009; 7: 105.

Chapter 8
Sea ice and icebergs

Wolfgang Dierking[1,2,3]

The major topic of this chapter is space-borne synthetic aperture radar (SAR) surveillance of ocean regions that are permanently or seasonally covered with ice. Surveys of sea ice conditions are required for production of ice charts, and sequences of SAR images are used for monitoring movement and deformation of the ice. Ice charts show the regional distribution of open water areas and different ice types, often supplemented by additional information such as areal fractions of ice types and floe sizes. Ice drift is displayed in maps of vector fields, which may be complemented with data quantifying their errors or reliability and the degree of deformation. A less extensive but nevertheless important topic of this chapter is the surveillance of icebergs. Ice charts and information on ice movements and iceberg occurrences are essential for the safety of ship traffic and offshore operations, and for several scientific questions. Issues addressed here concern methods for retrieving information about ice conditions, using different SAR systems and technologies.

8.1 Introduction

Many people intuitively link "maritime surveillance" with acquiring information about regional conditions in the *open* ocean, such as wave heights, speed and direction of currents, or location and spread of oil spills. However, ocean regions that are ice-covered during winter or even year-round are also continuously monitored. Sea ice is "any form of ice found at sea which has originated from the freezing of sea water" [1]. Wind, ocean currents, tides, air and water temperatures, and water salinity all influence the characteristics of local ice covers. Since the forces acting upon the ice and the driving meteorological and oceanic parameters vary spatially and over time, so do the ice conditions. This leads to the development of various ice types and structures that are introduced below. Sea ice is not only

[1]Centre for Integrated Remote Sensing and Forecasting for Arctic Operations (CIRFA), UiT The Arctic University of Norway, Tromsø, Norway
[2]Polar Meteorology, Alfred Wegener Institute Helmholtz Centre for Polar and Marine Research, Bremerhaven, Germany
[3]Member of International Ice Charting Working Group

common in the Arctic and around Antarctica but during the winter season, it is also prevalent in sub-polar Seas such as the Labrador, Greenland, Baltic, Bohai, and the Sea of Okhotsk. Sea ice is a hazard to ship traffic and offshore structures; hence there is a need to issue information on ice conditions and iceberg occurrences on a regular basis and on demand. Local and regional ice conditions are gathered by national ice services and presented as ice charts that are updated at regular intervals, mostly daily or weekly. Ice charts are produced mainly based on the manual analysis of satellite images, and information gathered through observations from ships and aerial reconnaissance. Such charts present ice conditions in terms of concentration, stage of development, and structural features, following the terminology of the World Meteorological Organization (WMO). Because of their general availability and comparatively high spatial resolution, C-Band synthetic aperture radar (SAR) systems are currently the "space eyes" of operational ice services.

Not only does maritime surveillance provide data for operational monitoring to safeguard operations in the ice, it also contributes to the documentation of environmental changes in the Arctic and Antarctic. The Polar Regions influence global climate and reveal alterations associated with climate changes. The loss of sea ice (i.e., the decrease of ice extent, concentration, and thickness) in the Arctic especially during summer months—well documented and repeatedly reported in the public media—is linked to rising surface temperature and atmospheric water vapor content [2]. Due to the decline of sea ice, the fraction of open water areas in the Arctic increases. Since water absorbs more solar energy than ice (ice has a much higher reflectance than water), the heat content of the upper ocean layer increases, causing earlier melt of the ice in spring and later ice growth in autumn. This is the ice-albedo feedback. The expectation of an easier access to Arctic regions has triggered considerations concerning the expansion of marine transport through the Northwest and Northeast Passages and the possibilities to exploit natural resources in the Arctic Ocean. Because topographical and bathymetric characteristics around Antarctica differ strongly from the Arctic, the change of the Antarctic sea ice extent has been small. Averaged over the entire Antarctic ocean area, a slight increase was observed for the period 1999–2009 [3]. The analysis of single regions (considering the time interval from 1979 to 2010) revealed an increase of sea ice extent in the Ross Sea but a decrease in the Bellingshausen and Amundsen Seas [4]. In the years from 2016 to 2019, the total ice extent during austral summer was lower than the 1981–2010 median (https//nsidc.org/data/seaice_index/compare_trends). Information on ice conditions in Antarctic waters is required for cargo vessels and tankers, which transport supplies to Antarctic stations, and for ship-borne science expeditions and tourist voyages.

Icebergs, even more so than sea ice, impose a risk on ship navigation and offshore structures. An iceberg is "a massive piece of ice of greatly varying shape, protruding more than 5 m above sea level, which has broken away from a glacier and which may be afloat or aground" [1]. Icebergs consist of freshwater ice. After the Titanic sank in 1912 because of a collision with an iceberg, maritime nations established the International Ice Patrol (IIP) in 1914. Since then the IIP has been responsible for detecting and tracking icebergs in the North Atlantic Ocean. Currently, the IIP is funded by 13 nations (among them Canada, Denmark, Norway,

Sweden, and Germany) and is operated by the US Coast Guard. Iceberg monitoring is mainly based on fixed-wing aircraft reconnaissance flights, but the use of satellite imagery is contemplated. For detected icebergs, their drift is forecast using a computer model that requires ocean current and wind data as input.

The European Union has addressed the Arctic (including the Arctic Ocean and adjacent seas) as a region of priority [5], highlighting its strategic, environmental, and socio-economic importance and its role as an indicator of climate change. The focus of an integrated EU Arctic policy is on (1) climate adaption in the Arctic regions, (2) sustainable development in and around the Arctic (e.g., exploitation of oil and gas) as well as safe and reliable navigation, and (3) international cooperation on Arctic issues (e.g., scientific research, fisheries management, and ecosystems protection). Considering the sparse population and the lack of transport links, the continuous monitoring of the Arctic with satellites is regarded as being essential.

Continuous long-term satellite observations are important sources for detecting effects of climate variations on the world's ocean on large spatial scales (such as the increase of sea surface temperature, sea level rise, and the decrease of Arctic sea ice extent) [6]. Data from altimeters and gravity missions are used to monitor regional changes in sea level. Sea surface temperatures are obtained from thermal channels of radiometers, and passive microwave radiometers and scatterometers deliver information on variations in the extent, areal fraction, and drift patterns of sea ice in the Arctic and Antarctic at spatial resolutions of tens of kilometers. Space-borne SAR, on the other hand, is typically operated with much higher resolutions between a few meters and 150 m, but it has not yet provided the spatial and temporal coverage required for studies of climate-induced changes of sea ice in the Polar Regions. However, this may change. The Sentinel-1 mission of the European Space Agency (ESA) currently consists of a constellation of two satellites, which cover the entire Earth's surface every six days in an exact repeat cycle. When acquiring images on ascending and descending orbits, the revisit frequency is 3 days at the equator and less than 1 day over the Arctic. With the addition of more satellites, the time interval between observations of the same point on earth is shortened correspondingly. Furthermore, SAR systems become more and more relevant as suppliers/sources of input data for computer models used to simulate and forecast ocean-ice-atmosphere interactions, since the spatial resolution of such models has been increased to reflect sea ice changes taking place over distances of a few kilometers [7]. SAR assumes also an important role for improving the understanding and analysis of data from altimeters, microwave radiometers, and scatterometers, and for validating information products derived therefrom.

What is this chapter about? It is not an overall review of technologies and methods used in SAR remote sensing of sea ice and icebergs—this would exceed the editorial limits. Sections 8.2 to 8.8 are based on a subjective choice of mostly recent articles that deal with major problems in space-borne SAR surveillance of sea ice and icebergs, and with interesting suggestions to solve these problems. In these articles, the reader will find hints to relevant older studies that could not be referenced here.

In the preceding paragraphs, the importance of monitoring sea ice and icebergs for operational ice charting and for recognizing changes in the harsh and vast marine regions of the Arctic and Antarctic was highlighted. The major questions now are: What information about sea ice can be retrieved from SAR imagery? Which SAR technologies, image processing methods, and retrieval algorithms are available for this purpose? This requires an understanding of SAR signatures gained by comparison with ground-based and airborne measurements on/over the ice. In the sections below, we deal with the interaction between radar waves and ice, mapping of sea ice conditions, advanced measurement technologies that are promising for sea ice monitoring, the retrieval of ice drift and detection of icebergs, and problems encountered in validating results and their workarounds.

8.2 Microwave response of ice

For making optimal use of SAR images of sea ice and icebergs, it is mandatory to understand the interactions between microwaves and ice. To begin with, SAR system parameters that are important in this context are shortly summarized before taking a closer look at different interaction mechanisms.

For remote sensing of sea ice with space-borne SAR, predominantly three frequency bands have been used to date: L-band (wavelength 15–30 cm, frequency 1–2 GHz), C-band (3.75–7.5 cm, 4–8 GHz), and X-band (2.4–3.8 cm, 8–12.5 GHz). In discussions of future SAR mission scenarios, also Ku-band (1.7–2.4 cm, 12.5–18 GHz) has been considered [8]. A SAR system transmits and receives electromagnetic waves at a predefined polarization. The polarizations of transmitted and received waves are indicated by the combination of two letters (e.g., "HH," "VV," "HV," and "VH" for linear polarized waves, where "H" means "horizontal" and "V" is "vertical"). SAR systems such as RADARSAT-2 and ALOS-2 PALSAR-2 provide "fully" polarimetric ("quad-pol") imaging modes. In a quad-pol system, the complex scattering coefficients S_{HH}, S_{VV}, S_{HV}, and S_{VH} (which comprise amplitude and phase of the scattered signal) are measured simultaneously at all four polarization combinations given above. In practice, the phase differences between channels are used rather than the direct phase information. In the case of sea ice, the phase difference between the HH- and the VV-polarized channel reveals distinct patterns linked to variations of sea ice properties, but the differences between polarizations HH and HV or VV and VH show random noise-like spatial variations [9]. The complex scattering coefficients S_{HV} and S_{VH} are equal. One limitation of fully polarimetric data acquisitions is that it can be implemented only at narrow swath widths (PALSAR-2: 30–50 km, RADARSAT-2: 50 km, and RADARSAT Constellation Mission RCM: 20 km). This restriction can be overcome in the so-called compact polarimetric option. On the RCM, compact polarimetry ("CP") is implemented such that a circular-polarized signal is transmitted ("R" right-hand or "L" left-hand), and the backscattered return is measured at H- and V-polarization, respectively ("dual-pol CTLR," where the latter means Circular Transmit and Linear Receive). The same principle was used for the Indian

RISAT-1. The advantage of CP is that it carries more information than the dual-pol mode, but it is not equivalent to quad-polarization since it transmits only one polarization [10]. The use of polarimetry and CP for sea ice classification is addressed further below.

If microwave radiation is impinging upon ice (be it sea or freshwater ice) at a given incidence angle, different processes take place. At the surface of the ice, a fraction of the incident energy may be reflected, another part may be scattered. This depends on the surface roughness at length scales between less than one radar wavelength and up to many radar wavelengths. A smooth surface acts like a mirror, which means that a specular reflection takes place. With increasing roughness, scattering of the incident intensity into different directions increases and the intensity of the mirror reflection decreases. The intensities of reflection and scattering contributions depend on the dielectric contrast between the ice and the air above, or—if radiation penetrates into layered ice—of the dielectric constants above and below an interface between two layers. The dielectric constant of sea ice is a function dominated by ice salinity and temperature, and, to a lesser extent, is dependent on ice density and geometrical properties of air and brine inclusions in the ice. Brine is a highly concentrated solution of salt that accumulates in small cavities between the ice crystals. The brine content is largest for newly formed sea ice. It decreases at later stages of ice growth and is zero for freshwater ice. This implies that the imaginary part of the dielectric constant—the loss factor, which quantifies the absorption of the electromagnetic waves in a medium—is high for young sea ice and very low for freshwater ice. As a consequence, the penetration depth of radar signals into saline young sea ice is small (roughly on the order of one radar wavelength) and much larger, namely about 100 times the radar wavelength, in pure ice. For example, at temperatures of $-10°C$ and less, the penetration depth into Arctic first-year ("FY") ice (for definitions of ice types see Table 8.1) with salinities between 5 and 12 psu (practical salinity units) is about 3–15 cm at X-band, 7–30 cm at C-band, and 0.15–1.0 m at L-band. Radar waves refracted into the ice (at an angle according to Snell's law) are scattered at inclusions such as air bubbles, interstices, and brine pockets and, in the case of layered ice, at interfaces between two layers. The intensity of scattering caused by inclusions depends on their volume fraction, size, and shape. A SAR system transmits and receives radar pulses at an oblique angle. Therefore, only contributions scattered from the ice surface and volume in the direction of the SAR determine the variations of the received intensity: at HH- and VV-polarization, for example, smooth high-salinity ice appears dark in radar images and rough low-salinity ice with air bubbles shows up as bright patches.

Regarding the characteristics of scattering from ice surfaces, it must be noted that the assumption of resonant Bragg scattering (small perturbation theory), which is typical for the ocean surface, is strictly speaking too simplistic for sea ice. The roughness spectrum of an ice surface, covering variations of surface elevation caused by centimeter-scale undulations as well as irregularly spaced ice blocks in ice ridges and rubble fields (Figure 8.1) with sizes of decimeters to meters, is spatially more inhomogeneous than the roughness of the ocean surface caused by

Table 8.1 Basic level ice types
Descriptions of ice types (second column) are based on [1]. It is assumed that the ice surface is homogeneous over a larger area of at least several to tens of meters, without signs of mechanical deformations such as ridges. Pancake ice, which appears as accumulations of ice plates often characterized by raised rims, is also listed here. Examples of scattering characteristics in SAR intensity images are given, assuming that the radar response is received at C-band linear like-polarization (HH, VV) or cross-polarization (HV,VH), and that data were acquired in freezing conditions. C-band was selected since it has been the standard for the operational ice services

Ice type	Description	Typical scattering characteristics
Open water	In sea ice classification, the separation of open water from ice is a basic requirement	A smooth water surface appears dark. If the surface is rough due to the influence of wind, the radar response is stronger at like-polarization but remains weak at cross-polarization.
Frazil ice/ grease ice	Ice crystals in the uppermost water layer forming during the initial phase of ice growth.	The accumulation of ice crystals dampens capillary and short gravity ocean waves, reducing the backscattered energy. In like-polarized intensity images areas of frazil and grease ice cause dark patches in brighter appearing open water zones.
Nilas	A smooth elastic ice crust up to 10 cm in thickness that grows under very calm ocean conditions. Nilas has a high salinity	Nilas often acts like a mirror, appearing dark in the SAR image. From laboratory measurements, there is evidence that brine inclusions also contribute to scattering. (From initial growth until 10 cm ice thickness, the radar backscattering increases by 5–10 dB at like- and cross-polarization since single brine pockets interconnect and grow in size [12]. Nevertheless, in satellite images the backscattered intensity remains at levels close to the noise floor)
Young ice	Later development stage replacing nilas on a smooth ocean surface, thickness 10–15 cm (gray ice), and 15–30 cm (gray-white ice)	The scattering behavior is in general similar to nilas
Pancake ice	Rounded ice plates of 0.3–3 m size and thickness of up to 10 cm, which develop on a rough ocean surface, often with rims raised above the surface, caused by collisions between single plates	The intensity backscattered from pancakes depends on their sizes, the fraction of water between them, and the height and width of the rims. This causes a variable appearance in SAR images that ranges from dark to bright

(Continues)

Table 8.1 (Continued)

Ice type	Description	Typical scattering characteristics
First-year ice (FY)	Closed ice cover of 30–200 cm thickness, resulting from increasing thickness of young ice and pancakes that are frozen together into solid sheets	Due to weathering and erosion effects, the centimeter-scale surface roughness is larger compared to younger ice, hence the intensity of surface scattering increases. Volume scattering from air bubbles may occur because of a lower salinity, which leads to a moderately bright appearance in SAR images at like- and cross-polarization
Multiyear ice (MY)	Ice that has survived one or several summer's melt, up to 3 m or more in thickness. Variations of surface elevation are smoother than on FY ice. In particular, in the upper part of MY ice, the salinity is often close to zero	Because of the very low salinity and an increased volume fraction of air bubbles, the backscattering from MY ice is larger than for the young and FY ice. (The intensity contrast to FY ice is significant at C-band but only modest at L-band)

waves. Processes corresponding to modulations of capillary and short gravity waves due to variations of wind, surface currents, or orbital motion of large gravity waves do not occur. While the roughness of a water surface can, in fact, be simulated as a function of meteorological and oceanic conditions, this is not possible for an ice surface. Instead, typical ranges of roughness parameters have been collected for various ice types from many field measurements. A general approach of modeling rough surface scattering used for the ocean and for ice assumes that the surface can be approximated by small planes or "facets." The scattered field can be considered as being the superposition of signals reflected from single facets, taking into account both amplitudes and phases of individual signals. The tilt and orientation angles of the facets are modeled as statistical distributions. The facets are larger than the radar wavelength and may be covered by smaller roughness elements with sizes comparable to or less than the radar wavelength. The interplay between smaller and larger roughness elements determines the characteristics of radar scattering.

Volume scattering has to be considered to varying degrees. For high salinity sea ice, it is in general weak. For freshwater ice, it may be dominant if inhomogeneities, such as air-filled cavities or layers of different dielectric properties, exist in the ice. In sea ice older than 1 year, air bubbles occupying volume fractions of up to 0.25 are common. Their sizes are on the order of millimeters. They may also be found in younger ice, usually with much lower volume fractions. Air bubbles are of variable shape but can be well approximated by spheres when evaluating their scattering efficiency theoretically. Brine pockets are another type of inclusion occurring in sea ice, as mentioned above. Their volume fraction is large during initial ice formation and growth. Their shape often resembles ellipsoids or needles.

Accumulated frazil ice with visible wave pattern (Beaufort Sea)

Nilas in the vicinity of a moving vessel (Bothnian Bay)

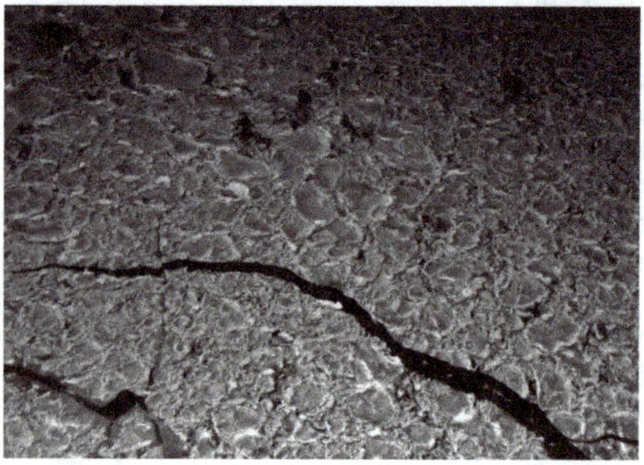

Pancake ice (Bothnian Bay)

Figure 8.1 Ice conditions.
Shown are examples of different growth stages and deformation features. Photographs were taken from an airplane or from a ship. More photographs can be found in [1]. Photos Courtesy: W. Dierking

Ice ridge in melting FY ice (Beaufort Sea)

Rubble field (Bothnian Bay)

Weathered ridges on MY ice after summer melt (Central Arctic)

Figure 8.1 (Continued)

Heavy rafting in grey and grey-white ice (Bay of Finland)

Lead in rough snow-covered FY ice (Weddell Sea)

Figure 8.1 (*Continued*)

Measurements reveal that the length of the major axis ranges from less than 1 millimeter up to a few millimeters, and that the ratio between minor and major axes varies between 3 and 8 [11]. The size of brine inclusions increases with temperature. At later stages of ice growth, the brine drains out, leaving pockets filled with air. If ice has survived the summer melt season, the salinity has decreased considerably because brine has been displaced by melt water from the ice surface [11].

These more general explanations will now be more closely linked to different sea ice types as they appear in the Arctic and Antarctic. The larger part of the sea ice cover in the Arctic and Antarctic is drifting pack ice that moves due to forces exerted by wind and ocean currents. Fast ice is stationary because it is fixed to the shoreline, grounded on the sea floor, or blocked by icebergs or islands. The most important ice types and structures discussed in this chapter, their properties, and their typical radar signatures are listed in Tables 8.1 and 8.2. The notation "important" refers here to the understanding of general radar signature characteristics and signature variations. The right columns of Tables 8.1 and 8.2 are only valid for radar signatures observed in intensity images acquired at C-band. Radar

Table 8.2 Ice deformation features and frost flowers
As for Table 8.1, typical scattering characteristics are provided for C-band linear like- and cross polarization, assuming freezing conditions. Single linear deformation features (rafting zones, ridges) are clearly visible only in images with high spatial resolutions ≤ 10 m (Note: L-band is better suited to detect zones created by compressive or shear forces.)

Ice feature	Properties	Typical radar signature
Rafted ice	A zone where pieces of thin ice slid over each other. Rafting mainly occurs in nilas and young ice	Rafting causes a net-like pattern of narrow strips of moderately high backscattering in a darker background, which often is stronger at cross-polarization
Ridge	A line or wall composed of smaller and larger ice blocks and fragments (sizes between several centimeters and a few decimeters) forced into piles by compressive forces or by shear	Ridges reveal spots of bright radar returns due to specular reflections from surface facets and volume scattering from air voids and cracks deeper down in the ridge
Rubble field	An extended area of broken pieces of ice that were compressed and partly piled up	A rubble field is characterized by bright spots of specular reflections and patterns of varying brightness due to surface and volume scattering
Lead	A fracture or opening in the ice of at least a few meters in width. Leads are exposing either an open water surface or are covered by thin ice, the latter sometimes with frost flowers on its surface	See Table 8.1 (open water, nilas, and young ice) and column "frost flowers" below
Frost flowers	Saline ice crystals of a few centimeters in diameter that grow on nilas and young ice under calm and cold conditions	Patches of frost flowers cause a strong scattering intensity at C-band since they increase the surface roughness at cm-scale considerably. (They are less effective scatterers at L-band.)

signatures of other frequency bands reveal different characteristics, as indicated for L-band. Examples of ice conditions observed in the field are shown in Figure 8.1.

Sea ice is mostly covered by snow, which is dry in freezing conditions. Dry snow is a mixture of ice crystals and air. In the frequency range from L- to X-band, it has usually only minor effects on the backscattering characteristics of the different ice types and deformation features. Radar scattering from the snow surface can often be neglected because the difference between the dielectric constants of air and dry snow is small. The intensity of snow volume scattering is in general much lower than the scattering intensity from the ice below. However, if at higher radar frequencies the ratio between radar wavelength and the length of ice crystals approaches values $\leq$ 10, the contribution of the snow volume to the total scattering may be distinct. Interaction mechanisms between snow and ice (e.g., upward wicking of brine from the ice into the lower snow layer) may change the ice surface roughness and vertical dielectric profile in the snow. Wet snow is a mixture of ice crystals, air, and liquid water. If the air is continuous throughout the pore space and liquid water occurs at isolated spots, the snow is in the "pendular" regime of liquid water concentration. If the water is continuous throughout the pore space and air is trapped in isolated bubbles, the regime is called "funicular." In wet snow, the penetration depth of the radar signal may be significantly reduced, depending on liquid water content, snow density, and radar frequency. If it is smaller than the snow depth, the backscattered radar signal is entirely dominated by the properties of the snow layer. Therefore, it is difficult or even impossible to separate ice types in SAR images acquired during the melt season, in particular at higher frequencies.

8.3 Operational sea ice mapping

This section deals with ice chart production in operational services, with studies toward possible automation strategies, and with some factors that complicate ice type classification (including ice-water separation). Table 8.3, which is based on

Table 8.3 Requirements for operational ice charting
Shown are spatial and temporal resolution requirements for monitoring of different ice parameters, based on [13] and [14]. Polynyas are openings in the ice with sizes between 10 km^2 and more than 10,000 km^2 that occur if strong katabatic winds from land push the ice away from the coast or if a spot of warm surface water hampers ice growth. For ridges and icebergs, only parameters such as length, width, and areal density can be retrieved from single SAR images, but not heights. Note that the temporal requirements are not associated with the latency for the availability of SAR data (which is between less than 1 h and 2 h; D. Flett, CIS, personal communication, 2018)

	Strategic		**Tactical**	
Parameter	**Spatial**	**Temporal**	**Spatial**	**Temporal**
Ice edge location	5 km	Daily	<1 km	6 h
Ice concentration	<100 m (±10% accuracy)	Daily	<25m (±5% accuracy)	6 h
Ice types	50–100 m	Daily	<20 m	6 h
Leads/polynyas	50–100 m	Daily	<20 m	6 h
Ridges	<50 m	Daily	<10 m	6 h
Ice decay stage	20 km	Weekly	5 km	Daily
Iceberg location	<50 m	Daily	<5 m	Hourly

[13] and [14], lists the requirements for operational monitoring considering different ice parameters, distinguishing between *strategic* and *tactical* information. The former refers to the regional survey as provided in the daily ice charts and is, for example, useful for route planning. The latter requires a detailed local analysis in the vicinity of a vessel to support immediate operations.

Since the operational ice services preferably use wide swath images, the spatial resolution is between 50 and 100 meters. This means that some of the requirements for tactical information listed in Table 8.3 are not met. Besides the spatial resolution, the radar frequency, polarization, and incidence angle determine how well ice types and water can be separated in a SAR scene. Today ice analysts combine images acquired at like- and cross-polarization (usually HH, HV), and they take into account the noise level of the radar system and the incidence-angle sensitivity of the backscattered intensity. Ice types can be separated by their radar intensities only to a certain extent. Therefore, ice analysts turn their attention to another important attribute, namely the texture: does the ice area appear featureless or does it reveal structural variations caused by, for example, clusters of ice ridges, presence of leads, or edges of floes? Are ice floes rounded or angular? In the development of automated procedures, variations of intensities at different polarizations or of polarimetric parameters (if available) are the first choice in most cases, but many studies also consider various texture parameters. Not yet included in automated procedures but a common practice in manual ice mapping is to consider the temporal history of the ice development for the regime/region being analyzed.

8.3.1 Manual generation of ice charts

The major challenge of ice charting is to reliably provide information on the state of the sea ice cover in a given region as soon as possible after SAR images are acquired, and with sufficiently small time intervals to capture changes in ice conditions. The general term "ice classification" includes ice–water separation as well as the distinction of different ice types. The WMO standard for sea ice and iceberg nomenclature provides a large number of terms that help to describe ice conditions in great detail (e.g., stage of development, forms of landfast ice, occurrence of floating ice, and features related to motion and deformation processes). The nomenclature is mainly based on visual observations in the field, for example, from the bridge of a ship or from an airplane. Imaging radar sensors receive their information at much longer wavelengths than the human eye. Hence, any ice classification obtained from SAR images requires linking the observed spatial radar signature and texture variations to the WMO ice terminology and to assess to which detail ice conditions can be characterized with the given SAR image as the data source. The separability of ice types and discernibility of ice structures depends not only on the radar properties as mentioned above but also on the relative influence of cm-scale ice features (Section 8.2), which do not differ consistently between different ice types and may cause different "appearances" (signatures) of a single ice type in a SAR image.

At the Canadian Ice Service (CIS), for example, ice charts are produced primarily based on RADARSAT-2 dual-pol ScanSAR Wide imagery, supplemented by Sentinel-1 IW (Interferometric Wide swath) and EW (Extra Wide swath) images. In addition, data from AMSR-2 (Advanced Microwave Scanning Radiometer 2), VIIRS (Visible Infrared Imaging Radiometer Suite), MODIS (Moderate Resolution Imaging Spectroradiometer), and GOES (Geostationary Operational Environmental Satellite) are used (D. Flett, CIS, personal communication, 2018). These instruments provide data from different spectral ranges (visible, near and thermal infrared, and microwaves) at different spatial resolution (10 m–25 km) and coverage (250 km–2000 km). However, in the visible range, only acquisitions during daylight are useful, and in the visible and infrared range, almost cloud-free conditions are needed. The ice charts are generated manually. Trained ice analysts, who use a geographic information system (GIS), manipulate the available images such that they can optimally interpret them within a short time (CIS: 1 h). Additionally, they consider meteorological conditions and information from ships and airplanes, and align and crosscheck their result with preceding ice charts. The information on sea ice conditions is then sent to ship navigators who use it for strategic or tactical route planning. Figure 8.2 depicts an example of an ice chart as distributed by the Canadian Ice Service, together with the SAR images that were taken as the basis for its production.

8.3.2 Toward automated segmentation and classification

Numerous studies have been published that propose techniques for automated ice classification. One major argument for switching to (at least partly) automated ice charting in the near future is the increasing number of operational satellites, which means on the one hand that a larger number of scenes is available per day and on the other hand that it is possible to reduce ambiguities in ice type separation by combining data of different SAR imaging options, modes, and frequencies. But what are the requirements for an automated procedure? Most importantly, the procedure has to be fast, reliable, and robust, and the necessary space in computer memory should be small. These are criteria valid for any algorithm to be employed operationally. The computational speed depends on the size of the original image. For accelerating the processing it may be necessary to down-sample the image, which means that the spatial resolution is reduced and the detection of structural details is more difficult. An algorithm is said to be reliable if it produces consistent results for similar ice conditions, and robustness is related to the algorithm's capability of trapping errors during execution.

Other, more specific criteria aim at making use of the full information content of a digital image. This means to take into account variations of image brightness as well as the spatial arrangement of those variations. For a single sea ice scene, separate images can be generated, for example, for different polarizations. In the case of quad-pol data acquisitions, not only intensities at like- and cross-polarizations but also different *polarimetric parameters* such as polarization ratios or the phase difference between different polarizations can be depicted separately or as overlays in an RGB format. In such cases, the question is which polarizations or polarimetric parameters are optimal for separating different ice types? Several studies deal with this problem but only one example can be given here. For the detection of leads in sea ice in Sentinel-1 dual-polarized (HH, HV) images, Murashkin *et al.* [15] suggested using the product HH × HV and the ratio HH/HV. Four cases can occur: open water leads under (1) calm and (2) strong wind conditions, and thin ice covered leads (3) with and (4) without frost flowers (for a definition of "frost flowers" see Table 8.2). At C-band, the HH-polarized intensity is strong for cases (2) and (3) and low for (1) and (4). At HV polarization, leads appear dark for cases (1), (2), and (4) and bright for (3). Hence, the magnitude of HH × HV can be expected to be low for (1) and (4), mid-range for (2), and large for (3), and the magnitude of HH/HV is much larger than one for (2) and closer to one for the other cases. It is important to keep in mind that intensities and polarimetric parameters are not constant for a given ice type because of natural "in-type" variations in, for example, temperature, salinity, small-scale surface and volume properties, and because of the influence of speckle. The latter is recognized as a grainy appearance of SAR images caused by the interference of elementary scattering returns of single sources within an area defined by the SAR's spatial resolution. Hence for each ice type, one has to deal with statistical distributions of intensities and polarimetric parameters. The respective distributions usually overlap between different ice types.

Image brightness is mostly represented by gray tones. The spatial arrangement of gray tones in a given region, that is, its texture, is quantified by different parameters. One popular method is to calculate gray-level co-occurrence matrices (GLCMs), but other approaches for computing texture parameters exist. In one study dealing with ice type separation and significance level testing, the GLCM achieved the highest accuracy in classifying sea ice types compared to two other methods for feature extraction (Gabor filters and Markov random fields) [16]. The elements of GLCM matrices reflect brightness changes between pairs of pixels in an image. To be more specific, the number of pairs with a certain combination of gray tones is counted as a function of offset (in number of pixels) and direction. The elements of the GLCM are hence measures of the probability that a certain change from gray level i to gray level j occurs at a particular distance and direction. Note that one GLCM is required for each offset and direction. The number of quantization levels in the image determines the size of the GLCM (which is a square matrix). It is immediately clear that, for example, 16 bit data result in a much too large matrix (65536 rows and columns). Therefore, the first step usually is to rescale the image to sizes between typically 4 bit (16 × 16 matrix) to 7 bit (128 × 128) [16–19]. Further steps include fixing the size of the sliding window for calculating the GLCM (e.g., 32–128 [17], 5–101 [18], and 9 [19] for square images with 2,500–5,000 pixels side length), the sliding distance (dependent on window size), the offset, and the direction. This means that in principle several GLCMs can be combined to characterize a given texture. Mostly, directions 0°, 45°, 90° and 135° are used, and the resulting GLCMs or texture parameters derived from the GLCMs are averaged to ensure rotational invariance. The window size is a balance between catching small local variations of texture (requiring small windows) and computational stability (which is higher when using large windows). The offset (which, in the different studies selected, was between 1 and 64 pixels) determines the spatial scale of the investigated texture parameters. The motivation for explaining the GLCM more in depth here was to give an example of the relatively extensive calculation procedure that may be necessary when quantifying texture, and of the tuning parameters needed. It is emphasized that useful textural information can also be obtained with less complicated processing.

From the GLCM, many different texture parameters, also denoted as texture features or GLCM features, can be calculated. However, several of them are correlated or even redundant. Maillard *et al.* [19] used only two parameters, contrast and entropy, which they regarded as most successful for sea ice classification. Zakhvatkina *et al.* [17] employed eight, Leigh *et al.* [18] nine, and Murashkin *et al.* [15] twelve features. Contrast quantifies the magnitude of local brightness variations. The entropy characterizes the degree of disorder among pixels in the test window, so a large magnitude indicates variable image content, for example, noise or structures. The correlation measures the linear dependency between the brightness levels of the reference pixel and its neighbors. The angular second moment (ASM) is related to the homogeneity of an area. The ASM is close to zero if distinct structures are visible and equals one if all pixels reveal the same brightness. Murashkin *et al.* [15] and Leigh *et al.* [18], who applied feature selection algorithms (to be introduced in Section 8.4.1), listed entropy among the least important features for lead identification and ice-water discrimination, respectively, whereas the other three parameters mentioned are among the more useful ones. The relative importance of various texture features may vary between different selection algorithms and depends on the type of input data. Leigh *et al.* [18], for example, used local averages and local maxima of intensities from 5×5 and 25×25 windows besides intensities at HH- and HV-polarizations on a pixel level as input images. Murashkin *et al.* [15] employed the intensity at HH-polarization, the band ratio HH/HV, and the band product HH $\times$ HV.

The retrieval of some texture parameters seems to be only weakly sensitive to the incidence angle, but a systematic investigation of different texture parameters as a function of incidence angle is still lacking. In general, the inclusion of texture parameters in sea ice classification is regarded as an advantage in automated sea ice classification. Yu and Clausi [20], however, noted that reliable texture features specific to a single ice type are difficult to extract. In fact, systematic comparison of texture features and their relative importance are rare. Graphs and images comparing several texture features for different ice types are, for example, shown in Zakhvatkina *et al.* [17], which provide at least a qualitative idea of the usefulness of single features in the classification process.

Other methods used for ice classification comprise, for example, class separation through predefinition of thresholds, neural networks, the Support Vector Machine (SVM), and the Random Forest Classifier. All these methods require training data to tune them for separating different ice types. They belong to the group of *supervised* classification schemes and display image segments that are labeled with a class name. Among the *unsupervised* algorithms, which are run without any training data, one finds clustering techniques such as k-Means and Isodata, which seek to identify pixels with similar statistical properties. In this case, the final product is a segmented image without any class labels.

A few classification approaches were published that are very closely related to operational generation of ice charts. The first fully automated, near real time, operational classification scheme was ARKTOS (Advanced Reasoning using Knowledge for Typing Of Sea ice), which was developed in collaboration with the U.S. National Ice Center. The basic idea of this system is to mimic the decision process of ice analysts when separating ice conditions [21]. Besides SAR, ARKTOS uses other sources of information, such as sea ice concentration from passive microwave radiometry and ice climatology data. (The notation "ice climatology" refers to the long-term properties of sea ice in a given region. It can be used to identify averages, trends, and extreme conditions.) The SAR scenes are segmented using "watershed merging." First, an initial set of separate regions is created through the application of a watershed algorithm. For the decision whether adjacent regions shall be merged, their sizes, average intensity difference, and boundary gradient are used as criteria. For each of the final segments, a set of attributes is calculated. The definition of these attributes is based on clues considered in the visual ice classification. The attributes comprise parameters derived from intensity, texture, geometric properties, and boundary characteristics of each region. Attributes are then used to build a set of rules that are applied for the final ice type labeling.

Another approach built in close co-operation with the CIS is the MAGIC system (MAp-Guided Ice Classification). MAGIC uses the polygons drawn by the ice analysts as input. For each polygon, a *pixel-wise* classification is performed on the basis of unsupervised segmentation and a subsequent automated labeling procedure. It is important to note that a manually drawn ice chart only provides information on average ice conditions for each polygon as a whole. With MAGIC, ice conditions can be characterized on a pixel-level, and it is possible to quantify spatial variations over the area of a single polygon. The segmentation module includes different techniques (see [22] for a list of related articles). Among these techniques, the Iterative Region Growing with Semantic (IRGS) is regarded as state-of-the-art. It combines segmentation with classification using a Markov Random Field (MRF) model. The latter allows incorporation of the effect of local spatial dependencies between pixels. The underlying idea of the IRGS is to closely link classification with the region-growing process. Details can be found in [20]. The problem of automatic labeling is discussed in [19] and [23]. So-called egg codes providing information about different ice parameters (see Figure 8.2) are used as input for classifying the segments in each polygon but can be changed by the user. MAGIC can be adapted for different tasks. For example, Leigh *et al.* [18] describe a procedure to separate ice and open water in dual-polarized (HH, HV) RADARSAT-2 images. In this case, two methods are combined: the IRGS approach is applied to the HV-polarized image and, in parallel, a pixel-based SVM classification is carried out, using 28 images of texture and backscatter features calculated from the HH- and HV-polarized scenes.

Not only conventional parameters derived from intensity and texture can be employed for ice classification, for the separation of ice and water, Komarov and Buehner [24] proposed using the difference between wind speeds estimated from

SAR (as described in [25]) and obtained from numerical weather prediction (NWP), the spatial standard deviation of the SAR wind speed, and the spatial correlation between the HH- and HV-polarized intensity. Hence, they directly link measurements of meteorological parameters and radar intensities. The aim is to assimilate the ice concentration information thus produced into prediction systems for forecasts of ice conditions. For open water areas, the difference between wind speeds derived from SAR and predicted by the weather model is close to zero, but over ice, it is larger since here the SAR wind retrieval (designed for open water) fails. As input to forecast models, the level of confidence for the identification of ice and water is needed. The authors analyzed more than 15000 RADARSAT-2 scenes acquired from 2010 to 2016 over Canadian and Greenland waters to determine the probability distribution functions of the three variables mentioned above and subsequently the level of confidence of ice/water retrievals.

8.3.3 Incidence angle sensitivity

In wide-swath SAR imaging modes, physical antenna beams covering adjacent swaths are combined to achieve large ground-range coverage (for RADARSAT-2 between 300 and 500 km, for Sentinel-1 between 250 and 400 km). The incidence angle θ increases from the near- to the far-range edge of an image. The angle difference between near- and far-range edges is a function of the swath width and the altitude/height from which the image is acquired. The difference is larger if the swath width is large and the flight or orbit altitude is low. At the CIS, for example, RADARSAT-2 ScanSAR Wide images (available in two combinations of adjacent beams with incidence angles ranging from 20° to 46° and 20° to 49°) and Sentinel-1 IW and EW modes (29°–46° and 19°–47°, respectively) are used for ice chart production. As the incidence angle increases, the backscattered intensities of a given ice type at like-polarization (HH, VV) decrease. Theoretical models show that the incidence angle sensitivity of the radar intensity at like-polarization mainly depends on the ice surface roughness spectrum at scales from centimeters to meters. Volume scattering, which may occur at low ice salinities, also shows an incidence angle sensitivity in particular at larger values of θ. The sensitivity is largest for scattering from smooth surfaces, and moderate or low for scattering from rough surfaces [26]. In practice, sensitivities are linked to different ice classes (in [26], for example, to level FY and deformed FY ice). In a few cases, the cross-polarized intensities (HV, VH) appear to be almost invariant to incidence angle changes, but other studies demonstrated that this is not valid in general. Polarimetric features (e.g., co- and cross-polarization ratios) and texture parameters may also reveal more or less pronounced changes of their magnitudes across the swath width, and these changes have to be considered in manual and automatic classification of sea ice (see, e.g., Section 8.4.1). One possibility to alleviate the incidence angle effect on the radar intensity is to use "gamma" $\gamma = \sigma^0/\cos\theta_I$, where σ^0 is the backscattering coefficient (in linear scale), and θ_I is the local incidence angle at a given range. However, γ still reveals a not negligible decrease with increasing θ [27]. Another possibility to reduce the incidence angle effect is to determine linear trends for σ^0 (in dB) as a function of θ, as reported in several studies that are listed in Table 1 of [26]. Depending on the ice type and season (freezing or melting conditions), an increase of $\Delta\theta = 10°$ causes a decrease of, for example, σ^0 between 1.3 dB and 3.4 dB at HH-polarization. In most cases, a SAR scene covers different sea ice types with different incidence angle sensitivities. Hence, one needs to know the ice type for each position in the image before removing the incidence angle effect. On the other hand, the removal is required before automated ice classification can be started. A solution is to use an average trend determined over the entire image, which, however, means that for some ice types the incidence angle effect is not fully compensated, while for other ice types it is overcompensated. This calls for a more sophisticated method. Doulgeris and Cristea [28] proposed to consider the incidence angle effect directly in the statistical model for segmentation of intensity images. In a first study, they assume a linear decrease of the intensity I for each ice class k: $I_k(dB) = a_k + b_k\theta$ where a_k is the mean intensity at a reference incidence angle, and b_k is the slope term. In the image segmentation process, the intensity distribution of each ice type is not only described by its mean a_k and standard deviation σ_k but also by the slope term b_k. This approach can be extended to multivariate analysis, if, for example, image data are acquired at different polarizations or in polarimetric mode. It has also to be taken into account that the incidence angle dependencies of intensities and polarimetric/texture parameters are usually not linear over the incidence angle ranges mentioned above. The effect of considering the incidence angle sensitivity in segmentation is visualized in Figure 8.3. Lohse *et al.* [29] demonstrated the implementation of the incidence angle sensitivity as an ice class property into a Bayesian classifier.

8.3.4 Melting conditions

During the summer season, ice charting based on C-band SAR is more challenging than for freezing conditions. In this context, the use of L-band SAR images as an additional data source has been discussed and investigated [13,30,31]. The study by Casey *et al.* [31] provides the most detailed and valuable analysis to date. Their findings reveal that signature contrasts between FY and MY (multiyear) ice are larger at L-band during the beginning and end of the melting season but can be very low at both C- and L-band when melting is strong. Information on presence and areal fractions of MY ice is important since it is thicker and harder than FY ice and presents a more severe hazard to ship traffic and offshore platforms. It also has to be considered that ship traffic increases in summer. The melt season can be divided into different stages, which reveal different characteristics of the backscattered intensity [32]. In winter, the radar signature of MY ice is dominated by volume scattering from air bubbles in the upper low-salinity layer, whereas the major mechanism for the more saline but less bubbly FY ice is scattering from the snow-ice interface. The dry snow layer has only a minor influence on the scattering characteristics (see Section 8.2 above). FY and MY ice can be well separated at C-band but less clearly so at L-band. In comparison to the winter situation, major changes taking place successively in summer are that (1) volume scattering from

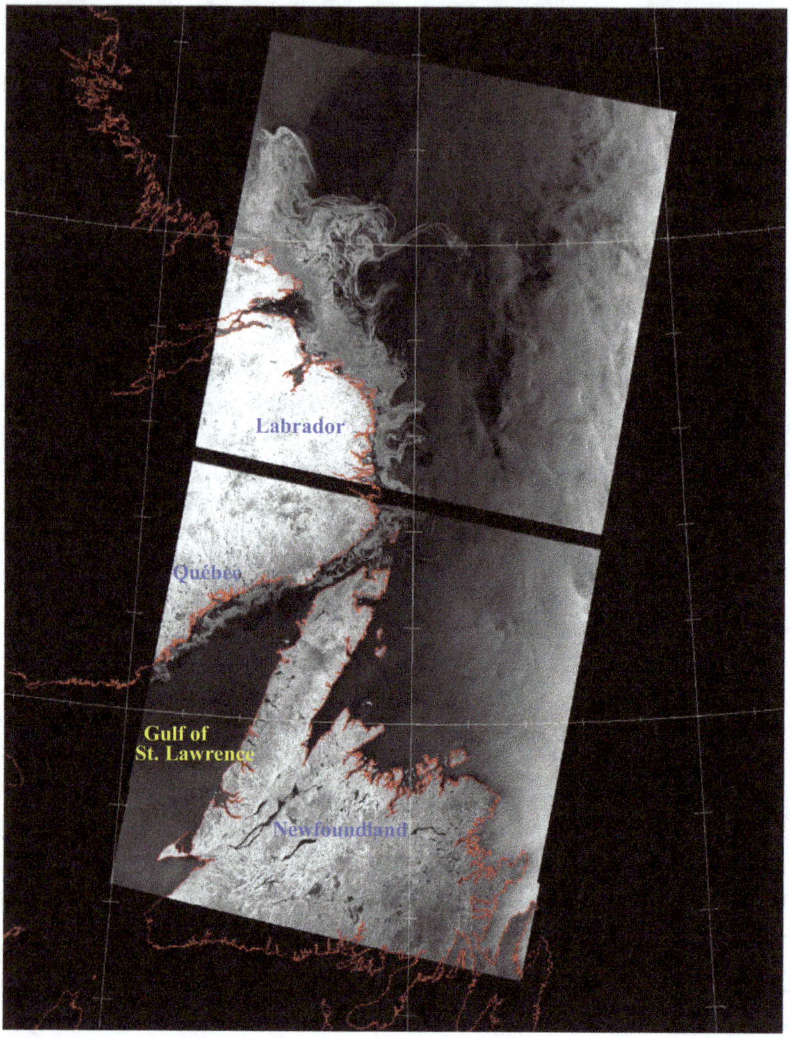

(a)

Figure 8.2(a) SAR scenes from Radarsat-2.
Here, two ScanSAR images acquired at HH-polarization in mid-January 2019 are shown. Land areas are delimited from ocean and sea ice by red-drawn coastlines. Sea ice is observed adjacent to the coasts of Québec and Labrador. Ice with a rough surface (e.g., ridged or pancake ice) appears bright. The dark zone at the seaward edge of the pack ice in the Gulf of Lawrence indicates the presence of frazil or grease ice. Also the open water surface reveals areas of darker and brighter gray tones. This is caused by the changing radar incidence angle from near range (at the right margin) to far range and by varying surface roughness due to wind and currents. RADARSAT-2 Data and Products. © MDA Geospatial Services Inc. (2019). All rights reserved. RADARSAT is an official mark of the Canadian Space Agency.

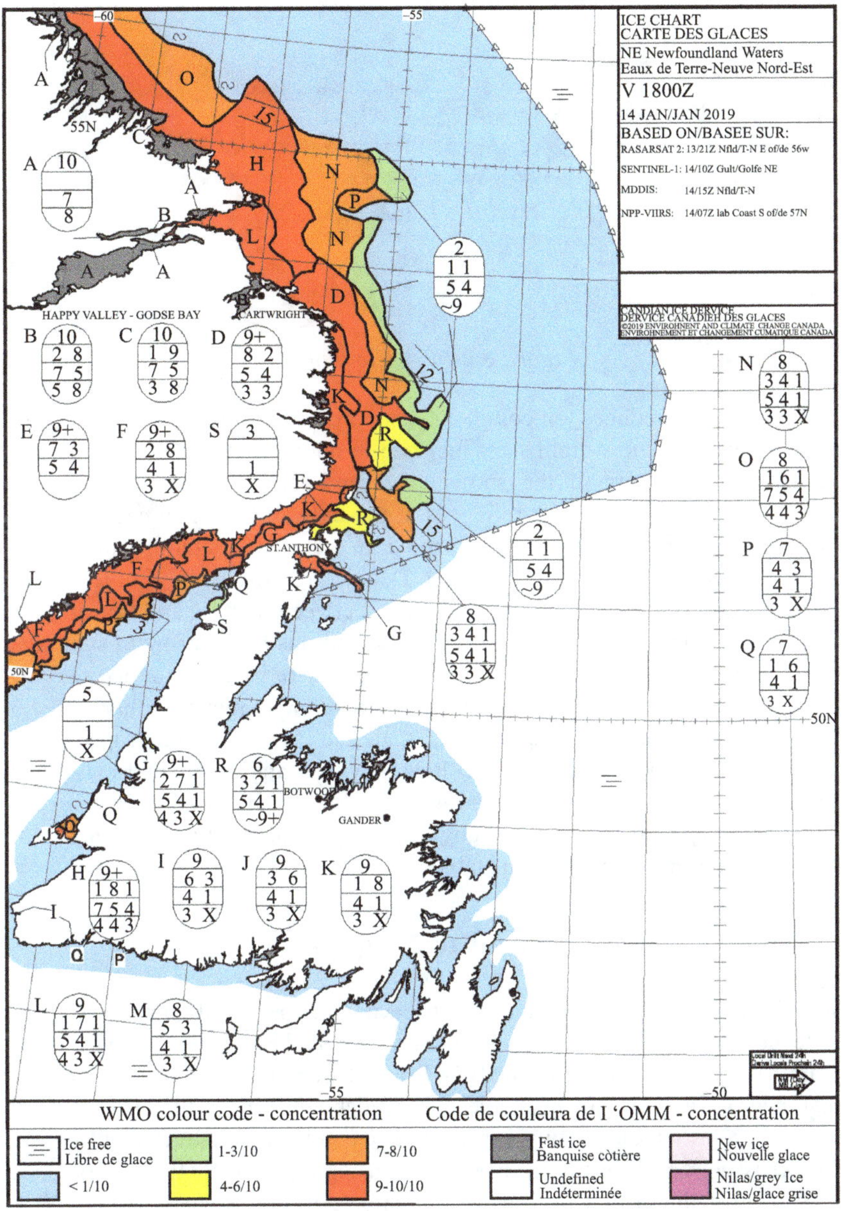

Figure 8.2(b) Ice chart as distributed by the CIS.
This ice chart was generated on the basis of the SAR images shown in Figure 8.2(a). The ice-covered area is delineated by coloured polygons. Different colors are linked to different concentrations as shown in the legend. The indicated polygons separate zones of distinct ice conditions, which are described using the "egg code" (shown on the next page). This code specifies ice concentration, ice types present in the polygon, their areal fractions, and their relative floe sizes. Only the dominant ice types are considered, and usually only three are listed for each polygon. Detailed rules for the egg code can be found in [1]

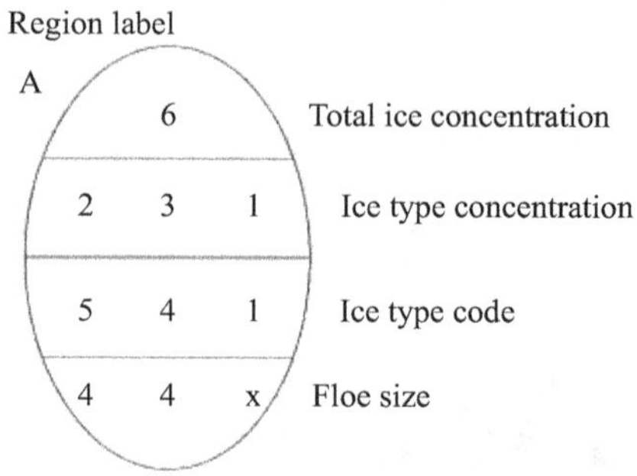

Figure 8.2(b) (Continued)

the ice volume is reduced or completely suppressed as the snow cover and ice surface get wet, (2) snow grain sizes increase through processes of metamorphism, (3) liquid water occurs in the snow cover and later on the ice surface (see Figure 8.1), and (4) melt water refreezes at the base of the snow pack, increasing the snow-ice interface roughness or forming an ice layer containing numerous air bubbles (denoted as "superimposed ice"). Details are provided in Table 8.4.

When meteorological conditions alternate between melting and freezing within short time intervals (in particular during the transitions from spring to summer and from summer to autumn), the radar signature characteristics of the ice may differ significantly from one to the next freezing interval due to metamorphism processes in the snow cover, changes of centimeter-scale ice roughness and properties of air bubbles in the ice. This causes difficulties in visual and automated classification since the brightness patterns observed in temporal sequences of SAR images may reveal varying appearances, even though the spatial distribution and relative fractions of different ice types remain stable.

8.4 Advanced measurement techniques

Quad-pol measurement modes have already been implemented in a number of different satellite missions (RADARSAT-2, ALOS PALSAR, and PALSAR-2 on ALOS-2), and CP-data have been available from RISAT-1 and from PALSAR-2 (here as experimental mode). Therefore, many studies are available dealing with various aspects of polarimetric SAR remote sensing of sea ice. The combination of different frequencies is another issue, which has been investigated using airborne SAR imagery and to a lesser extent satellite data. Only very few studies focus on the combination of images acquired at different spatial resolutions (this item is not examined here), and on the use of interferometric SAR for the retrieval of sea ice topography and deformation.

8.4.1 Polarimetry

In multipolarization radar imaging, quad-pol measurements provide the most comprehensive information since they involve all complex scattering coefficients

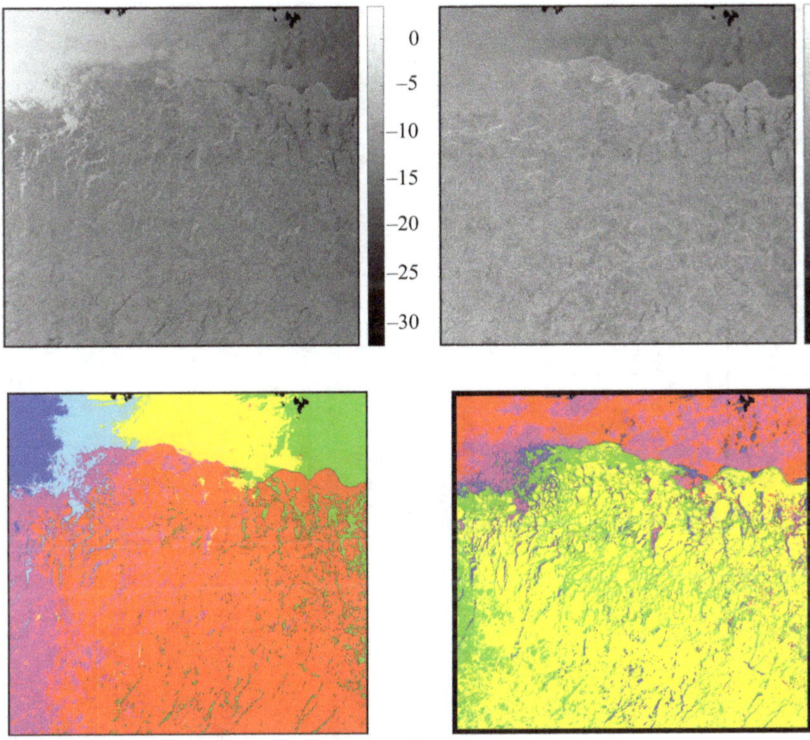

Figure 8.3 Incidence-angle guided image segmentation.
The images show HH-polarized intensity (upper left), HV-polarized intensity (upper right), the result of segmentation (based on both HH- and HV-polarization) without considering the incidence angle effect (lower left), and with considering it (lower right). The intensity images are from Sentinel-1, acquired in Extra-Wide swath mode over Fram Strait. The colors of segments were selected arbitrarily and do not match between the lower images. The water surface, in the uppermost part of the intensity images, shows decreasing brightness from left to right (near-range to far-range) at HH-polarization that reveals the systematic incidence angle effect, which is less obvious at HV. Brighter and darker patches are likely due to varying wind speeds over open water or due to new ice formation. Some curvilinear structural details are easier to recognize at HV. Courtesy: Anthony Doulgeris, UiT The Arctic University of Norway

(see Section 8.2). However, for operational sea ice mapping, the major drawbacks of quad-pol images are that they can only be acquired over very narrow swath widths, the related data volumes and processing times are large, and the calibration procedure is more demanding. The practical gain of quad-pol measurements for sea ice classification consists of two additional parameters besides the like- and cross-polarization intensities: the correlation $\rho_{VV\text{-}HH}$ and phase difference $\delta_{VV\text{-}HH}$ between the VV- and HH-polarized channels (Section 8.2). The phase difference

Table 8.4 Melt stages and related radar signature characteristics at C- and L-band
Descriptions are based on observations in the Arctic [31,32]. Since only few observations are available until now for the intensity contrast, the results may not be valid in any case. In the Antarctic, periods of melting are shorter than in the Arctic and are predominantly observed in the Weddell and Bellingshausen Seas around the Antarctic Peninsula [33]. Here, "second year" is abbreviated as "SY."

Melt stages	Changes in snow and ice	σ^0 FY ice	σ^0 MY ice	Intensity contrast $\sigma^0_{MY}-\sigma^0_{FY}$
Early melt	Rapid growth of snow grains due to increased air and snow temperatures; liquid water and brine occur in basal snow layer during day, refreezing during night	Diurnal variations may occur due to scattering from large grains in the snow in the daytime and from the ice surface at night; scattering remains at winter level if surface roughness is large	Close to typical winter level, stable	Remains close to the winter level at C- and L-band, increases at larger incidence angles
Melt onset	Liquid water in interstices of the snow (pendular regime); At water volume fractions > 7% breaking of bonds between snow grains and percolation of water (funicular regime)	Significant diurnal fluctuation; increase in the pendular regime due to scattering from (brine) wetted snow grains; continuous decrease during funicular phase	Strong decrease in particular at C-band due to reduction in radar penetration depths and suppression of the volume contribution; instead, scattering contributions are from wet air-snow and snow-ice interfaces	In the pendular regime very low at C-band, for $\theta > 25°$ still distinct at L-band; changing from positive to negative values during transition to funicular regime, for $\theta > 25°$ larger (negative) contrast at C-band than at L-band
Advanced melt	Melt ponds and flooding due to melting of snow; during the *ponding* phase extensive but shallow ponds occur on FY ice, and smaller but deeper ponds on MY ice; the surface is a mixture of melt ponds, wet snow patches, and bare wet ice in the *drainage* phase	Highly variable dependent on areal surface water fraction; for low wind, scattering is weak, larger wind speeds roughen the water surface which causes a high level of backscattering; contributions of ice surface roughness (e.g., hummocks) have also to be considered	Same as for FY ice	During ponding phase only weak and variable (positive and negative); during drainage phase, weak at C-band but distinct at L-band; incidence angle sensitivity negligible at L-band, weak at C-band

(Continues)

Melt-freeze transition	Refreezing of liquid water in snow and on ice; FY ice that survived melting turns into SY	Increase of melt-freeze cycles; during freezing periods, the penetration depth of the radar signal increases and backscattering approaches typical MY ice levels	Same as for FY ice	Starts to grow again, approaching typical winter magnitudes

$\delta_{VV\text{-}HH}$, for example is useful for separating nilas and young ice from thicker ice [34]. The ellipsoid- or needle-shaped brine inclusions are preferentially oriented in the vertical direction. Due to this alignment, the overall dielectric constant of nilas and young ice is uniaxially anisotropic. This means that the vertically and horizontally polarized radar waves propagate with different speeds causing a non-zero phase difference between the HH- and the VV-polarized signal. Older ice behaves like an isotropic medium, that is, the phase difference is zero. An example of visualizing polarimetric parameters as images is shown in Figure 8.4.

From the directly measured parameters (the scattering matrix elements, from which covariance and coherency matrices are obtained), several polarimetric features can be calculated, as indicated in Section 8.3. Examples are the co- and cross-polarization ratios, the degree of polarization, geometric brightness, or decompositions such as entropy, anisotropy, and alpha angle ("H/A/α"). Definitions and more examples of polarimetric features can be found in [14,34,35] and references

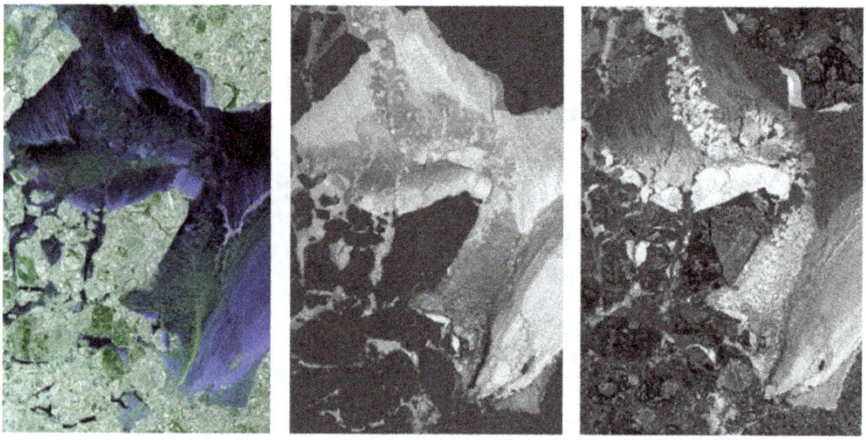

Figure 8.4 Polarimetric parameters.
These images show an RGB-intensity image (pixel spacing 5 m), with R: HV, G: HH, B: VV intensity (left), the copolarization ratio VV/HH (middle), and the phase difference HH-VV (right). In the intensity image, open water and frazil ice appear in blue (low backscattering, VV $\gg$ HH), dark green are zones of nilas, gray or gray-white ice (VV>HH, still relatively low backscattering level), and bright green are FY ice floes with ridges (similar magnitudes at HH and VV, high level of backscattering). The copolarization ratio is around 1.0 for FY ice (dark gray tones), and > 1 for the thinner ice types and for open water (light gray and white). The phase difference is close to zero for FY ice and open water, and between 30° and 60° for the thin ice types. Images were acquired by the airborne Danish EMISAR system in C-band quad-pol mode over the Greenland Sea. For investigating the potential of advanced technologies, airborne or high-resolution SAR images are preferred since it is easier to distinguish single ice types and to analyze their radar signature in detail

cited therein. Combinations of different polarimetric parameters that are related to typical scattering mechanisms are valuable for enhancing the visual and automated ice type separation. One example is the Pauli-basis, which, expressed using the elements of the scattering matrix, is defined as $|S_{HH}\text{-}S_{VV}|$, $2|S_{HV}|$, $|S_{HH}+S_{VV}|$. These elements (in their given order) represent double-bounce scattering, depolarization, and single bounce scattering and can be assigned to the RGB color scheme [34]. Another set of parameters was suggested by Geldsetzer *et al.* [35] for ice type separation based on CP: $P_{depol} = \sigma^0{}_{RL}(1\text{-}m)$, $P_{phase} = \sigma^0{}_{RL}\ \sin[(\delta_{RV\text{-}RH}+\pi)/4]$, and $P_{copol} = 0.5\sigma^0{}_{RL}\sigma^0{}_{RV}/\sigma^0{}_{RH}$. Here, $\sigma^0{}_{RL}$ is the right-circular-transmit left-circular-receive scattering coefficient, RV and RH denote right-circular transmit and vertical or horizontal linear-receive, respectively, m is the degree of polarization, and $\delta_{RV\text{-}RH}$ is the phase difference between RV- and RH-polarization. The first term accentuates contributions from volume and multiple scattering (typical for MY ice and ridges under freezing conditions), the second emphasizes phase differences $> 0°$ (observed for thin ice), and the third highlights polarization diversity (which young ice, FY ice, and open water may exhibit). These parameters were processed as RGB images and assessed by a CIS analyst in comparison to an HH/HV data product, in preparation for the RCM [36]. The analyst found that the CP-product provided more visual information than its dual-pol equivalent in a variety of ice conditions (e.g., better discrimination between wind-roughened open water and young ice, floe boundaries appear more distinct). However, this assessment was regarded as preliminary, and other polarimetric parameters may even improve the classification accuracy [35]. An important result of Geldsetzer *et al.* [35] is that the selection of the optimal parameters depends on the ice-type pair to be separated, and on the incidence angle range. Like the radar intensities, polarimetric features reveal sensitivities to the incidence angle θ that are dependent on ice type and differ between features. The magnitudes of polarimetric features as a function of θ, however, show more complex shapes than the intensity.

Although all studies published hitherto agree that polarimetric or CP-data improve the overall ice classification capability, there is yet no consensus with regard to the choice of polarimetric parameters/features to be included in the classification process. Considering the large number of features (e.g., in [35], 26 parameters were used), the question arises as to how to select the ones that maximize information content and computational efficiency. Some features are extracted directly from the covariance matrix. Others are calculated from its eigenvalues (which are identical to the ones of the coherency matrix) or derived using decomposition techniques (such as $H/A/\alpha$ [37] or the Target Scattering Vector Model [38]). Also, statistical quantities such as the relative kurtosis [39] are useful. For some classification methods, the shape of the parameter value distributions is also important (e.g., Gaussian versus non-Gaussian). Besides the computational burden required for the inclusion of many features, too large a feature set decreases the classification accuracy (Hughes phenomenon). Different techniques for reducing the number of features are available. Features can be judged based on how well they separate ice types, for example, by calculating the distance between the respective value distributions for each ice type. To determine a rank order, different methods can be employed, for example, the sequential forward feature selection or the recursive feature elimination. The forward feature selection algorithm starts with the parameter that achieves the best classification result and step-by-step adds the next parameter that reveals the largest classification improvement. Feature elimination as applied, for example, in [15] means to step-wise remove the features with the lowest influence on the classification performance. Since the algorithm needs to be fast, not all possible feature combinations can be assessed, and a feature already included in the subset can usually not be removed or re-added at a later stage.

8.4.2 Multifrequency

From the optical domain, the principle of multispectral imaging is well known, that is, the enhancement of information through the combination of images measured at different wavelengths. Because of technical reasons, this strategy is much more difficult to implement for imaging radars. While space-borne optical multispectral data acquisitions can be carried out using a single instrument, recent space-borne multifrequency radar measurements require combining data from different satellite missions. Hence it is often necessary to coordinate the respective data takes for a given area and time through negotiations with different space agencies. By identifying matching areas in images acquired at different frequencies and extracting the radar signatures of different ice types, it is possible to analyze pros and cons of the corresponding radar bands relative to one other, wherein possible differences in spatial resolution, polarization, incidence angle, and number of looks (i.e., the level of speckle reduction) must also be considered. Because C-band is the standard frequency for operational ice monitoring, it is often used as a reference to assess the usefulness of other frequency bands. Most investigations cover the comparison of two or three bands within the range from P- to X-band. The major finding is that the gain for sea ice classification is larger with multifrequency dual-pol data than with single-frequency quad-pol data. The different appearance of sea ice at C- and L-band is demonstrated in Figure 8.5.

What now are the specific results? Lee *et al.* [40] investigated an airborne data set from the Beaufort Sea, covering P-, L-, and C-band quad-pol images. They found that the discrimination between leads and FY ice was difficult at L-band, as was the discrimination between MY and ridged FY ice areas at C-band. The overall classification accuracy was highest at L-band. Based on studies of airborne dual-pol (HH/HV and VV/VH) C- and L-band images complemented by aerial photography from three sea ice regimes around Svalbard, Dierking [41] came to partly similar conclusions: L-band imagery was in general better suited to distinguish brash ice, rubble fields, and ridged ice from smooth level ice. For two regimes, L-band was worse for separating new from older and thicker level ice. A combination of L- and C-band images is required to separate lead ice covered with frost flowers from uncovered lead ice. Dierking and Dall [42] demonstrated by means of aircraft

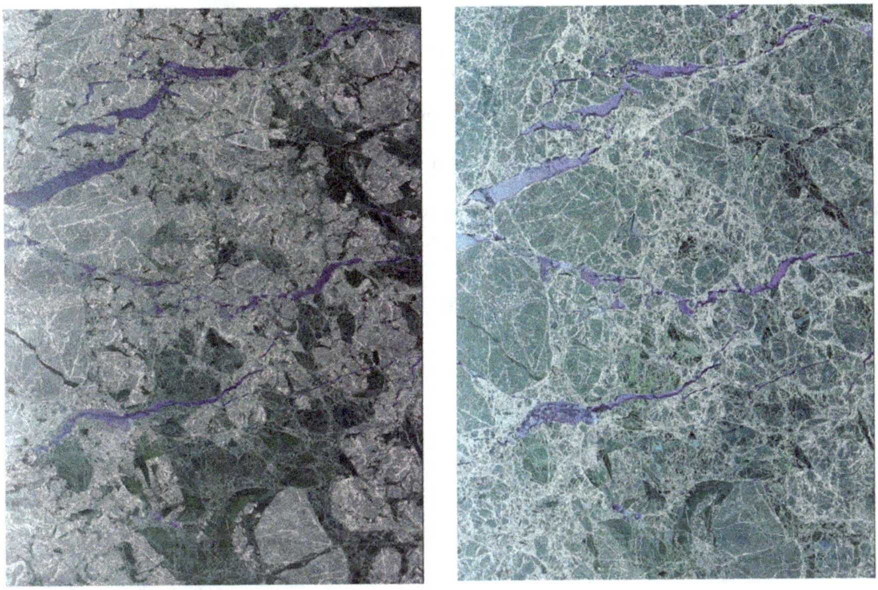

Figure 8.5 Difference between C- and L-band signatures of sea ice. C- (left) and L-band (right) intensity images (pixel spacing 5 m, swath width 12 km), with R: HV, G: HH, B: VV. They were acquired by the Danish EMISAR system north of Svalbard. At C-band, gray and gray-white ice (dark green) can be well separated from thicker FY ice (light green). This is more difficult at L-band, where, on the other hand, many deformation structures show up more clearly

quad-pol SAR data acquired over the Baltic Sea and north of Svalbard that L-band is much better suited for the detection and characterization of ice ridging on scales of meters to several 100 meters than C-band. X-, C-, and L-band satellite data were compared in different studies, for example [43–46]. Eriksson *et al.* [43] investigated ice regimes of the Baltic Sea and confirmed that the detection of ice ridges is least difficult in L-band images. They found that the information content of X- and C-band images is largely equivalent. L-band SAR is less sensitive to wet snow on the ice (the significance of L-band for ice classification in melting conditions was elucidated above). Dierking [44] analyzed images from Fram Strait. He concluded that thin ice radar intensities are sometimes much more difficult to distinguish from those of thicker ice at C- and X-band than at L-band. This is inconsistent with the observations of Lee *et al.* [40] but supports the result in [41] that, in some cases, L-band may be better to separate new and older ice. Johansson *et al.* [45] used quad-pol satellite images acquired north of Svalbard and found that X- and L-band complement C-band in the identification of leads and new ice. They suppose that, in their case, the higher spatial resolution at X- and L-band may have contributed to this result. Johansson *et al.* [46] examined quad-pol (L, C) and dual-pol (X) images from the same region with the objective to accurately detect new ice. They found that the combination of entropy and co-polarization ratio is well suited to separate new ice from open water and thicker sea ice at all three radar bands throughout the winter and spring season.

The conclusions that can be drawn from these and other studies are that lower frequencies (L, P) are superior for detecting deformation structures such as ridges and rubble fields, and fragmented ice such as brash, which is a relevant result for operational ice mapping considering that deformed thicker ice poses a hazard to marine operations. When it comes to separating new and thin ice types from one another and from the thicker surrounding ice, sometimes L-band reveals better results while, in other cases, C- or X-band are superior. The reason behind these apparently contradicting observations is yet not clear but is probably linked with differences in small-scale ice properties and related scattering mechanisms. The mapping of thin ice and open water areas, in particular leads, is also of importance for ice charting, since they are favorable routes for ships.

8.4.3 Interferometry

Information on variations of sea surface elevation, in particular ridge heights and inter-ridge distances, are important for operations in the pack ice or on landfast ice. Ridge heights need to be measured with accuracies of ±20% (at 50 m spatial resolution), and for tactical purposes as much as ±10% accuracy at 10 m resolution [13]. Sea surface topography data are also useful for studies on atmosphere–ice interactions or sea ice mechanics. In principle, interferometric SAR may provide the necessary data [47]. However, at present, this technique can only be applied over landfast ice or requires special orbit configurations.

For interferometric height retrieval, it is necessary to acquire two radar images from slightly different sensor positions. This can be achieved using only one satellite by combining images from repeated orbits. (Repeat orbits do not match perfectly in space, but differ typically by a few hundred meters.) Another possibility is to use data from two satellites operated in a pair configuration, such as TanDEM-X. The geometric distance between the two sensor positions is denoted as baseline, which is usually separated into across- and along-track components oriented perpendicular and parallel to the satellite velocity vector, respectively. The former determines the sensitivity to height variations, the latter to ground displacements along the line-of-sight (LoS) between radar and ground area. However, the distance cannot be selected arbitrarily long, since the correlation between the two received signals, which decreases with increasing baseline, has to be sufficiently high for height or displacement retrievals. The magnitude of the correlation (i.e., the "interferometric coherence") depends on the time difference between image acquisitions and decreases if the ice and snow properties change between data acquisitions due to wind, melting, and refreezing. For repeat-pass interferometry over drifting pack ice, the coherence is too low for any useful retrieval of ridge heights and displacements. The phase difference between two (correlated) radar signals received from the same ground area element consists of contributions related to both ice surface topography and ice movement plus components due to the influence of the atmopshere and ionosphere, scattering out of the ice volume, and radar system noise. The presentation of spatial variations of the phase difference in an image is called an interferogram. For further details, the reader is referred to Dierking *et al.* [47] and references cited therein.

The potential of InSAR measurements for ridge height retrieval is examined in Dierking *et al.* [47]. They investigated satellite tandem configurations operated at different radar frequencies from K_a to L-band, considering statistics of ridge heights and widths as well as possible magnitudes of ice drift. They found that with favorable satellite mission configurations (i.e., tandems with a large across-track baseline and spatial resolution of ~10 m) surface height variations can be retrieved with relative errors $\leq$ 0.5 m. However, the along-track baseline cannot be constantly kept at zero because of larger risks of collisions between the two satellites. Non-zero along-track baselines complicate the height retrieval as ice movements on the order of 0.05 m/s result in significant contributions to the measured interferometric phase. A potential (but yet not tested) solution proposed in [47] is the independent determination of the LoS velocity using the Doppler technique described in Section 8.5 below, and the subsequent removal of the motion contribution from the interferometric phase. Another issue is the lack of near-future plans for satellite configurations that are optimized for operational generation of ridge height maps (such as the cross-track pendulum). Yitayew *et al.* [48] used interferometric TanDEM-X data to retrieve heights of ridges on temporarily stationary ice in Fram Strait. Additional measurements of the sea-ice surface topography with a stereo camera and a laser altimeter served as references for comparison. Ridge heights $\geq$ 0.5 m could be estimated with a root- mean-square error of 0.3 m or less.

The capability of interferometric SAR to reveal temporal coherence and to monitor small displacements on centimeter to decimeter scale has been considered useful to determine the extent of landfast ice and assess its stability by detecting zones of movements and deformation. Landfast ice is sea ice that forms and remains temporally stationary along the coast [1]. It may be attached, for example, to the shore, shoals, or grounded icebergs and extends a few meters to several hundred kilometers from the coast. It is traversed by people of coastal communities for fishing and hunting, may carry a network of ice roads for transportation, and reduces coastal erosion during winter storms. The time interval during which ice has to remain stationary until it is called landfast is not defined, but often a duration of at least 20 days is set as a limiting minimum [49]. Recent publications on applications of InSAR for studies of landfast ice deal with X-, C-, and L-band [49–52]. The shorter the wavelength, the smaller are the displacements that can be retrieved. Berg *et al.* [50] used two COSMO SkyMED image pairs, each acquired with 24 h's time difference between the single images, to present two case studies of deformation patterns in the north-eastern part of the Bay of Bothnia. They were able to retrieve ice displacements on the order of 1 cm and related the deformation to compressive forces caused by pack ice at the seaward edge of the landfast ice zone. Marbouti *et al.* [52] based their investigations in the Gulf of Bothnia on two images from Sentinel-1 that were taken with a 12-days' time lag. Over this period, the surface of the landfast ice remained stable enough to preserve coherence. The advantage of a larger temporal gap is that landfast and slowly drifting ice can be distinguished more reliably. The authors observed displacement of landfast ice on the order of 40 cm across the 20 km wide study area and attributed it to forces caused by sea level tilt, currents, thermal expansion, and pressure exerted by drift ice. Meyer *et al.* [49] employed ALOS PALSAR imagery acquired over coastal regions of Alaska. L-band interferometry is less sensitive to displacements of the ice than C- or X-band but reveals a higher coherence over time spans of tens of days. For ALOS PALSAR, the time lag between orbits with repeating ground tracks amounted to 46 days. After application of a phase filter, Meyer *et al.* [49] found that the difference in the 46-days-coherence between landfast and drifting ice was large enough to use coherence as a first step to separate both ice classes and to generate a landfast ice mask. They suggested a subsequent processing chain that comprises morphological filtering to fill holes in the landfast ice mask and segmentation to isolate patches erroneously classified as landfast in the drifting ice zone. A comparison to a complementary method for landfast ice detection revealed that their method is of high robustness and accuracy for a wide range of environmental conditions. Dammann *et al.* [51] focus specifically on the assessment of landfast ice stability for support of strategic planning of coastal ice traffic and related tactical decisions using interferograms obtained from Sentinel-1 imagery.

8.5 Ice displacement and deformation

The drift of sea ice influences the spatial and temporal distribution of heat and salinity in ice-covered ocean regions. Heat is released at locations where ice forms and is absorbed where it melts. Due to brine rejection during ice growth, the salinity of the water layer below the ice increases. Where old ice melts, the upper water layer becomes less saline. Deformation processes are related to local variations and discontinuities in the drift field. "Compressive" deformation of sea ice, which shows up as ridges, hummocks, and rubble fields, is a potential threat to ship navigation and offshore structures. "Divergent" ice deformation, on the other hand, results in openings and leads, which may serve as possible shipping routes. Wind, ocean currents, and internal ice stress affect ice drift patterns. On larger spatial scales ($\geq$ 100 km), Coriolis force and ocean surface tilt have to be considered. Discontinuities of the drift field occur at the edges of stationary landfast ice, at bulges of the coastline, at grounding points, and near icebergs, or are caused by cyclonic weather events and converging currents. CIS has run an automated algorithm to extract sea ice motion from SAR imagery. The resulting products are used as additional source of information for the production of ice charts (Dean Flett, CIS, personal communication, 2018). For the near future, one objective at CIS is to develop techniques for assimilating ice drift information derived from RCM SAR into ice forecast models [53].

Ice drift is determined from pairs of SAR images. In contrast to the requirement that pairs collected by a single radar sensor can only be used for interferometry if they are acquired from repeat orbits, the only constraint for drift retrieval is the need for a sufficiently large spatial overlap between image acquisitions from different orbits within a time interval dependent on the observed ice drift regime. Strictly speaking it is only possible to determine *the direct distance* between the positions of a distinct ice feature visible in images 1 and 2. Irregular patterns of ice drift within the temporal gap between image acquisitions, for example, back and forth movements due to tidal currents or curvilinear motion, cannot be reconstructed. Larger temporal gaps of several hours to a few days occur, in particular, in image sequences of a single-satellite, acquired over narrow swaths. Gaps can be reduced when combining data from different satellites. Lehtiranta *et al.* [54] showed that it is also possible to estimate sea ice motion from a pair consisting of an L- and a C-band image, although this is less reliable than combinations L-L or C-C. Because of the relatively narrow widths of overlaps between SAR images even at ScanSAR or wide-swath modes (often considerably less than 500 km), drift fields can usually be determined only on local and regional scales. With a systematic data acquisition strategy such as the RADARSAT Geophysical Processor System [55], however, it is possible to cover large parts of the Arctic Ocean. Sea ice drift vectors can be provided at spatial resolutions of a few tens to hundreds of meters with good accuracy under favorable conditions [56,57]. The achievable accuracy depends on ice conditions and initialization of the retrieval algorithm [58].

For the retrieval of ice drift patterns from SAR image pairs, pattern matching and feature tracking can be used. The former is based on either the normalized cross correlation [54] or the phase correlation [59], or a combination of both [56,60]. For robust cross or phase correlations, distinct and stable spatial radar intensity variations (e.g., caused by the presence of leads, ridges, and ice floe edges) are required. This means that any deformation of the ice between acquisitions of images 1 and 2 (i.e., the occurrence of *new* deformation features) disturbs the matching process. The correlation is computed for windows of predefined size. Larger windows provide more reliable measures of consistency between stationary patterns but cause larger errors if translational and rotational movements vary within the covered area [58]. Smaller windows preserve local drift variations but may lead to lower accuracies of drift retrievals. The normalized correlation function is more robust against noise but is computationally less efficient and requires repeated calculations for different orientations of the window to capture rotation of ice floes [57]. Lehtiranta *et al.* [54] describe a method to hardware-accelerate cross-correlation calculations. The phase correlation approach is much faster and can be extended to detect rotational movements. However, the magnitude of phase correlation is not a reliable measure of similarity [60]. Komarov and Barber [60] and Korosov and Rampal [57] use the cross-correlation coefficient to judge the quality of a pattern match, whereas Hollands *et al.* [61] and Karvonen [59] added additional criteria (e.g., texture features, or consideration of lower phase correlation peaks besides the maximum). Feature tracking algorithms identify single distinct structures in an image, such as ridges or leads or parts of them. Dependent on the implementation, they can be very fast [62,63]. In a second step, the detected structures are compared to potentially matching features in subsequent images. With pattern matching, displacements can be retrieved on a fixed regular grid

(Eulerian approach), whereas the spatial distribution of drift vectors resulting from feature tracking is heterogeneous and changes from one image pair to the next (Lagrangian approach). The Eulerian approach is preferred for comparisons with results from numerical models for simulating and forecasting sea ice dynamics. The results of the Lagrangian approach are superior for visualizing sea ice deformation, which is related to shape and size changes of single grid cells defined by the positions of the drift vectors. Berg and Eriksson [58] proposed to start with pattern matching based on phase correlation. Only if the magnitude of phase correlation remains below a given threshold, a feature-tracking module is activated to track single ice floes. Korosov and Rampal [57] used the combination of both in reversed order: Feature tracking is first employed to provide a rapid first guess of magnitude and direction of drift vectors, followed by pattern matching, which in their case is carried out using cross-correlation.

Pattern matching is optimally conducted using resolution pyramids and cascades. One pyramid is formed by gradually low-pass filtering the original high-resolution image to generate a sequence of scenes with decreasing spatial resolution. Pattern matching is started on the coarsest resolution level. The results obtained at levels of coarser resolution are used to initialize the calculations on the next finer resolution level. The window size (in absolute length units) for the calculations of correlations remains constant within one pyramid, that is, a window contains more pixels at higher resolution. The cascade is a sequence of coupled pyramids in which the window size is decreased. The advantage of the cascading approach is that the robustness of the computation is increased and the result is a drift field of high spatial density.

Komarov and Barber [60] analyzed HH- and HV-polarized RADARSAT-2 ScanSAR images and found that the number of displacement vectors retrieved at HV was larger than at HH. Cross-polarized images record ice structures more clearly provided that the noise level is sufficiently low. Cross-polarization facilitates ice-water separation, estimation of floe size, separation of FY and MY ice, and identification of large-scale deformation features [64]. Lehtiranta *et al.* [54] and Howell *et al.* [65] compared drift retrievals based on C- and on L-band images. Both teams agree that L-band is superior in most cases. Lehtiranta *et al.* [54] focused on single image pairs from ENVISAT, RADARSAT, and ALOS PALSAR, respectively, acquired with 1–2 days' time difference over the Bay of Bothnia. Howell *et al.* [65] investigated nine image pairs (mostly with temporal differences of less than a day) from RADARSAT-2 and from PALSAR-1/PALSAR-2, respectively, that were taken in the Canadian Arctic over young, FY and MY ice under freezing conditions ("dry" ice), and over FY and MY ice under melting conditions. Except for dry MY ice, more drift vectors with, in most cases, higher cross-correlation coefficients were obtained using the L-band image pairs.

While the approaches described so far rely on pairs of images and map the two-dimensional displacement field valid for the time gap between the acquisitions of the two images, Kraemer *et al.* [66] introduced another method for which only a single SAR scene is required to estimate the instantaneous (sub-second) sea ice motion along the LoS between the radar and a surface element. For that purpose, they employed the Doppler frequency derived from Envisat stripmap mode SLC (single-look complex) data. In a second study, Kraemer *et al.* [67] compared Doppler LoS sea ice drift velocities derived from Sentinel-1A EW data with GPS tracks from a drifting ice station as well as with vector fields obtained using traditional cross correlation for a Sentinel-1/Radarsat-2 image pair acquired with a time difference of only 25 min. They found that in EW mode it is not possible to obtain Doppler estimates with the required precision for typical Arctic ice drift speeds, which vary mainly between 0.03 and 0.2 m/s. More precise measurements, however, may be possible by using higher resolution modes like IW or wave mode. Another possibility to measure instantaneous LoS velocity components of drifting ice is to use along-track interferometry. This, however, is only possible with a satellite tandem for which the along-track baseline is on the order of milliseconds and the across-track baseline is short enough to minimize the influence of ice topography on the interferometric phase [47]. The spatial resolution of this method is higher than for the Doppler approach. Both methods fail to measure the drift component directed perpendicular to the LoS. It is not yet clear whether sequences of single subsecond LoS velocity component measurements, which are usually separated by hours to days, are representative for an assessment of the real two-dimensional ice velocity field and its changes on temporal scales from several minutes to hours and days.

8.6 Icebergs

The threat that icebergs pose to ship traffic and offshore structures has been in the public awareness ever since the sinking of the Titanic. Although technologies and strategies for monitoring waters with frequent occurrences of icebergs have been improved considerably since then, iceberg encounters cannot be totally avoided. For example, 57 iceberg-ship incidents were reported for the northern hemisphere between 1980 and 2005 (BBC News, https://www.bbc.com/news/magazine-17257653). For science, icebergs are of interest for several reasons. In the Antarctic, iceberg calving is the largest source of freshwater flux from the ice sheet into the ocean, but the relative contribution of smaller icebergs is highly uncertain. Tracking of larger icebergs reveals patterns of mean ocean currents. Icebergs affect local water circulation, the growth and areal distribution of sea ice, and primary biological production.

About 7/8 of an iceberg is located below the water surface, as can easily be deduced by comparing the typical densities of seawater and freshwater ice (1025 and ~900 kg/m^3, respectively). Under water, the horizontal extension of icebergs is usually 20% to 30% larger than above the water surface. Icebergs can be categorized into classes of different sizes and shapes [68,69] (see Table 8.5 and Figures 8.6 and 8.7). For the iceberg that caused the sinking of the Titanic, eye-witnesses estimated its height above water at between 15–30 m and its length to be 120 m. Ocean currents are the dominating factor among the forces that determine the drift of large icebergs. The relative influence of wind depends on the shape and effective

Table 8.5 Size and shape categories of icebergs
Photos of iceberg shapes can be found in [68]

Naming	Lengths [m]	Shape
Growler	0–5	Tabular
Bergy bit	5–15	Non-tabular
Small berg	15–60	Domed
Medium berg	60–120	Wedge
Large berg	120 –220	Dry dock
Very large berg	>220	Pinnacle
		Blocky

(a)

Figure 8.6(a) Bergy bits, small domed icebergs, and medium iceberg of wedge shape close to Rothera Station, Antarctic Peninsula. Photo Courtesy: W. Dierking.

cross-section of the iceberg's sail (the portion above the water surface). Melting also influences the drift by changing the ratio between the volumes of sail and keel (the latter is the below-water portion). If the weight of the iceberg beneath the water surface has decreased to such an extent as to create a state of instability, the iceberg may roll over as gravity seeks to right the instability, bringing most of its weight beneath the sea surface again.

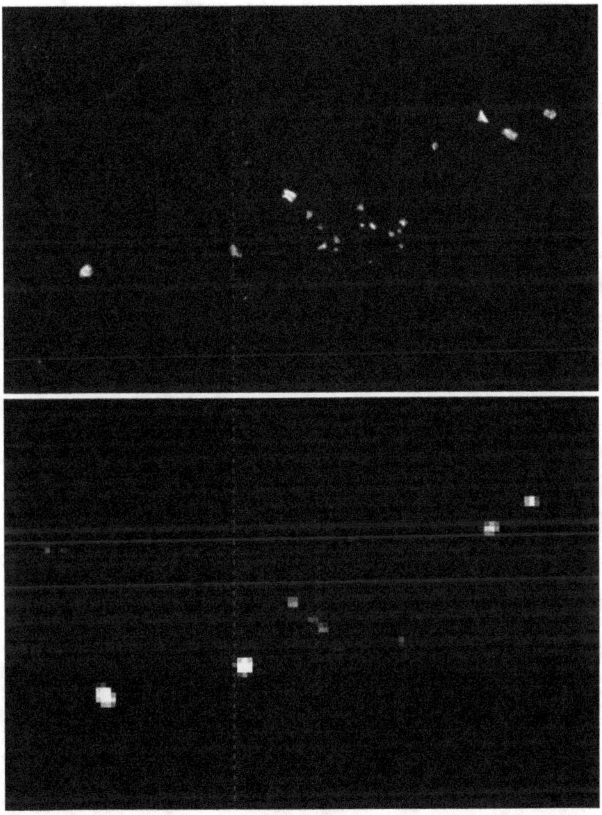

(b)

Figure 8.6(b) Small icebergs in optical and SAR images.
Icebergs in open water observed on the east coast of Wilczek Land, Franz Josef Land, in images of the Sentinel 2 Multi-Spectral Imager (top) and Sentinel-1 SAR (bottom). The former image was acquired in the visible range with a spatial resolution of 10 m, the latter in ScanSAR mode at 90 m resolution. The time difference between image acquisitions was 2 h. The width of the images is 4.7 km. This image pair demonstrates clearly the difficulties of detecting small icebergs in ScanSAR mode. Some icebergs are missed in this example. For the iceberg cluster in the middle this is an effect of the coarse spatial ScanSAR resolution. A strong radar echo is not necessarily linked with a large iceberg (e.g., the second bright return from the left in the SAR image). No echo is received from one iceberg in the group of 3 bergs to the right. Photo Courtesy: Ingri Halland Soldal, Nansen Environmental and Remote Sensing Centre, Norway

The primary information required for iceberg detection is the geographical position of an iceberg and the time of observation [69]. Similar to the production of ice charts, the preferred satellite SAR imaging mode for iceberg monitoring is ScanSAR or wide swath to cover larger areas during one overpass. The spatial

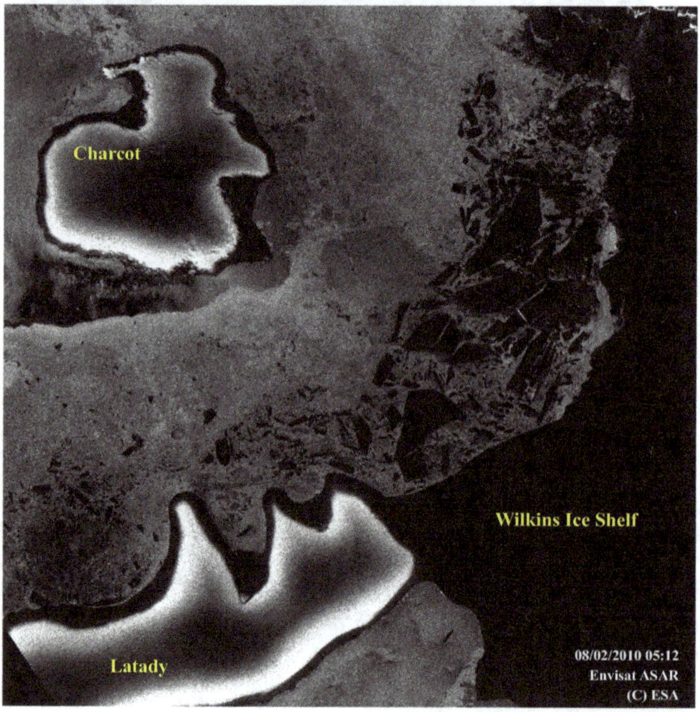

Figure 8.7 Tabular icebergs at Wilkins Ice Shelf, Antarctic Peninsula. Photos Courtesy: W. Dierking.

Top: large and very large tabular icebergs in front of the Wilkins Ice Shelf, observed during a measurement flight in 2010. Bottom: C-band SAR images taken over Wilkins. The ice shelf and the calved icebergs appear dark, the surrounding sea ice gray and textured. Charcot and Latady Island are covered by freshwater ice. The bright margins indicate strong backscattering. © European Space Agency resolutions of recent ScanSAR modes are between 20 m (TerraSAR-X) and 150 m (RADARSAT-2), at swath widths between 100 and 500 km. Another factor that influences the identifiability of single icebergs is speckle, the strength of which is expressed as effective number of looks (ENL), where a lower number means a stronger influence of speckle. The ENL for ScanSAR modes ranges typically from 3 to 12. In view of these numbers, the detection of growlers, bergy bits, and small icebergs is challenging and, in many cases, not possible with ScanSAR modes (Figure 8.6). TerraSAR Stripmap (SM) mode, for example, offers a better spatial resolution of 7–8 m with ENL = 6 but a swath of only 30 km (single-polarization). The corresponding numbers for Sentinel-1 SM are 9 m with ENL = 3.7 and a swath of 80 km.

Both surface and volume reflections and scattering contribute to the radar response of an iceberg [70]. Its surface may be covered by snow or firn, or it may be pure ice if it has recently rolled over. Reflections or scattering may occur from sidewalls, dependent on the shape and orientation of the iceberg (e.g., double bounce reflections from sides of tabular bergs). Since icebergs consist mainly of freshwater ice, the penetration depth of radar waves under freezing conditions is large (e.g., 1–5 m at X-band and 3–14 m at C-band). Radar waves are scattered at air voids, lenses, pipes, cracks, and other volume impurities. Multiple reflections are possible between bottom, sidewalls, surface, and larger volume structures.

Iceberg detection is often carried out using a CFAR (Constant False Alarm Rate) detector, which is designed to detect pixels with large differences of intensity relative to the surrounding pixels (the "background"). To achieve this, the local probability density function (PDF) of the background intensity needs to be known. An intensity threshold is fixed such that only a small number of background pixels will be classified as icebergs. A certain "probability of false alarms" cannot be avoided since usually the intensity-PDFs of icebergs and background overlap. In a study of ENVISAT ASAR data (Wide Swath and Image Mode) and ERS-2 images from the Weddell Sea, Wesche and Dierking [71] found that the observed intensity distributions of icebergs, water, and sea ice can be well matched by a K-distribution for intensities at linear scale. The comparison of iceberg and background intensity distributions revealed varying overlaps dependent on season and pixel size of the image product. Under freezing conditions, the mean radar intensities of icebergs are significantly larger than they are for smooth sea ice or water surfaces at low to moderate wind speeds [71–73] (see Figure 8.6(b)). The intensity contrast between iceberg and background is low in rough sea states and deformed sea ice. When air temperatures are above zero degrees, wet snow and liquid water on the surface obstruct scattering contributions from the volume, and the iceberg may appear dark in a brighter background of rough water under high-wind conditions or in rough sea ice (Figure 8.7).

In practical applications, the detection configuration consists of a square-shaped test window, which is surrounded by a training window [72]. In the detection process, this configuration is moved over the image. The ideal situation is that the test window is located over an iceberg and the training window covers an area of open water or sea ice (the background) around the iceberg. However, with a predefined window size, the center window may only cover a fraction of larger icebergs, and, if icebergs are small, some pixels in the center window may belong to the background. In particular, if the areal density of icebergs is large, parts of the training window may overlap adjacent icebergs, which means that the background statistics are biased. The problem can be alleviated by repeating the detection process one or more times, considering positions of already identified icebergs and excluding corresponding pixels in the training windows [72]. For the pixel-wise detection approaches, a recommended second step is to merge pixels belonging to one iceberg and to identify false alarms caused, for example, by wave crests on the ocean surface. For false-alarm detection, a window is placed over a detected object such that it copies the object's shape but is larger. If the average pixel brightness obtained from this window (not considering the pixels belonging to the detected object) differs considerably less from the object brightness than in the training/test window case, the detection is regarded as false alarm. Applying this procedure, Frost *et al.* [72] achieved a detection rate of 89.5% using TerraSAR-X Stripmap mode HH-polarized images from the west coast of Greenland and from the Grand Banks.

While Frost *et al.* [72] and Wesche and Dierking [71] focused on single-polarization SAR images, Marino *et al.* [74] proposed a new detector that combines the HH- and HV-polarized channels. It is defined as $(\langle HV \rangle_c - \langle HV \rangle_t) / \langle HH \rangle_t$, where $\langle \, \rangle$ are spatial averages of intensities at linear scale over the center (c) and training (t) window. This detector is applied over an image in a pixel-wise fashion and multiplied with the intensity value of the respective pixel at HV-polarization. The effect is a significantly larger contrast between icebergs and background than when using only single-polarization intensities. Akbari and Brekke [75] investigated the potential of iceberg detection in polarimetric images and defined a feature vector extracted from the covariance matrix, consisting of backscattered intensities at HH-, VV-, and HV-polarization, geometric brightness, cross-polarization ratio, and relative kurtosis. The feature vector is input to an unsupervised algorithm that generates a segmented image. For identifying segments as icebergs, Akbari and Brekke [75] employed a rule classifier that adopts brightness, segment size, eccentricity, and shape compactness as criteria. Hence, instead of focusing on each pixel individually, this procedure tests the SAR feature and shape properties of each segment. Also, Mazur *et al.* [76] based their detection scheme on identification of objects that are obtained from image segmentation. They applied the object based image analysis (OBIA), which performs object identification at multiple scale levels, and started with large segments that were split into smaller segments at five different levels. This process is steered through default values for scale, shape, and compactness. To distinguish between icebergs and background at each level, Mazur *et al.* [76] used several criteria: intensity given as a function of incidence angle (expressed as γ, see Section 8.3.3); the contrast between γ of the center object and the mean γ of all objects in the immediate neighborhood; the contrast between γ of the center object and the mean γ of all objects within a range of 50 pixels; and in addition a set of geometric parameters, for example, area and compactness. They analyzed ENVISAT ASAR imagery from the Amundsen Sea. Based on visual inspection and comparisons with aerial photos, the average detection accuracy was 96.2%, varying between 84.6% and 100%, dependent on iceberg size, shape category, and season. The overall false alarm rate was 7%. The best results were achieved for spring and winter and for the smallest icebergs. Mazur *et al.* [76] explained this intuitively unexpected result by the use of small image segments at the final step of processing and the consideration of background characteristics relative to identified objects.

The gain of using polarimetric C-band images was also investigated in [70] and [77]. Dierking and Wesche [70] focused on icebergs in sea-ice-covered ocean regions. The combined use of selected polarimetric parameters together with intensity improved the discrimination performance between icebergs and sea ice in many, but not all, cases. Denbina and Collins [77] included simulations of compact polarimetry in their analysis. They proposed to reconstruct the quad-pol covariance matrix from CTLR data and use "pseudo-quad-pol" parameters for detection. For this, they had to assume relationships between the linear-polarization HH-VV coherence and the cross-polarization ratio. As Rainey [10] pointed out, it is difficult to judge whether such assumptions, which are necessary for the reconstruction, are valid. From a practical point of view, such an approach may be acceptable since Denbina and Collins [77] could demonstrate that, for iceberg detection, the performance of pseudo-quad-pol data obtained at larger incidence angles ($>25°$) is better than with HH-HV and VV-VH combinations.

Another important aspect is the discrimination of ships and icebergs [73]. While the radar intensity received from icebergs at HV-polarization is much smaller than at HH-polarization, the HV- and HH-polarized signals reflected or scattered from ships are of similar magnitudes. Surface and volume scattering dominate the radar response from icebergs, whereas the main response from ships reveals large contributions of dihedral and trihedral scattering. Considering the geometrical properties of the identified object is also helpful, but may fail in cases when icebergs resemble ships in shape and size, or when objects are too small to reliably extract their shape from the radar image. Howell *et al.* [73] used a CFAR approach on ENVISAT ASAR images and found that icebergs and ships were easier to detect at HH-polarization than at HV. According to Kerbaol and Collard [69], who based their analysis on ERS-2, RADARSAT-1, and ENVISAT images, the ideal situation for detection of ships is a large incidence angle, low wind speed, high spatial resolution, and HH-polarization.

8.7 Validation

The validation of results obtained from algorithms designed for ice classification, iceberg detection, and retrieval of drift and deformation parameters is difficult and rarely satisfying. For sea ice classification, two main reasons can be named: Firstly, satellite image acquisitions cover a large region within a time interval of a minute (e.g., for Sentinel EWS, an area of $410\ \text{km} \times 6.7\ \text{km/s} \times 60\text{s} = 164820\ \text{km}^2$), which obviously is not possible for field parties. The possibilities for carrying out validation campaigns are limited because of the challenging and expensive logistics. Secondly, the radar signatures depend on a multitude of spatially varying ice parameters ranging from centimeter to kilometer scales. In particular, properties of small-scale structural elements, such as ice surface undulations at centimeter range or air bubbles in the ice, have a strong impact on the radar response. While large-scale ice conditions such as spatial distributions of open water patches in ice or ridge density may be obtained from air (or helicopter)-borne measurements which cover larger areas in a short time, this is not possible for small-scale properties. Often, even the required large-scale information is not retrievable. For example, if the ice is covered with snow, it is not possible to distinguish ice types in optical images. Ice ridges are difficult to recognize in images acquired under low light conditions. Thickness, as a criterion for ice type separation, can be estimated from thermal IR imagery only for thinner ice. For the assessment of ice drift retrievals, data of drifting buoys or beacons are optimal. Mostly, however, the number of buoys covered in a SAR scene is very low or even zero. For validating estimates of deformation parameters, arrays of several buoys are needed, which are usually deployed during field campaigns. Hence only few regional SAR image sequences can be directly compared with results from buoy arrays. For judging the quality of iceberg detections, reports from reconnaissance flights and ships as well as data from coastal radars are useful but their availability is limited spatially and temporally. Hence the question arises: which means do researchers use to analyze and quantify the results of their algorithms?

For sea ice classification, many researchers use operational ice charts or consult sea ice experts for an assessment of their results [17,19,21,23,35,78]. Ice charts can often be regarded the best available source for evaluation of automated classification results, if they are compiled by an experienced ice analyst who integrates information that comes from external sources such as data from ships and aircraft, reports from shore, complementary optical, thermal IR, and passive microwave imagery, the ice history as known from past analyses, and meteorological data. Karvonen *et al.* [79] described an experiment in which five groups consisting of four to five ice analysts had to estimate the ice concentration in two RADARSAT-2 ScanSAR Wide Mode images at HH-HV combination from the Gulf of Bothnia. The first scene revealed lower contrasts between water and ice types because of melting and refreezing events before data acquisition. The concentration estimates had to be provided for fixed polygons in steps of 10%. The standard deviations between groups ranged from 0% to 29% in the first image and from 0% to 5% (with one exception of 26%) in the second. In parallel, an automated ice-water separation was performed, using σ_{HV}, entropy, and autocorrelation at HH- and HV-polarization. Compared to the average result of the ice analysts, the automated algorithm underestimated the concentration, in the first image by 16% and by 1% in the second. Moen *et al.* [34] reported on the manual interpretation of a RADARSAT-2 fine quad-pol scene taken over the sea ice north of Svalbard. The SAR scene was provided to two ice analysts as separate VV, HH, and HV layers and as Pauli-decomposition, complemented by photographs and ice logs from a field campaign that took place during image acquisition. The ice cover had to be separated into 13(!) stages of ice growth. The two independently generated ice charts revealed considerable differences. This was mainly attributed to the fact that in particular the discrimination of five different stages of FY ice in radar images is extremely difficult in practice. This example demonstrates that the reliability and robustness of ice classification decrease if ice types or stages of ice growth have to be separated based on a classification scheme that is too detailed.

If comparison to buoys or beacons, for example, as carried out in studies [57,60] is not possible, the judgment of ice drift/displacement patterns and corresponding deformation parameters automatically retrieved from SAR images can be achieved by manually determining displacements from image pairs [56,58]. Buoys deployed on the ice provide drift data only for single floes, but at high temporal resolution on the order of minutes. To be able to automatically provide a preliminary judgment of the retrieval quality independent of reference data, Hollands *et al.* [61] proposed to use a "confidence factor" (CFA) that is based on a combination of the cross-correlation coefficient, image texture parameters, and an intensity threshold to exclude the effect of single strong reflectors that only show up for certain incidence and orientation angles. The magnitude of the CFA is used to identify regions of less reliable displacement information, in which pattern matching is difficult due to the lack of distinct ice structures or due to very low intensity contrasts. Another possibility is back-matching [60,61]. The displacement field is calculated twice, the first by using image 1 as reference and the second with image 2 as reference. If drift vectors obtained from image constellations 1–2 and 2–1 reveal larger differences in magnitude and direction, they are regarded as less or not reliable. The main point is that the calculations from pair 1–2 are carried out on a grid slightly different from pair 2–1; otherwise, the difference is always zero. For comparison, however, one of the resulting displacement fields has to be resampled to the grid of the second field. Statistical errors in retrievals of displacement and deformation were theoretically derived in [80]. The displacement error consists of the positioning errors in either of the two images, and the tracking error, which is related to ambiguities in pattern matching or feature tracking. Estimates of the displacement error can be obtained by comparing results automatically generated from SAR images with tracks of buoys. Since buoys remain fixed to the ice floe on which they are deployed, their tracking error is zero. Displacement errors, in turn, affect the estimates of deformation parameters. The latter are calculated from line integrals around the boundary of grid cells and reveal so-called boundary-definition errors when the grid cell boundaries and deformation zones (the latter showing up as discontinuities in the displacement field) are unfavorable oriented to each other. Solutions to reduce this effect are proposed in [81] and [82].

The assessment of automatic iceberg detection needs to answer the questions: How many icebergs are correctly identified ("true positives," TP)? How many are missed ("false negatives," FN)? How many objects are marked incorrectly as icebergs ("false positives," FP)? Often, the only possibility is to carry out a visual identification on the SAR images that are also used for automatic detection [72,76]. This approach can be successful if icebergs reveal a sufficient intensity contrast relative to the background and cover a couple of pixels in the SAR image. It fails when the size of an iceberg is similar to the pixel dimension or when the radar signal of an iceberg is too weak to be visually recognized in the SAR image. Optimal is a combination of optical and SAR images acquired with only short time gaps (typically less than 1 h) between their acquisitions. This situation is rare, and usually one must take into account the drift of icebergs when comparing optical and SAR images acquired at different times, except for icebergs that are grounded or captured in landfast ice. In a few cases, ground information is available locally [72].

While the classification accuracy is often quantified in a confusion matrix [35], other metrics are more useful for cases in which an ice class or detected icebergs occupy only a small areal fraction of an image. Accuracy is the number of correctly classified pixels over the sum of all pixels, defined by (TP+TN)/(TP+FP+TN+FN), where TN (true negatives) are pixels correctly identified as not belonging to a given class (e.g., all noniceberg pixels recognized as such). For iceberg detections, TN is usually very large compared to TP, FP, and FN. More useful metrics are therefore the probability of missed detections, PMD = FN/(TP+FN), and the probability of false alarms, PFA = FP/(FP+TN). Plots of PMD versus PFA help to assess the balance between both parameters. Small PFAs are linked with high PMDs and vice versa [77]. A similar assessment scheme was applied in [15] for detection of leads. Murashkin *et al.* [15] evaluated "precision" (defined as TP/(TP+FP)) versus "recall" (TP/(TP+FN)), with TP being the number of pixels correctly classified as lead, FP the number of nonlead pixels erroneously classified as lead, and FN the number of lead pixels that are erroneously classified as nonlead. A higher precision of classification means to accept a lower probability of detection (i.e., a lower recall). Precision-recall or PMD-PFA curves are useful to determine detection thresholds. An introduction to these concepts can be found in [83].

8.8 Conclusions

What are the major points to keep in mind? One goal of sea ice and iceberg SAR-surveillance is to develop and apply methods for retrieving information about ice conditions and iceberg occurrences from imaging radar data. Hence, two major questions are: *which methods are optimal? And why?* In the sections above, several classification and retrieval methods were described that produced good results for the ice and environmental conditions under which they were tested. Criteria for judging those methods are (or should be) robustness, reliability, use of computer memory, and computational speed (see Section 8.3). In addition, their complexity has to be taken into account, for example, the number of required tuning parameters and potential difficulties to determine their optimal values or the difficulty in tracing intermediate results in chains of different algorithms. Another factor is "global applicability"—do they always deliver results as good as for the conditions under which they were tested?

An important issue in this context is that it is *not possible to prove that a method is perfect* even if it is. As discussed in Section 8.7, reference data for an assessment of classification and retrieval results that are sufficiently extensive and detailed are impossible to obtain, considering the required logistical efforts for field and airborne campaigns and the problems to obtain data from different satellite systems for a given area within very short time intervals. Another issue is that the radar response of ice is strongly dependent on small-scale properties as explained in Section 8.2. Small-scale properties can change within short distances of a few to tens of meters. Their spatial variations are only weakly, or not at all, correlated with ice types. They may be responsible for distinct patterns recognized in SAR images but may change within short time intervals (e.g., due to melt and refreeze cycles). These changes complicate ice type separation, ice displacement retrievals and iceberg detections since they are not predictable and can hence not be considered in the development of the method. Despite these inherent problems in development and application of automatic classification and retrieval algorithms, the achieved results are encouraging and have a great potential to support (but not to replace) analysts in operational services and in science. An interesting experiment not yet performed would be the systematic comparison of different methods and implementations for classification, drift and deformation retrieval, and iceberg detection for test sites with different conditions.

The assessment of methods for ice chart production requires a comprehensive knowledge of various image-processing techniques, including algorithms for clustering, segmentation, classification, decomposition, feature selection or reduction, and for quantifying separability of ice types. One of the more specific items is, for example, the *dramatically increased number of parameters that can potentially be used in sea ice classification and iceberg detection*. This includes a multitude of polarimetric and textural features. Special polarimetric parameters reflect certain scattering mechanisms, which in turn depend on small and large-scale sea ice properties. The separation of scattering mechanisms can improve ice classification and iceberg detection at least to a certain degree, as demonstrated in various studies (Sections 8.4 and 8.6). Ice reveals structures due to spatial distributions of ridges, leads, and broken floes. Therefore, the additional consideration of texture may further increase the classification accuracy, but, dependent on the method, calculations of texture can be relatively extensive and may require tuning of input parameters (Section 8.3). In addition, results depend on the spatial scale over which texture is calculated. So far, different studies recommend different sets of parameters for classification. Some studies showed or indicated that the optimal composition of parameters depends on the specific objective of the classification (e.g., only ice-lead separation or complete ice type discrimination), radar frequency, radar incidence angle, and ice conditions. The usefulness of polarimetric and texture parameters can be assessed in specific case studies, considering criteria such as complexity and speed of computation, correlations among single parameters, and how strongly they are linked to certain ice properties. Based on the case studies, a pre-selection of the most promising parameters should be possible.

Comparative analyses with different automated ranking/reduction methods would also be valuable.

For sea ice classification and ice drift retrieval, it was demonstrated that the use of data from *SAR systems operating at different frequencies* is beneficial. For classification, however, one requirement is that images at different frequencies must be acquired within short time intervals, ideally a few seconds to minutes. If the different SAR images show exactly the same spatial arrangement of ice floes, ice structures, and open water areas, it is possible to directly combine them into a layer stack for automated classification, retrieval, or detection. While it is in principle possible to compensate for the effect of uniformly continuous ice drift and Earth rotation, layer stacking is more complicated if deformation of the ice took place between data acquisitions by the different SAR systems and if this deformation is strong enough to significantly change the spatial arrangement in a scene. In this context, experiments on using ice drift and deformation retrieval algorithms for matching temporally separated multifrequency images are useful. For the combinations of SAR with other sensor types (optical and thermal imagers, passive microwave radiometers, and altimeters), similar considerations have to be made.

A challenging task is *to provide useful data on accuracies, errors, or corresponding metrics for every single classification, retrieval, and detection product.* When algorithms are developed, it is mandatory to "validate" them based on independently measured and sufficiently detailed reference data. However, the latter are often difficult to obtain from regions of the polar oceans (Section 8.7). If they are available, they are often only valid for specific ice and environmental conditions. In particular when results are produced without having matching reference data available, it is necessary to provide the user with a measure of the suitability of the data and algorithms that were used for generating the respective end product. Data may have been acquired under unfavorable imaging conditions, and the algorithm may not work optimally for the specific ice conditions. The intercomparison of results from different algorithms may provide a relative measure of uncertainties.

Acknowledgments

For helpful discussions and for providing comments to this chapter, I would like to thank Prof. Torbjørn Eltoft, UiT The Arctic University of Norway, Assoc. Prof. Leif Eriksson, Chalmers University of Technology, Sweden, Dean Flett, Manager of Applied Science Section/Canadian Ice Service, and Nick Hughes, Leader of Norwegian Ice Service. I am in particular grateful to John Falkingham, Secretariat of the International Ice Charting Working Group, and Assoc. Prof. Anthony Doulgeris, UiT The Arctic University of Norway, who reviewed and edited this chapter in detail. Finally, I am thankful to Tom Zagon, Physical Sciences Specialist, Canadian Ice Service, for providing Figure 8.2, and Ingri Halland-Soldal, a PhD student at Nansen Environmental and Remote Sensing Center, Norway, for preparing Figure 8.6(b).

References

[1] WMO Sea-Ice Nomenclature, Volumes I, II and III, working document WMO-No.259. Publications, Manuals and Guides. Published 02 April 2015. https://www.jcomm.info.

[2] Vihma T. "Effects of Arctic sea ice decline on weather and climate: A review." *Surveys in Geophysics.* 2014, vol. 35, no. 5, pp. 1175–1214. doi:10.1007/s10712-014-9284-0

[3] Nghiem S. V., Rigor I. G., Clemente-Colón P., Neumann G., and Li P. P. "Geophysical constraints on the Antarctic sea ice cover." *Remote Sensing of Environment.* 2016, 181, pp. 281–292. doi:10.1016/j.rse.2016.04.005

[4] Parkinson C. L., and Cavalieri D. J. "Antarctic sea ice variability and trends, 1979–2010." *The Cryosphere.* 2012, vol. 6, pp. 871–880. doi:10.5194/tc-6-871-2012

[5] Joint Communication to the European Parliament and the Council: An integrated European Union policy for the Arctic, Document 52016JC0021, JOIN/2016/021. October 2018 https://eur-lex.europa.eu/legal-content/EN/ALL/?uri=JOIN:2016:0021:FIN

[6] Garcia-Soto C., Vazquez-Cuervo J., Clemente-Colón P., and Hernandez F. "Satellite oceanography and climate change." *Deep Sea Res. Part II: Topic. Stud. Oceanogr.* 2012, vol. 77–80, pp. 1–9. doi:10.1016/j.dsr2.2012.07.004

[7] Rampal P., Bouillon S., Ólason E., and Morlighem M. "neXtSIM: A new Lagrangian sea ice model." *The Cryosphere.* 2016, vol. 10, pp. 1055–1073. doi:10.5194/tc-10-1055-2016

[8] Rott H., Cline D., Duguay C., *et al.* "CoReH2O - A Ku- and X-Band SAR mission for snow and ice monitoring." *Proceedings of EUSAR 2008*, Paper 5.2.2, 7th European Conference on Synthetic Aperture Radar, June 2–5, 2008, Friedrichshafen, Germany.

[9] Dierking W. "Sea ice monitoring by synthetic aperture radar." *Oceanography* 26(2), pp. 100–111, 2013, doi:10.5670/oceanog.2013.33.

[10] Rainey R. K. "Comparing compact and quadrature polarimetric SAR performance." *IEEE Geosci Remote Sens. Lett.* 2016, vol. 13, no. 6, pp. 861–864. doi:10.1109/LGRS.2016.2550863

[11] Eicken H. "From the microscopic, to the macroscopic, to the regional scale: Growth, microstructure and properties of sea ice." in D. N. Thomas and G. S. Dieckmann (eds.), *Sea ice: An introduction to its physics, chemistry, biology and geology*. Blackwell Science; 2003, Chapter 2.

[12] Nghiem S. V., Kwok R., Yueh S. H., *et al.* "Evolution in polarimetric signatures of thin saline ice under constant growth." *Radio Sci.* 1997, vol. 32, no. 1, pp. 127–151. doi:10.1029/96RS03051

[13] Flett D. G. "Operational use of SAR at the Canadian ice service: Present operations and a look to the future." *Proceedings of the Second Workshop on Coastal and Marine Applications of SAR*, 8–12 September 2003, Svalbard, Norway, ESA SP-565, pp. 183–198, 2004

[14] Scheuchl B., Flett D., Caves R., and Cumming I. "Potential of RADARSAT-2 data for operational sea ice monitoring." *Can. J. Remote Sens.* 2004, vol. 30, no. 3, pp. 448–461. doi:10.5589/m04-011

[15] Murashkin D., Spreen G., Huntemann M., and Dierking W. "Method for detection of leads from Sentinel-1 images." *Annal. Glaciol.* 2018, pp. 124–136. doi:10.1017/aog.2018.6

[16] Clausi D. A. "Comparison and fusion of co-occurrence, gabor and MRF texture features for classification of SAR sea-ice imagery." *Atmosphere-Ocean.* 2001, vol. 39, no. 3, pp. 183–194. doi:10.1080/07055900.2001.9649675

[17] Zakhvatkina N., Korosov A., Muckenhuber S., Sandven S., and Babiker M. "Operational algorithm for ice-water classification on dual-polarized RADARSAT-2 images." *The Cryosphere.* 2017, vol. 11, pp. 33–46. doi:10.5194/tc-11-33-2017

[18] Leigh S., Wang Z., and Clausi D. A. "Automated ice-water classification using dual-pol SAR satellite imagery." *IEEE Trans. Geosci. Remote Sens.* 2014, vol. 52, no. 9, pp. 5529–5539. doi:10.1109/TGRS.2013.2290231

[19] Maillard P., Clausi D. A., and Deng H. "Operational map-guided classification of SAR sea ice imagery." *IEEE Trans. Geosci. Remote Sens.* 2005, vol. 43, no. 12, pp. 2940–2951. doi:10.1109/TGRS.2005.857897

[20] Yu Q., and Clausi D. A. "SAR sea-ice image analysis based on iterative region growing using semantics." *IEEE Trans. Geosci. Remote Sens.* 2007, vol. 45, no. 12, pp. 3919–3931. doi:10.1109/TGRS.2004.839589

[21] Soh L. K., Tsatsoulis C., Gineris D., and Bertoia C. "ARKTOS: An intelligent system for SAR sea ice image classification." *IEEE Trans. Geosci. Remote Sens.* 2004, vol. 42, no. 1, pp. 229–248. doi:10.1109/TGRS.2003.817819

[22] Clausi D. A., Qin A. K., Chowdhury M. S., Yu P., and Maillard P. "MAGIC: MAp-guided ice classification system." *Can. J. Remote Sensing.* 2010, vol. 36, suppl. 1, pp. S13-S25. doi:10.5589/m10-008

[23] Ochilov S., and Clausi D. A. "Operational SAR sea-ice image classification." *IEEE Trans. Geosci. Remote Sens.* 2012, vol. 50, no. 11, 4397–4408. doi:10.1109/TGRS.2012.2192278

[24] Komarov A. S., and Buehner M. "Automated detection of ice and open water from dual-polarization RADARSAT-2 images for data assimilation, *IEEE Trans. Geosci. Remote Sens.* 2017, vol. 55, no. 10, pp. 5755–5769. doi10.1109/TGRS.2017.2713987

[25] Komarov A. S., Zabeline V., and Barber D. G. "Ocean surface wind speed retrieval from C-band SAR images without wind direction input." *IEEE Trans. Geosci. Remote Sens.* 2014, vol. 52, no. 2, pp. 980–990. doi:10.1109/TGRS.2013.2246171

[26] Mäkynen M., and Karvonen J. "Incidence angle dependence of first-year sea ice backscattering coefficients in Sentinel-1 SAR imagery over the Kara Sea." *IEEE Trans. Geosci. Remote Sens.* 2017, vol. 55, no. 11, pp. 6170–6181. doi:10.1109/TGRS.2017.2721981

[27] Dierking W. "Technical assistance for the deployment of airborne SAR and geophysical measurements during the ICESAR 2007," Final Report Part 2: Sea Ice, ESA-ESTEC Contract No. 20655/07/NL/CB, 2008.

[28] Doulgeris A. P., and Cristea A. "Incorporating incidence angle variation into SAR image segmentation." Geoscience and Remote Sensing Symposium, IGARSS 2018, Valencia, Spain, IEEE International. doi: 10.1109/IGARSS.2018.8519043

[29] Lohse J., Doulgeris A. P., and Dierking W. "Mapping sea ice types from Sentinel-1 considering the surface-type dependent effect of incidence angle." *Annal. Glaciol.* 2020

[30] Arkett M., Flett D., De Abreu R., Clemente-Colón P., Woods J., and Melchior B. "Evaluating ALOS-PALSAR for ice monitoring – what can L-band do for the North American Ice Service?" *2008 IEEE Geoscience and Remote Sensing Symposium Proceedings* (IGARSS 2008). IEEE, Boston, MA, USA, 2008. Vol. 5, pp. 188–191. doi:10.1109/IGARSS.2008.4780059

[31] Casey J. A., Howell S. E. L., Tivy A., and Haas C. "Separability of sea ice types from wide-swath C- and L-band synthetic aperture radar imagery acquired during the melt season." *Remote Sens. Environ.* 2016, vol. 174, pp. 314–328. doi:10.1016/j.rse.2015.12.021

[32] Yackel J. J., Barber D. G., Papakyriakou T. N., and Breneman C. "First-year sea ice spring melt transitions in the Canadian Arctic Archipelago from time series synthetic aperture radar data, 1992–2002." *Hydrol. Process.* 2007, vol. 21, pp. 253–265. doi:10.1002/hyp.6240

[33] Drinkwater M. R., and Liu X. "Seasonal to interannual variability in Antarctic sea surface melt." *IEEE Trans. Geosci. Remote Sens.* 2000, vol. 38, no. 4, pp. 1827–1842. doi:10.1109/36.851767

[34] Moen M.-A. N., Doulgeris A. P., Anfinsen S. N., *et al.* "Comparison of feature based segmentation of full polarimetric SAR satellite sea ice images with manually drawn ice charts." *The Cryosphere.* 2013, vol. 7, pp. 1693–1705. doi:10.5194/tc-7-1693-2013

[35] Geldsetzer T., Arkett M., Zagon T., Charbonneau F., Yackel J. J., and Scharien R. K. "All-season compact-polarimetric C-band SAR observations of sea ice." *Can. J. Remote Sensing.* 2015, vol. 41, pp. 485–504. doi:10.1080/07038992.2015.1120661

[36] Drouin H., Geldsetzer T., Arkett M., and Pegitzer M. "Visual Assessment of RCM compact-polarimetry modes for Canadian Ice Service operational implementation." Report Canadian Ice Service, March 2015.

[37] Cloude S. R. and Pottier E. "An entropy based classification scheme for land applications or polarimetric SAR." *IEEE Trans. Geosci. Remote Sens.* 1997, vol. 35, no.1, pp. 68–78. doi:10.1109/36.551935

[38] Touzi R. "Target scattering decomposition in terms of roll-invariant target parameters." *IEEE Trans. Geosci. Remote Sens.* 2007, vol. 45, no. 1, pp. 73–84. doi:10.1109/TGRS.2006.886176

[39] Doulgeris A. P. "A simple and extendable segmentation method for multi-polarization SAR images." Proc. POLinSAR 2013, 6th Workshop Sci.

Applicat. SAR Polarimetry and Polarimetric Interferometry, ESA SP-713, Frascati, Italy, 2013

[40] Lee J. S., Grunes M. R., and Kwok R. "Classification of multi-look polarimetric SAR imagery based on complex Wishart distribution." *Int. J. Remote Sens.* 1994, 15(11), pp. 2299–2311. doi:10.1080/01431169408954244

[41] Dierking W. "Mapping of different sea ice regimes using images from Sentinel-1 and ALOS synthetic aperture radar." *IEEE Trans. Geosci. Remote Sens.* 2010, vol. 48, no. 3, pp. 1045 – 1058. doi:10.1109/TGRS.2009.2031806

[42] Dierking W., and Dall J. "Sea ice deformation state from synthetic aperture radar imagery—Part 1: Comparison of C- and L-band and different polarizations." *IEEE Trans. Geosci. Remote Sens.* 2007, vol. 45, no. 11, pp. 3610–3622. doi:10.1109/TGRS.2007.903711

[43] Eriksson L. E. B., Borenäs K., Dierking W., *et al.* "Evaluation of new spaceborne SAR sensors for sea-ice monitoring in the Baltic Sea." *Can. J. Remote Sensing*. 2010, vol. 36, suppl. 1, pp. S56–S73. doi:10.5589/m10-020

[44] Dierking W. "Thin ice classification and thickness estimation using ASAR, PALSAR, RADARSAT-2 and TerraSAR-X data: First results." Proceedings of ESA Living Planet Symposium 2010, Bergen, Norway, ESA Special Publication SP-686, 2010.

[45] Johansson A. M., King J. A., Doulgeris A. P., *et al.* "Combined observations of Arctic sea ice with near-coincident colocated X-band, C-band, and L-band SAR satellite remote sensing and helicopter-borne measurements." *J. Geophys. Res. Oceans*. 2017, vol. 122, pp. 669–691. doi:10.1002/2016JC012273.

[46] Johansson A. M., Brekke C., Spreen G., and King J. A. " X-, C-, and L-band SAR signatures of newly formed sea ice in Arctic leads during winter and spring." *Remote Sens. Environ.* 2018, vol. 204, pp. 162–180. doi:10.1016/j.rse.2017.10.032

[47] Dierking W., Lang O., and Busche T. "Sea ice local surface topography from single-pass satellite InSAR measurements: A feasibility study," *The Cryosphere*. 2017, vol. 11, pp. 1967–1985. doi: 105194/ tc-2017-40.

[48] Yitayew T. G., Dierking W, Divine D. V., *et al.* "Validation of sea-ice topographic heights derived from TanDEM-X interferometric SAR data with results from laser profiler and photogrammetry." *IEEE Trans. Geosci. Remote Sens*. 2018, vol. 56, no. 11, pp. 6504–6520. doi:10.1109/TGRS.2018.2839590

[49] Meyer F. J., Mahoney A. R., Eicken H., Denny C. L., Druckenmiller H. C., and Hendricks S. "Mapping arctic landfast ice extent using L-band synthetic aperture radar interferometry." *Remote Sens. Environ.* 2011, vol. 115, pp. 3029–3043. doi:10.1016/j.rse.2011.06.006

[50] Berg A., Dammert P., and Eriksson L. E. B "X-band interferometric SAR observations of Baltic fast ice." *IEEE Trans. Geosci. Remote Sens*. 2015, vol. 53, no. 3, pp. 1248–1256. doi:10.1109/TGRS.2014.2336752

[51] Dammann D. O., Eriksson L. E. B., Mahoney A. R., Eicken H., and Meyer F. J. "Mapping pan-Arctic landfast sea ice stability using Sentinel-1 interferometry." *The Cryosphere*. 2019, vol. 13, pp. 557–577. doi:10.5194/tc-13-557-2019

[52] Marbouti M., Praks J., Antropov O., Rinne E., and Leppäranta M. "A study of landfast ice with Sentinel-1 repeat-pass interferometry over the Baltic Sea." *Remote Sens*. 9, 833, 2017, doi:10.3390/rs9080833

[53] Arkett M., Braithwaite L., Pestieau P., *et al.* "Preparation by the Canadian Ice Service for the operational use of the RADARSAT Constellation Mission in their ice and oil spill monitoring programs." *Can. J. Remote Sens.* 2015, vol. 41, pp. 380–389. doi:10.1080/07038992.2015.1104631

[54] Lehtiranta J., Siiriä S., and Karvonen J. "Comparing C- and L-band SAR images for sea ice motion estimation." *The Cryosphere*. 2015, vol. 9, pp. 357–366. doi:10.5194/tc-9-357-2015

[55] Kwok R., Hunke E. C., Maslowski W., Menemenlis D., and Zhang J. "Variability of sea ice simulations assessed with RGPS kinematics." *J. Geophys. Res.* 2008, vol. 113, C11012. doi: 10.1029/2008JC004783

[56] Hollands T., and Dierking W. "Performance of a multi-scale correlation algorithm for the estimation of sea ice drift from SAR images: Initial results." *Annal. Glaciol.* 2011, vol. 52, no. 57, pp. 311–317. doi:10.3189/172756411795931462

[57] Korosov A. A., and Rampal P. "A combination of feature tracking and pattern matching with optimal parametrization for sea ice drift retrieval from SAR data." *Remote Sens.* 2017, vol. 9, pp. 258. doi:10.3390/rs9030258

[58] Berg A., and Eriksson L. E. B. "Investigations of a hybrid algorithm for sea icedrift measurements using synthetic aperture radar images." *IEEE Trans. Geosci. Remote Sens*. 2014, vol. 52, no. 8, pp. 5023–5033. doi:10.1109/TGRS.2013.2286500

[59] Karvonen J. "Operational SAR-based sea ice drift monitoring over the Baltic Sea." *Ocean Sci.* 2012, vol. 8, pp. 473–483. doi:10.5194/os-8-473-2012

[60] Komarov A. S., and Barber D. G. "Sea ice motion tracking from sequential dual-pol RADARSAT-2 images." *IEEE Trans. Geosci. Remote Sens*. 2014, vol. 52, no. 1, pp. 121–136. doi:10.1109/TGRS.2012.2236845

[61] Hollands T., Linow S., and Dierking W. "Reliability measures for sea ice motion retrieval from synthetic aperture radar images." *IEEE J. Selected Topics in Applied Earth Observations and Remote Sensing*. 2015, vol. 8, no. 1, pp. 67–75. doi: 10.1109/JSTARS.2014.2340572

[62] Muckenhuber S., Korosov A. A., and Sandven S. "Open source feature-tracking algorithm for sea ice drift retrieval from Sentinel-1 SAR imagery." *The Cryosphere*. 2016, vol. 10, 913–925. doi:10.5194/tc-10-913-2016

[63] Demchev D., Volkov V., Kazakov E., Alcantarilla P. F., Sandven S., and Khmeleva V. "Sea ice drift tracking from sequential SAR images using accelerated-KAZE features." *IEEE Trans. Geosc. Remote Sens*. 2017, vol. 55, no. 9, pp. 5174–5184. doi:10.1109/TGRS.2017.2703084

[64] Arkett M., Flett D., and De Abreu R. "C-band multiple-polarization SAR for sea ice monitoring – What can it do for the Canadian Ice Service." Proc. ENVISAT Symp. Montreaux, Switzerland, 2007.

[65] Howell S. E. L., Komarov A. S., Dabboor M., *et al.* "Comparing L- and C-band synthetic aperture radar estimates of sea ice motion over different ice regimes." *Remote Sens. Environ.* 2018, vol. 204, pp. 380–391. doi:/10.1016/j.rse.2017.10.017

[66] Kræmer T., Johnsen H., and Brekke C. "Emulating Sentinel-1 Doppler radial ice drift measurements using ENVISAT ASAR data." *IEEE Trans. Geosc. Remote Sens.* 2015, vol. 53, no. 12, pp. 6407–6418. doi:10.1109/TGRS.2015.2439044

[67] Kræmer T., Johnsen H., Brekke C., and Engen G., Comparing SAR-based short time-lag cross correlation and Doppler-derived sea ice drift velocities, *IEEE Trans. Geosc. Remote Sens.* 2018, vol. 56, no. 9, pp. 1898–1908. doi:10.1109/TGRS.2017.2769222

[68] Turnbull I. D., Fournier N., Stolwijk M., Fosnaes T., and McGonigal D. "Operational iceberg drift forecasting in Northwest Greenland." *Cold Regions Sci. Technol.* 2015, vol. 110 pp. 1–18. doi:10.1016/j.coldregions.2014.10.006

[69] Kerbaol V., and Collard F. "SAR-derived coastal and marine applications: From research to operational products." *IEEE J. Oceanic Eng.* 2005, vol. 30, no. 3, pp. 472–486. doi 20.1109/JOE.2005.857505

[70] Dierking W., and Wesche C. "C-band radar polarimetry: Useful for detection of icebergs in sea ice?." *IEEE Trans. Geosc. Remote Sens.* 2014, vol. 52, no. 1, pp. 25–37. doi:10.1109/TGRS.2012.2234756

[71] Wesche C., and Dierking W. "Iceberg signatures and detection in SAR images in two test regions of the Weddell Sea, Antarctica." *J. Glaciology.* 2012, vol. 58, no. 208. doi:10.3189/2012JOG11J020

[72] Frost A., Ressel R., and Lehner S. "Automated iceberg detection using high-resolution X-band SAR images." *Can. J. Remote Sens.* 2016, vol. 42, pp. 354–366. doi:10.1080/07038992.2016.1177451

[73] Howell C., Youden J., Lane K., Power D., Randell C., and Flett D. "Iceberg and ship discrimination with ENVISAT multi-polarization ASAR." Geoscience and Remote Sensing Symposium, IGARSS 2004, Anchorage, AK, USA, IEEE, pp. 113–116. doi:10.1109/IGARSS.2004.1368922

[74] Marino A., Dierking W., and Wesche C. "A depolarization ratio anomaly detector to identify icebergs in sea ice using dual-pol SAR images." *IEEE Trans. Geosc. Remote Sens.* 2016, vol. 54, no. 9, pp. 5602–5615. doi:10.1109/TGRS.2016.2569450

[75] Akbari V., and Brekke C. "Iceberg detection in open and ice-infested waters using C-band polarimetric synthetic aperture radar." *IEEE Trans. Geosc. Remote Sens.* 2018, vol. 56, no. 1. doi:10.1109/TGRS.2017.2748394

[76] Mazur A. K., Wåhlin A. K., and Krężel A. "An object-based SAR image iceberg detection algorithm applied to the Amundsen Sea." *Remote Sens. Environ.* 2017, vol. 189, pp. 67–83. doi:10.1016/j.rse.2016.11.013

[77] Denbina M., and Collins M. J. "Iceberg detection using compact polarimetric synthetic aperture radar." *Atmosphere-Ocean*. 2012, vol. *50*, no. 4, pp. 437–446. doi:10.1080/07055900.2012.733307

[78] Geldsetzer T., and Yackel J. J. "Sea ice type and open water discrimination using dual co-polarized C-band SAR." *Can. J. Remote Sens*. 2009, vol. 35, no. 1, pp. 73–84. doi:10.5589/m08-075

[79] Karvonen J., Vainio J., Marnela M., Eriksson P., and Niskanen T. "A comparison between high-resolution EO-based and ice analyst-assigned sea ice concentrations." *IEEE J. Sel. Topics Appl. Earth Observ. Remote Sens*. 2015, vol. 8, no. 4, pp. 1799–1807. doi:10.1109/JSTARS.2015.2426414

[80] Lindsay R. W., and Stern H. L. "The RADARSAT geophysical processor system: Quality of sea ice trajectory and deformation estimates." *J. Atmos. Ocean. Technol*. 2003, vol. 20, pp. 1333–1347. doi:10.1175/1520-0426 (2003)

[81] Bouillon S. and Rampal P. "On producing sea ice deformation data sets from SAR-derived sea ice motion." *The Cryosphere*. 2015, vol. 9, pp. 663–667. doi:10.5194/tc-9-663-2015

[82] Griebel J., and Dierking W. "Impact of sea ice drift retrieval errors, discretization, and grid type on calculations of ice deformation." *Remote Sens*. 2018, vol. 10, pp. 393. doi:10.3390/rs10030393

[83] Fawcett T. "An introduction to ROC analysis" *Pattern Recognit. Lett*. 2006, vol. 27, pp. 861–874. doi10.1016/j.patrec.2005.10.010

Chapter 9

SAR oil spill imaging, interpretation and information retrieval techniques

Camilla Brekke[1] *and Cathleen E. Jones*[1,2]

Oil spills in the ocean reduce the small-scale surface roughness, leading to increased forward scattering and reduced backscatter (Figure 9.1).

This causes a slick to be darker than surrounding clean water in radar images and, conversely, brighter in optical images given suitable sunlight conditions. Environmental monitoring of vast ocean areas to identify accidental or intentional spills is a challenging and costly task, especially in Arctic regions that lack sunlight during the polar night and commonly experience harsh weather conditions. Remote sensing by synthetic aperture radar (SAR) is a key solution to this problem, allowing imaging of the sea surface independent of sunlight and cloud coverage. SAR-based marine petroleum hydrocarbon spill (hereinafter referred to as *oil spill*) monitoring services support two main needs for information. First, regular surveillance of the seas is needed to detect the presence of accidental and illegal man-made oil spills. Second, during oil spill response operations, authorities need to know the characteristics of the spill, such as its extent, a forecast of how it will spread and be transported in the marine environment, and the location of the combatable thicker oil within the slick.

This chapter addresses the physical mechanisms behind SAR imaging of oil spills in the open ocean and proceeds with a discussion on the emerging SAR information retrieval techniques for detecting and characterizing these slicks. Section 9.1 covers oil spill information items possibly derived from SAR and limitations of existing methods. Challenges and trade-offs faced by operational service providers in retrieving these items are discussed in Section 9.2. Section 9.3 is about SAR image interpretation: the main contrast drivers for oil spills are explained and surface scattering models aiding interpretation are reviewed. In Section 9.4, the main techniques for oil spill detection and characterization are illustrated and current research questions are discussed. Concluding remarks and notes on further readings are given in Section 9.5.

[1]Department of Physics and Technology, UiT The Arctic University of Norway, Tromsø, Norway
[2]Jet Propulsion Laboratory, California Institute of Technology, Pasadena, CA, USA

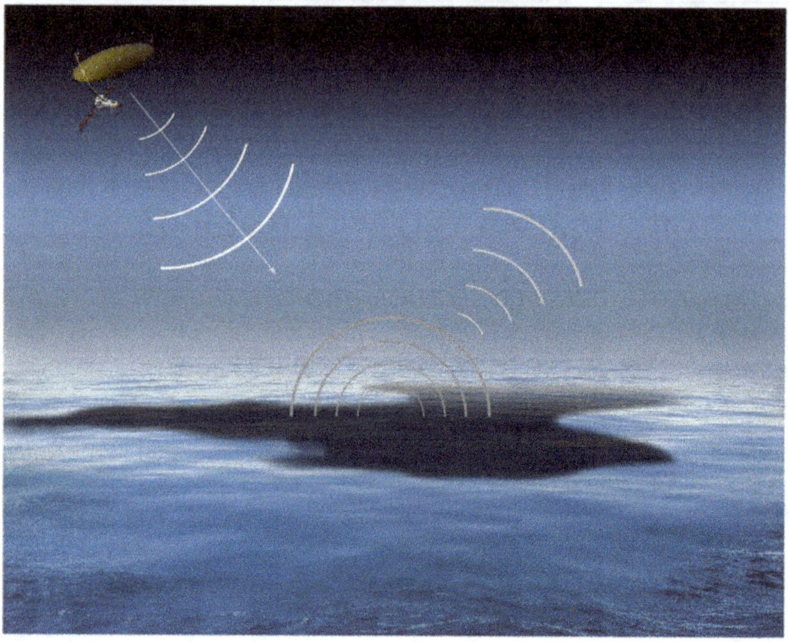

Figure 9.1 An oil spill causes increased forward scattering and reduced backscattering of the electromagnetic signal relative to nearby slick-free seawater

9.1 Information items requested and gaps

According to [1], there are six general oil spill-specific applications of remotely sensed data from any sensor: surveillance and detection, mapping for both tactical (short term) and strategic (long term) countermeasures, trajectory determination, direct support for oil spill countermeasures, gathering of legal evidence, and law enforcement regarding ship discharge of pollutants.

Table 9.1 summarizes some of the key information items available either directly or indirectly from SAR measurements. The phase of oil spill operations at which the information is needed is indicated. In practice, there is a large degree of latitude in acceptable latency depending on the specific situation because spills evolve differently depending on the meteorological and sea surface (metocean) conditions.

Current satellite SAR oil spill services aim to provide near real-time delivery of information after a scene is acquired and downlinked. For example, the European Maritime Safety Agency (EMSA) requires maximum delivery times of 20–120 min depending on the product (per Invitation to Tender No. EMSA/OP/6/2018), although specific operations depend upon the service organization, their access to satellite imagery, and the geographical location of the satellite downlink

Table 9.1 Key information items pertaining to oil spill surveillance, response, remediation, and preparedness involving SAR

Use	Information item	Comment
Detection and early warning	Position	Latitude/longitude information obtainable from SAR directly
	Source of origin	Oil spill source and trajectory prediction (hindcasting) models can be initialized by slick position in SAR
	Drift estimate (surface velocity)	Requires consecutive SAR acquisitions within minutes to hours to allow spatial feature correlation (nowcasting)
	Extent	Surface areal extent obtainable from SAR directly
	Slick type discrimination[a]	Only relative difference between oil slick types[b] (no label) obtainable from SAR
Tactical clean-up stage	Volumetric oil fraction[a]	Estimated directly from SAR for thick slicks, but requires assumptions on the dielectric constants of crude oil and sea water
	Thickness[a]	Only zones of relative larger thickness obtainable from SAR. Absolute thickness determination[a] requires calibration
Tactical clean-up stage and long-range monitoring	3D drift and spread prediction (surface velocity and vertical entrainment)	Numerical (forecasting) models can be initialized by slick contours provided from SAR
Strategic planning: Environmental impact analysis	Oil trajectory statistics	Numerical (forecasting) models can be initialized by slick contours provided from SAR and run for different meto-cean conditions
Strategic planning: Statistics of spill occurrence	Mapping	Position and contour obtainable from SAR
Remediation: Legal evidence	Volume[a]	Estimate from extent and thickness. Requires some calibration of SAR-derived relative thickness and modeling to estimate the entrained volume

[a]More studies are needed on this topic.
[b]Mineral versus biogenic oil slicks.

ground station. EMSA has a central role in monitoring and preventing pollution in European Union seas and ports. The agency uses satellite-based oil spill information from CleanSeaNet (CSN) to assist ship-source pollution response, as described for the North Sea region during 2007–2011 in [2]. During this period, three SAR satellites were used (ENVISAT, Radarsat-1 and -2) to cover 1,000 million km^2 sea area and 8,666 possible spills were detected. Out of these, 2,828 were checked on-site, and 50% within 3 h by aerial surveillance. Of the confirmed releases (745), 80% were mineral oil and 20% other substances. These statistics reveal one potential problem with the operational slick detection methodology: although rapid detection is important in the early warning phase, false alarms are a major concern. Even though satellite SAR provides a valuable synoptic overview of large ocean areas, natural slicks can trigger expensive aircraft or vessel operations. For comparison, the U.S. National Oceanic and Atmospheric Administration Satellite and Information Service (NOAA NESDIS) prepares Marine Pollution Surveillance Reports (MPSRs) [3] based on semiautomated surveys of remote sensing data by operational analysts. Potential releases are checked against the locations of known seeps, pipelines, rigs, well heads, and ship wrecks and local metocean conditions from buoys or models before reports are filed. Analysts use all available optical and SAR imagery, with Sentinel-1 now the dominant source. The latency between image acquisition and report submission ranges from 2 h to >24 h in exceptional cases, with typical delay of 8 h for Sentinel-1 scenes [Ellen Ramirez, private communication]. Between 200 and 300 reports are generated per year and human intervention in the image screening process eliminates many false alarms. In principle, the combination of latency and false alarms could be reduced through automation but currently there is no robust method to automatically differentiate accidental or intentional mineral oil spills from natural slicks using SAR. There is only an indication of SAR-based determination of relative difference between oil spill types (mineral vs. biogenic) reported in the literature [4] (also see Section 9.4.2). This is an outstanding knowledge gap between information needed by the authorities and other stakeholders and the information provided through satellite SAR alone.

During the tactical clean-up stage of a major oil spill combat situation, responders need to know the location of thicker layers of oil in the surface slick to initiate clean-up procedures such as in situ burning and mechanical or chemical dispersal or recovery [5]. A sea-surface oil spill is spatially nonuniform and typically contains zones of thicker and thinner oil layers, whether the oil is emulsified or not. It is established practice to separate areas within the slick between five distinct categories based on the visual appearance of the slick following the Bonn Agreement for Oil Appearance Code (BAOAC) [6]. The categories correspond to order-of-magnitude ranges of thickness and volume of the oil (see Table 9.2) and the classification is typically done by experienced observers from low-flying aircraft. Emulsions, which are mixtures of oil and water, increase the viscosity, volume, and thickness of the material and can be very stable [7].

A comparable SAR-based system to quantify thickness currently does not exist and its development is complicated by several factors. First, SAR does not measure the thickness directly, so a physical model must be developed relating to how oil layer thickness affects the parameters that influence the backscatter of waves in the microwave region of the electromagnetic spectrum. The main parameters are the ocean wave spectra and the dielectric constant. In the case of the latter,

Table 9.2 Oil classes and the corresponding thickness and volume ranges from the BAOAC [6]. Emulsified oil or "mousse" is oil-water mixtures that can be considerably thicker than a few millimeters and are characterized by color in shades of orange, brown, and/or red [7]

Code	Appearance description	Layer thickness (μm)	Layer volume (L/km^2)
1	Sheen (silvery/gray)	0.04–0.30	40–300
2	Rainbow	0.30–5.0	300–5000
3	Metallic	5.0–50	5000–50 000
4	Discontinuous true color	50–200	50 000–200 000
5	Continuous true color	$\geq$ 200	$\geq$ 200 000

emulsification of the oil, that is, the creation of stable or meta-stable oil-in-water or water-in-oil mixtures with altered dielectric properties from oil or seawater, is indirectly related to thickness because some forms of emulsions clump to form thick layers. Second, the small-scale ocean surface roughness is strongly influenced by instantaneous wind conditions and the large-scale wave structure depends on wind history and background swell. A valid model must account for these effects. The last complicating factor is a limitation of SAR instruments, namely a requirement that the SAR be sensitive to sufficiently low levels of backscatter to measure thickness variation. Because the backscatter intensity decreases as the thickness increases or dielectric constant decreases [8], instrument noise can dominate the measured signal over a slick, preventing an accurate determination of thickness for the thicker oil layers.

Nonetheless, the potential of SAR for measuring relative thickness [9] or the volumetric oil fraction of emulsified oil [10,11] has been demonstrated. With calibration using field observations of slick thickness, the SAR backscatter can be related to absolute thickness for a given slick and metocean conditions [12]. A relation between the BAOAC and a proposed oil/water mixing index (Mdex) to separate sheen from emulsions is proposed in [13]. Oil spill characterization is discussed further in Section 9.4.2.

The aim of oil spill modeling is to describe the slick's transport and fate in open seas. Upon initial detection, there is an immediate need for information on the oil spill source location, and both tactical and strategic planning relies on reliable predictions of the oil slick's future location. Therefore, both hindcasting and forecasting of the trajectory are important and, arguably, determining the source of origin can be more important in some situations. The drift velocity can be derived from two or more SAR images alone by feature correlation, but due to the time lag between two satellite SAR acquisitions, the speed is underestimated when the movement is not linear and the spatial shape of the spill, input to feature tracking, may not be recognizable if the lag is too long, hampering this method. Due to the limitation on the revisit time of space-borne SARs (discussed in Section 9.2.2), source identification by SAR is usually not reliable. The slick position and extent can be extracted from space-borne, air-borne, or ship-borne radar and used as input to oil spill hindcast trajectory models [14] or forecast models to predict the future trajectory. Oil on the ocean surface elongates in the direction of the wind while horizontal transport of oil droplets entrained in the water column is to a large extent determined by the currents. A reliable quantitative measure of slick volume cannot be derived from SAR images alone as the slick evolves because detection of oil particles within the water column is impossible due to the high conductivity of sea water. However, information about particle entrainment into the water column can be obtained by combining SAR and numerical models. Forecasting surface velocity and vertical particle entrainment is addressed in Section 9.4.3.

In general, it is clear that oil spill response requires rapid imaging and processing because slicks can move and change quickly on the sea surface, and that the longer the revisit time of the satellite missions, the less efficient satellite-based SAR approaches will be. Airborne SAR can provide very rapid repeat imaging to track the trajectory and fate of oil slicks [9] to support response and remediation, but would still rely upon satellite surveillance programs for initial detection. Airborne SAR has an advantage over spaceborne SAR in characterizing the oil thickness or volume because of the higher instrument signal-to-noise ratio (SNR) (discussed in Section 9.3.3).

9.2 Challenges

Vast ocean areas must be monitored continually by operational services to identify potential oil spills. Retrieving the information items discussed above involves several challenges and trade-offs that relate to factors both intrinsic and extrinsic to the methods and remote sensing instruments employed. The feasibility of executing oil spill clean-up operations depends to a large degree on the weather conditions, solar illumination, geography such as local topographic features that potentially influence wind fields and navigability, and tactical deployment issues like the local settlement and infrastructure and the accessibility of the polluted region [15]. SAR sensors operate at different frequencies and different polarizations, with a choice of operational modes that in some cases trade spatial resolution for image areal coverage. Oil slicks move with wind and currents and the physical and chemical properties evolve with time, changing not only their persistence and environmental impact but also their characteristic signatures for remote sensing. The major challenges that relate to the use of SAR for oil spill response are discussed in this section.

9.2.1 *Polarization diversity*

SAR antennas are designed to transmit and receive linearly polarized electromagnetic waves, with the orthogonal orientations denoted "vertical" (V) and "horizontal" (H). Circular polarization can be synthesized by simultaneously transmitting H and V pulses 90° out of phase in the time domain. Single-polarization (SP) systems transmit and receive the same polarization; dual-polarization (DP) systems transmit one polarization and can receive both the same and the orthogonal polarization; and quad-polarization (QP) systems transmit and receive two orthogonal polarizations. Electromagnetic scattering can be completely characterized by QP systems. Copolarized returns have the same transmit and receive polarization (e.g., HH or VV, where the first (second) letter indicates the transmit (receive) polarization), and cross-polarized returns have orthogonal transmit and receive orientations (e.g., HV or VH). In rare cases, a DP product will be available with the two copolarized returns (HH and VV), but typically DP modes are either horizontal (HH and HV) or vertical (VV and VH). A polarimetric system preserves and records the relative phase between and among the transmit and receive pulses, and systems that do not do this are called "incoherent." The term "compact polarization" (CP) refers to systems that do not maintain the DP collinear/orthogonal convention of the transmit and receive channels, and hybrid polarization (HP) [16] is one particular variant in which circular polarization is transmitted and linear polarization received. HP systems typically are not QP, that is, they transmit only right or only left circularly polarized waves.

SAR-based oil spill detection services have traditionally relied on single- and dual-polarization SAR systems with large spatial coverage. Over the last decade or so, several quad-polarimetric (QP) SAR systems have been launched into orbit, with Radarsat-2 (Canadian) launched in 2007 and Gaofen-3 (Chinese) in 2016. QP SAR allows improved interpretability and information retrieval regarding the polarization-dependent properties of a scattering surface; however, the increased information content usually comes at the cost of reduced spatial coverage. For example, Radarsat-2 Fine QP mode covers a 25 km swath [17], whereas Sentinel-1 extra wide (EW) SP and DP modes have 410 km swath width. Compact polarimetry was introduced to allow simultaneously high spatial coverage and increased polarimetric content [18], although compact polarimetry SAR still falls between conventional dual-polarization and quad-polarimetric SAR systems in terms of polarimetric information. The HP mode [16] has already been integrated into several missions, for example, RISAT-1 (Indian), ALOS-2 (Japanese), and the Radarsat Constellation Mission (RCM) (Canadian). High spatial resolution is not needed in many cases for oil spill monitoring. However, it can be useful for nowcasting of the drift using pixel offset tracking, for locating a point source on the surface like a rig or ship, for identification of internal slick zones, and for false alarm discrimination when the slick's shape or other spatial characteristics are used to differentiate natural phenomena from spills.

Table 9.3 gives some relevant operating parameters of current, recent past, and near-term future SAR instruments. Almost all provide SP or DP products, so the table includes information about other oil-spill relevant modes available as standard products. The table also includes the levels of noise equivalent sigma zero (NESZ) for comparison across the different missions and modes, and to inform further discussions in Section 9.3.3. The upcoming L-band NASA-ISRO SAR (NISAR) (USA/India collaboration) mission is designed with a new polarimetric mode specifically for ocean studies, called quasi-quad-polarimetry (QQP). NISAR will be able to operate with an 80 MHz bandwidth in its highest resolution mode, but more generally will acquire data at 20 MHz bandwidth to reduce the data

Table 9.3 Parameters of civilian SAR instruments operating currently, in the recent past, or under development. The "Remarks" contain information about polarization modes of particular relevance to oil spill surveillance. RISAT-1 deactivated in 2016. FQ, fine quad; SM, stripmap; LN, low noise; PP, PingPong; EW, extra wide-swath mode. Polarization modes: HH, H-transmit, H-receive; VV, V-transmit, V-receive; QP, quad-polarimetry; HP, hybrid polarity (circular-transmit, linear-receive); DP, dual-polarimetry; QQP, quasi-quad-polarimetry (DP HH/HV and DP VV/VH)

Band	Mission	Launch (year)	NESZ (mode/ channel)	Repeat (day)	Remarks
X	TerraSAR-X [19]	2007	−19 dB (SM)	11	DP HH/VV[a]
	COSMO-Skymed 1–4 [20,21]	2007–2010	−22 dB	1–8[b]	PP HH/VV (incoherent)
C	Radarsat-2 [17]	2007	−39 to -31 dB (QP)	24	QP
	RISAT-1 [22]	2012	−17 dB (FRS-1)	25	HP
	RCM [23]	2019 (plan)	−25 dB (LN)	4[c]	HP
	Sentinel 1A/B	2014/2016	−22 dB (EW)	6[d]	EW
	Gaofen-3 [24]	2016	−33 dB (QP[e])	29	QP
L	ALOS-2 [25]	2014	−36 dB (SM[e])	14	QP
	SAOCOM 1A/B [26]	2018 1A, 2019 1B (plan)	−35 dB (SM)	8[d]	HP
	NISAR	2022 (plan)	−25 dB[f]	12	QQP

[a]The dual-copolarization combination of HH/VV is preferred for oil spill monitoring [27].
[b]For constellation of four satellites.
[c]For constellation of three satellites.
[d]For constellation of two satellites.
[e]For HH channel.
[f]NISAR's requirement is that the NESZ be less than −25 dB across the full spectrum. Actual average value will be known after the hardware is built and will vary by mode.

volume. The QQP mode splits the full spectrum to obtain simultaneously-acquired HH/HV and VV/VH DP modes. One polarization will be transmitted with 20 MHz bandwidth in the lower half of the spectrum and the other with 5 MHz bandwidth in the upper half. This allows the preferred HH/VV combination to be exploited for oil spill observations, but at disjoint frequencies. The small shift in frequency will have little impact on ocean studies, most of which do not require coherent VV and HH data. The use and usefulness of polarization in oil spill studies will be discussed in Sections 9.3 and 9.4, respectively.

9.2.2 Imaging repeat interval

SAR satellites are usually launched in a polar orbit and image the ground with a fixed swath width that is mode-dependent. The satellite *repeat cycle* is the exact repetition of acquisition over the same location with the same imaging geometry. However, many locations can be imaged from different orbits because the spacing between the orbit tracks varies with latitude, and hence the revisit rate for imaging irrespective of geometry is much better at higher latitudes. The *temporal resolution* of a satellite is here defined as the time lag between two possible image acquisitions of the same area, either fully or partly overlapping and irrespective of imaging geometry. Table 9.3 shows the satellites' repeat cycle and Figure 9.2 illustrates the significant drift and evolution of the shape of various oil spills between two consecutive SAR acquisitions by Radarsat-2. The images were acquired 11.5 h apart in FQ mode and the radiometric changes are dominated by the difference in imaging geometry. Given the number of currently operational space-borne SAR missions (Table 9.3), higher temporal resolution is obtained when multiple sensors are exploited. However, uniform and consistent analysis across the different systems requires a thorough understanding of the differences between the various sensors and modes [18] and of the incidence angle-dependence and wind-dependence of the backscatter.

Airborne SAR instruments can reimage a slick from the same viewing geometry much more rapidly than satellite SARs, typically several times per hour. Figure 9.3 shows an example of a mineral oil seep imaged with the Uninhabited Aerial Vehicle Synthetic Aperture Radar (UAVSAR) L-band instrument six times in 2 h. The selected scenes shown are obtained 71 min apart. In this case, besides transport and spreading, the time evolution of the radiometry can provide information on weathering processes. With time series data, it is possible to measure transport of the thicker oil within the slick using the intensity as a proxy for thickness (darker returns in areas of relatively thicker or more emulsified oil).

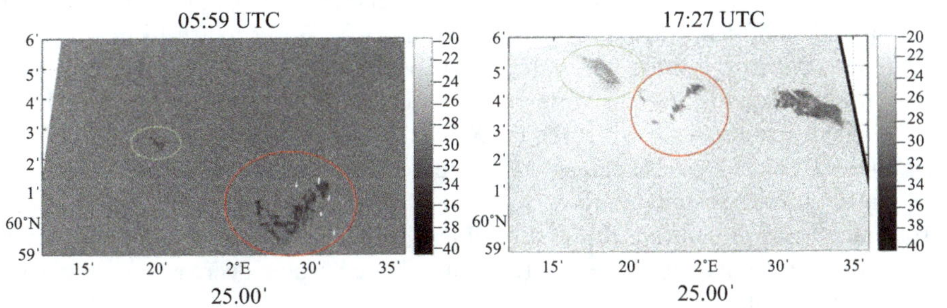

Figure 9.2 RADARSAT-2 VV intensity (dB) from two consecutive images of the same oil spills, showing both drift and change in size and shape of the slicks over an 11.5 h period. The increase in intensity from unslicked seawater arises from the change in incidence angle. Circled in green: plant oil spill. Circled in red: emulsion oil spill. Not circled in image on right: crude oil spill. RADARSAT-2 Data and Products. Copyright MDA LTD (2011)—All rights reserved

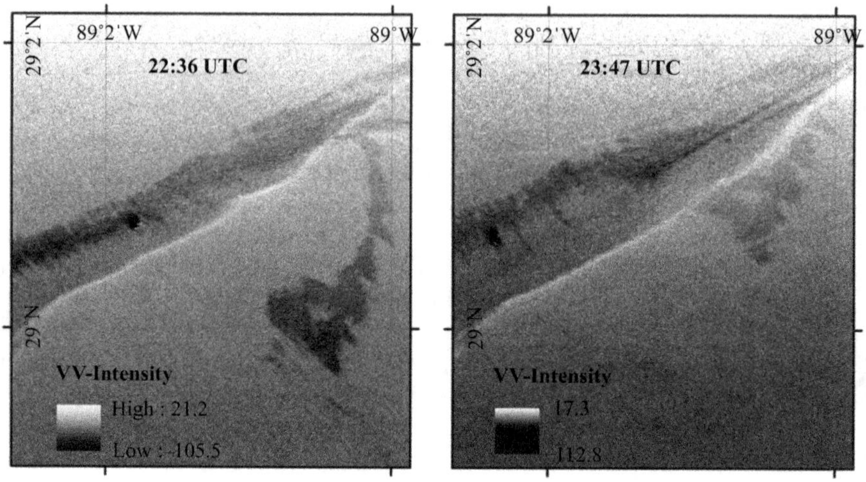

Figure 9.3 UAVSAR VV intensity (dB) from two images acquired 71 min apart of a slick originating from a seep. A mat of red-brown emulsion, identified visually from a low-flying aircraft and identifiable by the low radar return, is moving to the west-southwest

9.2.3 The weather window

The weather window is defined here as the wind speed range over which SAR is usable for imaging oil spills, and is related to the sea surface geometry at the scale of the radar wavelength. Much effort has been devoted to relating the normalized radar cross section to near-surface wind speeds, either empirically based on power laws [28] or theoretically based on physical models [29]. Oil spill detection by SAR is generally confined to wind speeds in the range of 2–3 m/s to 10–14 m/s [30, and references therein], but the minimum wind speed to generate Bragg waves producing measurable radar returns varies with frequency. Some approximate thresholds at 20° incidence angle are ~2.5 m/s for X-band, ~2.2 m/s for C-band and ~2.0 m/s for L-band [27,29]. These thresholds increase slightly with incidence angle and decrease with ocean temperature. At the high end of the range, the limit comes from mixing of the oil into the water column by breaking waves and the fact that the backscatter contrast between clean and slicked water decreases as the wind speed increases [30].

9.2.4 Transport and weathering of oil pollutants

After release, oil from spills at sea evolves through transport and weathering [31]. Slicks spread through the influence of gravity and winds, and are transported both on and below the sea surface. Oil released into the marine environment also develops over time through *weathering processes* acting on the pollutant and transforming its original physical and chemical properties. Some important transforming processes are evaporation, emulsification, dissolution, oxidation, and

Table 9.4 Description of some selected processes affecting the fate of oil from a spill that depend on sea-state and oil properties

Process	Description
Emulsification	Oil (immiscible) is dispersed by droplets of sea water. Contributes to the persistence of the spill due to an increase in viscosity and thickness with water content [32]. Stable emulsions contain 60%–85% water [31]
Horizontal spreading	Spreading of oil on the sea surface mainly in early phases after release [31]. The spread is not necessarily uniform and observations have revealed that 90% of the oil can be confined to an area of less than 10% of the slick extent [33]
Vertical dispersion and entrainment	Movement of oil droplets (size < 100 μm [31]) into the water column. A rule of thumb for the depth of mixing during entrainment says 1.5 times the wave height [31]. Wave breaking is required to initiate entrainment of a surface spill. Diffusion can mix the oil deeper into the water column [34]

biodegradation [31], some of which act rapidly. These processes determine the fate of an oil spill in the marine environment and they have implications for response and combat procedures. In the context of a clean-up operation, and taking the time-dependent weathering processes into account, the amount and type of the pollutant, the sea state, and the prevailing environmental and weather conditions will determine which active countermeasures should be made. Hence, information obtained from SAR measurements benefit from rapid sampling and, if possible, resampling of the radar images to maintain its relevance during field operations. Examples of oil spill development with time were shown in Figures 9.2 and 9.3, with examples from "rapid" satellite SAR imaging (< 1/2 day) and much more rapid (~1 h) airborne SAR repeat imaging performed in a single flight.

Table 9.4 gives a brief description of selected transport and weathering processes of particular relevance for this chapter. Geophysical parameters that can be affected by the presence of an oil slick on the ocean surface are the surface roughness, dielectric complex permittivity, surface tension, and surface viscosity. In Sections 9.3 and 9.4, we discuss how emulsification of oil changes the dielectric properties of the slick, which has an impact on its radar backscatter. In Section 9.4.3, we exhibit how SAR imagery can be integrated with modeling for improved understanding of the horizontal spreading and vertical entrainment and mixing.

9.2.5 False alarms

The accuracy of both manual and automatic oil spill detection services is reduced due to the presence of *look-alikes*, that is, other physical phenomena causing dark signatures in SAR imagery. There are a number of situations that give rise to radar-dark features in the ocean that are look-alikes for oil spills and can cause false alarms in SAR-based surveillance. To illustrate the difficulty, we compare mineral oil slicks with look-alikes on the sea surface. Figure 9.4 presents three examples of SAR scenes containing man-made oil slicks, which appear as dark spots on a brighter background. The left panel shows an example of a reoccurring produced water slick from an oil rig. Several examples of spilled emulsified oil are shown in the middle and right panels, and small plant oil spills (look-alikes) are visible in the upper right part of the middle panel and to the left of the large slick in the right panel. Examples of other look-alikes are low wind areas, grease ice, and natural biogenic films caused by, for example, algae, fish oil, and vegetable oil. If the interest is exclusively in detecting man-made oil spills or seeps, natural oil seeps also fall into the look-alike category. Figure 9.5 shows examples of SAR images of algae bloom and low wind, both naturally occurring radar dark features. In Figure 9.6, grease ice is seen as dark slicks in the SAR scenes.

A variety of descriptors such as spatial, contextual, and statistical features are often used to label and discriminate the slicks [35]. Hence, it is obvious that the SAR sensor's spatial and radiometric resolutions, along with the instrument noise floor, are of importance for the mineral oil slick discrimination capability.

9.3 Interpretation and modeling

To explain the contrast between an oil spill and the ambient sea water in an SAR image, we introduce the properties that characterize a material and dictate the scattering of electromagnetic waves. Next, we revisit and discuss how surface scattering models may aid interpretation of real SAR measurements. Finally, we discuss the implications of the signal's proximity to the noise floor, which can be critical to interpretation of radar-dark signals.

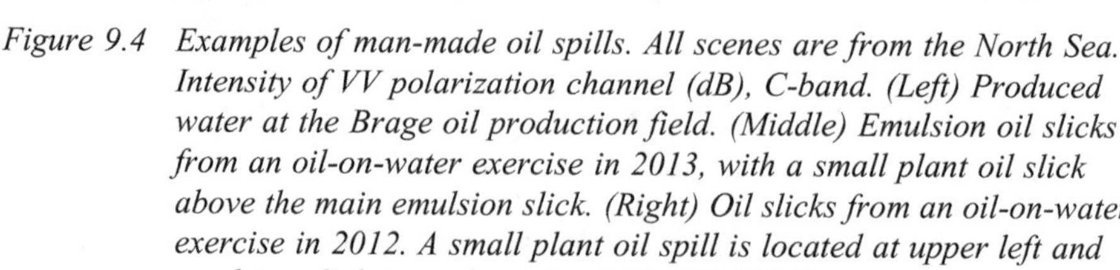

Figure 9.4 Examples of man-made oil spills. All scenes are from the North Sea. Intensity of VV polarization channel (dB), C-band. (Left) Produced water at the Brage oil production field. (Middle) Emulsion oil slicks from an oil-on-water exercise in 2013, with a small plant oil slick above the main emulsion slick. (Right) Oil slicks from an oil-on-water exercise in 2012. A small plant oil spill is located at upper left and emulsion slick is on the right. RADARSAT-2 Data and Products. Copyright MDA LTD (2017, 2013, 2012)—All rights reserved

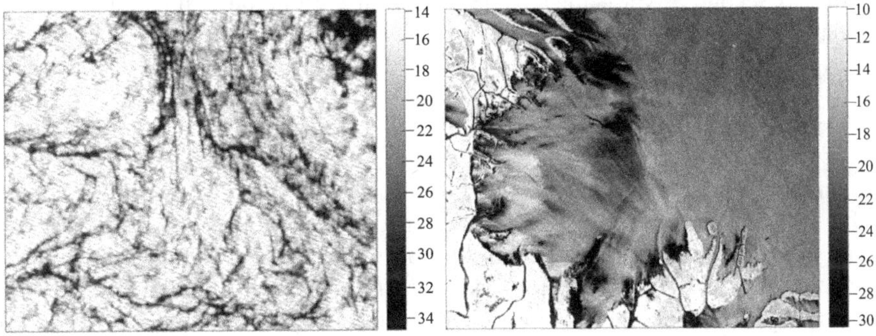

Figure 9.5 *Examples of look-alikes. Intensity of VV polarization channel (dB). (Left) RADARSAT-2 (C-band) scene of an algae bloom in the Baltic Sea. RADARSAT-2 Data and Products. Copyright MDA LTD. (2016)—All rights reserved. (Right) UAVSAR (L-band) scene of low wind condition and calm water over subaqueous vegetation within islands in the Wax Lake Delta, Louisiana, USA*

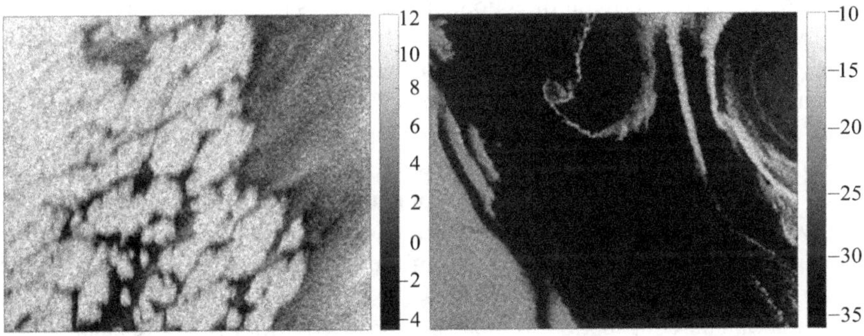

Figure 9.6 *Examples of look-alikes. Intensity of VV polarization channel (dB), C-band. Both panels show newly frozen sea ice causing dark slicks. (Left) Open water and grease ice in the Barents Sea. Interpretation from comparing with timeseries of Sentinel-1 data. (Right) Grease ice along the ice margin near Hopen island. RADARSAT-2 Data and Products. Copyright MDA LTD (2017, 2014)—All rights reserved*

9.3.1 Contrast drivers

In SAR imaging, there are two main drivers of contrast, that is, a change in backscatter power, between an oil slick and the surrounding clean sea surface. One driver is the slick's capability of damping the small-scale roughness, causing a smaller portion of the reflected radar energy to be scattered back to the radar antenna. A second driver is the change in dielectric properties within an oil spill as compared to uncontaminated sea water. Both are discussed below.

9.3.1.1 Surface roughness

All of the SAR instruments listed in Table 9.3 are side-looking and operate in a monostatic configuration (transmitter and receiver collocated). Therefore, they measure the backscattered signal, which has larger amplitude for rougher surfaces. If the surface was perfectly smooth, the angular pattern of the reflected scatters would be a delta function centered on the specular direction, and the sensor would detect no backscatter. When the surface is slightly rough the angular radiation pattern has two components, the coherent scattering component, which is the reflection component in the specular direction, and the diffuse scattering component, which represents scattering in all directions. The coherent part becomes negligible when the roughness increases and the radiation pattern approaches Lambertian scattering for very rough surfaces [36].

Two parameters are commonly used to characterize a naturally random surface: the spread of heights about the average value (vertical) and the variation of these heights across the surface (horizontal). The standard deviation of the surface height variation, σ_z, is referred to as the root mean square (rms) height of the surface. To model the smoothness of a surface, that is, the correlation of the random height variations in the lateral direction, an idealized surface correlation function is typically used together with the correlation length, l [37]. For two points on the surface, the surface correlation describes their statistical independence and the correlation length increases with the distance between two correlated points (for a smooth surface $l = \infty$). Given a specific correlation function, the spatial roughness of a surface is fully described by the rms height, σ_z, and the surface correlation length, l [37]. The impact of surface roughness on radar backscatter depends strongly on the wavelength of the radar, λ_r. This is accounted by scaling the roughness parameters by the wavelength of the radar, yielding $k_r\sigma_z$ and $k_r l$, where $k_r = 2\pi/\lambda_r$ is the radar wavenumber [38].

The dominant contribution to radar backscatter is from Bragg resonance scattering, namely resonant scattering from spatially correlated, wind-generated ocean wave components that are in phase with the incident radar waves. These components fall within the range from small capillary ripples to short gravity waves, depending on the radar frequency [27]. The Bragg wavelength of ocean waves resulting in resonance with the radar waves is given by $\lambda_B = n\lambda_r/(2\sin\theta)$, where θ is the incidence angle and $n = 1, 2, \ldots$ is the order of resonance with $n = 1$ producing the dominant return. For a given λ_r, the resonant waves are shorter at more oblique θ, and at a given incidence angle λ_B increases with λ_r. In general, the Bragg scattering model is incomplete to describe the backscatter from the ocean surface because longer waves also interact with the Bragg waves and affect the radar return through tilt modulation (changing the local surface orientation or slope), hydrodynamic modulation (changing the spatial distribution of ocean wave spectral components, for example, as short waves pile up in crests and spread out in troughs), and velocity bunching (nonlinear modulation caused by the SAR processing, which translates the surface velocity component in the look direction into a position shift in azimuth direction due to the Doppler shift) [39,40].

9.3.1.2 Relative dielectric constant

The dielectric properties of the scattering surface affect the electromagnetic interaction between the electromagnetic wave and the surface, and hence the backscattered signal. The *relative permittivity*, ϵ_r, of a material is a complex number given by

$$\epsilon_r(\omega) = \left(\epsilon^{'}(\omega) - i\epsilon^{''}(\omega)\right)/\epsilon_0 \tag{9.1}$$

where $\epsilon^{'}(\omega)$ is the real part of the material's permittivity, $\epsilon^{''}(\omega)$ is the imaginary part, ϵ_0 is the vacuum permittivity (permittivity of free space), $i = \sqrt{-1}$, and $\omega = 2\pi\nu$ is the angular frequency (also referred to as radian frequency) of the incident radar wave with frequency, ν. The term *dielectric constant* has been used interchangeably with the electric permittivity and also to refer to the real part of the relative permittivity. The term *relative dielectric constant* is here used to describe the complex ϵ_r.

A reduction in the effective relative dielectric constant means less total energy being reflected by the surface, and hence, the surface will appear darker. As discussed in Section 9.2.4, oil mixes with sea water through interaction with waves and weathering to form oil/water emulsification. Oil has a relative dielectric constant around $\epsilon_r^O = 2.3 - i0.02$ [10], much lower in both real and imaginary components than that of sea water, ϵ_r^{SW}, which is a function of temperature, T, and salinity, S, but exceeds ϵ_r^O under almost all open water ocean conditions. For example, calculations based on [36] for 1.26 GHz (L-band) frequency give $\epsilon_r^{\mathrm{SW}} = 70.1 - i82.7$ for typical Gulf of Mexico conditions ($T = 27°\mathrm{C}$, $S = 35$ ppt) and $\epsilon_r^{\mathrm{SW}} = 75.4 - i59.4$ for typical North Sea conditions (8°C, $S = 35.25$ ppt). Oil/water emulsions have relative dielectric constant lying between $\epsilon_r^O$and ϵ_r^{SW}. Recent work at L-band has suggested that it is possible to quantify the mixing of oil and sea water from SAR, based on work using a simple linear mixing model for the emulsion relative dielectric constant [10,13] or the more accurate Bruggeman formula [11].

Figure 9.7 depicts the impact of oil mixing with sea water on the relative dielectric constant. Radar backscatter is only sensitive to the slick's dielectric properties if the backscatter occurs from the oil, not the underlying sea water interface. This is caused by a thick layer of oil or oil emulsion on the sea surface [10,13]. Typically, most of a slick is not thick, with the general rule being that 90% of the oil is contained in 10% of the slick area [33]. Because of oil's low loss factor (imaginary part of ϵ_r^{O}), electromagnetic waves pass through very thin layers of oil without interacting, and so see an effective relative dielectric constant of sea water, same as for the clean water. This is certainly true for sheen layers (Table 9.2). Although studies have inferred the oil volumetric fraction from backscatter [10,11,13], none have yet been validated to determine the thickness threshold at which the oil layer's relative dielectric constant impacts backscatter. However, it is

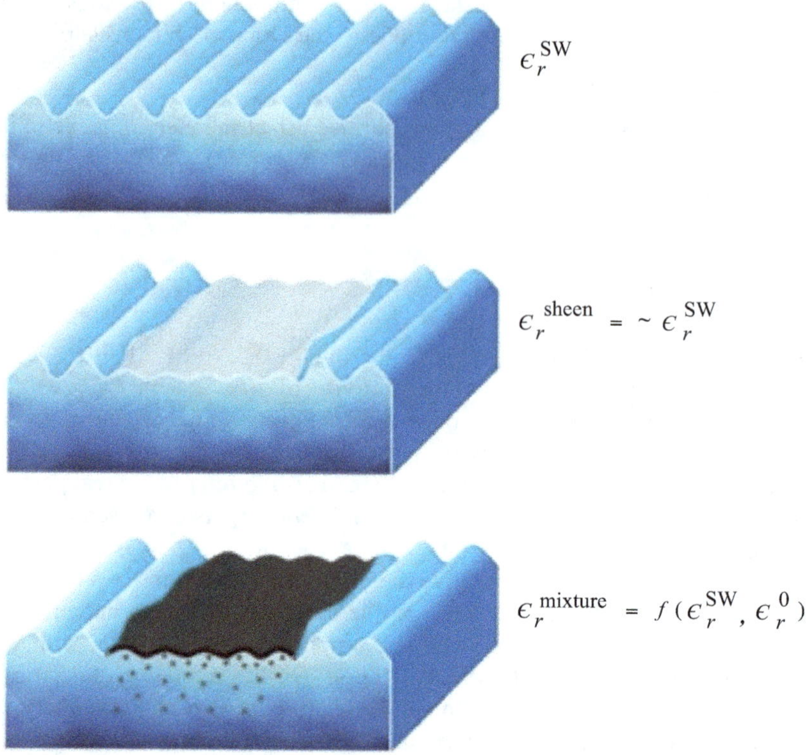

Figure 9.7 The impact of sheen and thicker mineral oil slicks on the relative dielectric constant: (top) clean sea water, (middle) sheen slick, and (bottom) thicker oil slick

known that emulsions often form thicker layers [7] and observed that red-brown emulsions within slicks have much lower backscatter than the rest of the slick [9,41]. This is a topic of active study. Although it is known from electromagnetic scattering theory that the layer thickness for interaction must be comparable to the wavelength within the medium, determination of the thresholds for both thickness and oil volumetric fraction at which the backscatter becomes sensitive and experimental confirmation for an oil spill in the open ocean remains an outstanding issue.

9.3.2 Surface scattering models

For the convenience of the reader, we proceed with a brief review of selected ocean surface scattering models adopted in the existing oil spill SAR literature, namely the small perturbation model (SPM), also referred to as the Bragg scattering model [42]; the two-scale model, also referred to as the composite model [42]; the

Table 9.5 Overview of selected ocean surface scattering models. Copolarization ratio: $\mathrm{PR} \equiv \sigma_0^{\mathrm{HH}}/\sigma_0^{\mathrm{VV}}$

Model	**Surface properties taken into account**	**NRCS components represented**	**PR function of**
SPM [42]	Ocean waves of wave-number $2k_r \sin\theta$ (Bragg wavenumber)	$\sigma_0^{\mathrm{pp}} = \sigma_{0B}^{(1)\mathrm{pp}}$, where $\mathrm{pp} \in \{\mathrm{HH}, \mathrm{VV}\}$ and (1) indicates first order. Depolarization effects not included and $\sigma_{0B}^{(1)\mathrm{pq}} = 0$, where $\mathrm{pq} \in \{\mathrm{HV}, \mathrm{VH}\}$	θ, ϵ_r
Two-scale model [42]	Adds tilt modulation by longer gravity waves to SPM	$\sigma_0^{\mathrm{pp}} = \sigma_{0B}^{\mathrm{pp}}$ and $\sigma_0^{\mathrm{pq}} = \sigma_{0B}^{\mathrm{pq}}$ Cross-polarization term included	$\theta, \epsilon_r, \psi, \xi$
Kudryavtsev model [43]	Adds depolarized scattering to two-scale copolarized NRCS, attributed to wave breaking from "steep" and "rough" patches	$\sigma_0^{\mathrm{pp}} = \sigma_{0B}^{\mathrm{pp}} + \sigma_n$ Nonpolarized scattering (non-Bragg scattering) term σ_n added ($\sigma_{0\mathrm{B}}^{\mathrm{pq}}$ not included)	$\theta, \epsilon_r, \psi, \xi, \sigma_n$

Kudryavtsev model [43], which includes a depolarizing scattering mechanism; and the extended Bragg (X-Bragg) model, a polarimetric surface scattering model originally developed for soil surface retrieval [38], but also applied to marine oil spill studies [44]. Table 9.5 summarizes which surface properties are taken into account and which normalized radar cross-section (NRCS) components, σ_0, are represented in the ocean scattering models considered here.

The scattering of electromagnetic waves from the ocean surface is often modeled through approximate solutions with assumptions on the dimensions of the scattering elements [38,45]. The SPM assumes the variation of the surface roughness to be small compared to λ_r, $k_r\sigma_z \ll 0.3$[38], and has a limited range of validity in terms of the ocean surface. The first-order NRCS components are given by

$$\sigma_{0B}^{(1)\mathrm{pp}} = 4\pi {k_r}^4 \cos^4(\theta)\left|g_{\mathrm{pp}}^{(1)}(\theta, \epsilon_r)\right|^2 W(0, 2k_r \sin\theta) \tag{9.2}$$

where $W(\cdot)$ is the 2D wavenumber spectral density of the surface roughness and $g_{\mathrm{pp}}^{(1)}(\theta, \epsilon_r)$ are the first-order scattering coefficients, which only depend on θ and ϵ_r [42]. For an explicit expression of $g_{\mathrm{pp}}^{(1)}(\theta, \epsilon_r)$ see [42]. The SPM model is popular [37], but does not include nonzero cross-polarized backscatter or depolarization effects. Note that in a polarimetric radar system, cross-polarization is non-null cross-polarized covariance matrix elements and depolarization is a nonunity normalized copolarization correlation coefficient or polarimetric coherence coefficient [38,46].

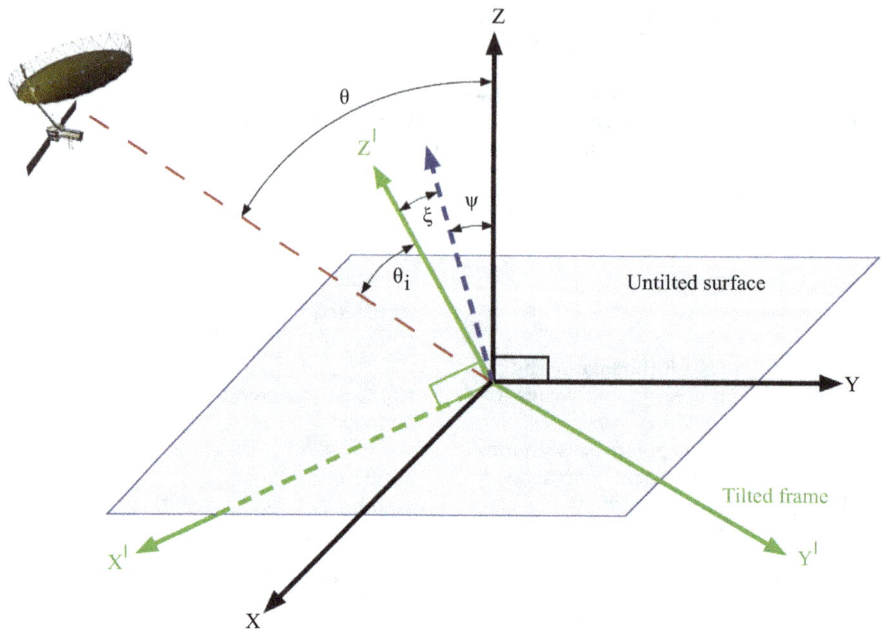

Figure 9.8 The SPM model assumes scattering from an untilted surface with incidence angle, θ, in the plane of incidence (the y–z plane as drawn). The two-scale model introduces tilted facets with in-plane tilt angle, ψ, and out-of-plane tilt angle, ξ, that characterize the rotation between the tilted surface normal, z', and the untilted surface normal, z

Figure 9.8 illustrates the more realistic sea surface geometry introduced by the two-scale model, a modification of SPM that leads to a non-null cross-polarization term [42]:

$$\sigma_{0B}^{\text{pp}} = 4\pi k_r{}^4 \cos^4(\theta_i) \left| \left(\frac{\sin(\theta+\psi)\cos(\xi)}{\sin(\theta_i)} \right)^2 g_{\text{pp}}^{(1)}(\theta_i, \epsilon_r) + \left(\frac{\sin(\xi)}{\sin(\theta_i)} \right)^2 g_{\text{qq}}^{(1)}(\theta_i, \epsilon_r) \right|^2$$
$$\times W(2k_r \cos(\theta+\psi)\sin(\xi), 2k_r \sin(\theta+\psi))$$

and

$$\sigma_{0B}^{\text{pq}} = 4\pi k_r{}^2 \cos^4(\theta_i) \left(\frac{\sin(\theta+\psi)\sin(\xi)\cos(\xi)}{\sin^2(\theta_i)} \right)^2 \left| g_{VV}^{(1)}(\theta_i, \epsilon_r) - g_{\text{HH}}^{(1)}(\theta_i, \epsilon_r) \right|^2$$
$$\times W(2k_r \cos(\theta+\psi)\sin(\xi), 2k_r \sin(\theta+\psi)) \tag{9.3}$$

where $\sigma_{0B}^{\text{HV}} = \sigma_{0B}^{\text{VH}}$, θ is the local incidence angle relative to the untilted horizontal plane, and $\theta_i = \cos^{-1}[\cos(\theta+\psi)\cos(\xi)]$ is the incidence angle for the tilted facet.* Now the first-order scattering coefficients depend upon the facet geometry in addition to the dielectric properties of the scattering surface [42]. Within the two-scale model, due to long gravity waves, the normal to the facet deviates from the vertical by an angle ψ in the plane of incidence and by an angle ξ in the plane perpendicular to the plane of incidence. Note that in this model $\sigma_{0B}^{\mathrm{pq}} = 0$ if $\xi = 0$, that is, if there is no out-of-plane tilt of the facet.

Building on the two-scale model, Kudryavtsev *et al.* [43] represented each copolarized NRCS as the sum of one $\sigma_{0B}^{\mathrm{pp}}$ component associated with conventional two-scale Bragg scattering and one nonpolarized scattering component, σ_n, due to non-Bragg scattering. The non-Bragg component is assumed to be the same for VV and HH, and is interpreted as being caused by wave breaking from steep and rough patches on the sea surfaces and other phenomena that can cause non-Bragg scattering [43,45]. A cross-polarization term is not included in this model. The non-Bragg contribution can in theory be removed using the polarization difference PD:

$$\mathrm{PD} \equiv \sigma_0^{\mathrm{VV}} - \sigma_0^{\mathrm{HH}} = \sigma_{0B}^{\mathrm{VV}} - \sigma_{0B}^{\mathrm{HH}} \tag{9.4}$$

The PD has been explored in oil spill studies and is reported to have good detection capability [47, and references therein].

Considering the robustness of the SMP model within its validity range [38], the X-Bragg model is a polarimetric model that attempts to widen the SPM model's validity range by allowing a random tilt of the facet, with the width of the distribution reflecting the roughness variability. It accounts for both cross-polarization and depolarization effects introduced through roughness-induced rotation of the Bragg coherency matrix (***T***-matrix, see, e.g., [10]) about an angle β in the plane perpendicular to the scattering plane. Assuming $P(\beta)$ to be a uniform distribution about zero with half-width β_1, the coherency matrix of the X-Bragg model is represented as [38]:

$$\boldsymbol{T} = \begin{bmatrix} T_{11} & T_{12} & T_{13} \\ T_{21} & T_{22} & T_{23} \\ T_{31} & T_{32} & T_{33} \end{bmatrix}$$

$$= \begin{bmatrix} C_1 & C_2 \mathrm{sinc}\,(2\beta_1) & 0 \\ C_2 \mathrm{sinc}\,(2\beta_1) & C_3(1 + \mathrm{sinc}\,(4\beta_1)) & 0 \\ 0 & 0 & C_3(1 - \mathrm{sinc}\,(4\beta_1)) \end{bmatrix}$$

where $C_1 = \left|g_{\mathrm{HH}}^{(1)} + g_{\mathrm{VV}}^{(1)}\right|^2$, $C_2 = \left(g_{\mathrm{HH}}^{(1)} + g_{\mathrm{VV}}^{(1)}\right)\left(g_{\mathrm{HH}}^{(1)*} - g_{\mathrm{VV}}^{(1)*}\right)$, and $C_3 = 1/2\left|g_{\mathrm{HH}}^{(1)} - g_{\mathrm{VV}}^{(1)}\right|^2$. A couple of interesting features could be derived from the extended Bragg model:

$$R(\beta_1) = (T_{22} - T_{33})/(T_{22} + T_{33}) \text{ and } M(\theta, \epsilon_r) = (T_{22} + T_{33})/T_{11} \tag{9.5}$$

*See "Notes on further reading" at the end of the chapter for more information about this topic.

where R is a roughness indicator (depending only on the orientation angle), and M is a material indicator (depending on the incidence angle and the dielectric constant). These features have been explored in oil spill studies and varying sensitivity to different oil spill types has been reported [47,48]. Note that both of these features involve cross-polarization terms and, hence, caution should be used when applied to low signal-to-noise ratio (SNR) systems (see Section 9.3.3).

According to Gade *et al.* [30], in the Marangoni damping theory, monomolecular slicks,[†] such as natural biogenic slicks, show a resonance damping behavior that is absent for mineral oil slicks. Hence, the theory of Marangoni damping is left out of this discussion, as the primary focus here is on crude oil or emulsions with a thickness larger than monomolecular films and different damping mechanisms acting on the ocean surface waves.

9.3.2.1 Interpreting the copolarization ratio from real SAR imagery

An appealing property of the SPM is that the roughness cancels in the copolarization ratio and it only depends on the local incidence angle (relative to the untilted horizontal plane) and the dielectric properties of the media:

$$\frac{\sigma_0^{\mathrm{HH}}}{\sigma_0^{\mathrm{VV}}} = \frac{\left|g_{\mathrm{HH}}^{(1)}(\theta, \epsilon_r)\right|^2}{\left|g_{\mathrm{VV}}^{(1)}(\theta, \epsilon_r)\right|^2} \tag{9.6}$$

In the two-scale model, the copolarization ratio depends on the large-scale roughness (through the tilt angles), the local incidence angle, and the dielectric properties of the media:

$$\frac{\sigma_0^{\mathrm{HH}}}{\sigma_0^{\mathrm{VV}}} = \frac{\left|\left(\frac{\sin(\theta+\psi)\cos(\xi)}{\sin(\theta_i)}\right)^2 g_{\mathrm{HH}}^{(1)}(\theta_i, \epsilon_r) + \left(\frac{\sin(\xi)}{\sin(\theta_i)}\right)^2 g_{\mathrm{VV}}^{(1)}(\theta_i, \epsilon_r)\right|^2}{\left|\left(\frac{\sin(\theta+\psi)\cos(\xi)}{\sin(\theta_i)}\right)^2 g_{\mathrm{VV}}^{(1)}(\theta_i, \epsilon_r) + \left(\frac{\sin(\xi)}{\sin(\theta_i)}\right)^2 g_{\mathrm{HH}}^{(1)}(\theta_i, \epsilon_r)\right|^2} \tag{9.7}$$

If the HH, VV, and HV NRCS values are known for the clean sea surface, the two ratios of the polarization-dependent cross-section terms, $\sigma_0^{\mathrm{HH}}/\sigma_0^{\mathrm{VV}}$ and $\sigma_0^{\mathrm{HV}}/\sigma_0^{\mathrm{VV}}$, measured in the clean water can be used to solve for the two tilt angles. Under the assumption that the longer wavelengths on the sea surface are unaffected by the presence of a slick, the angles derived from the clean water can be used as an estimate for the slick covered surface [10]. Then, assuming the two-scale model is a good estimate for the surface scattering for both oil and water, the copolarization ratio can be used to estimate ϵ_r for the oil slick.

[†]See "Notes on further reading" at the end of the chapter for more information about this topic.

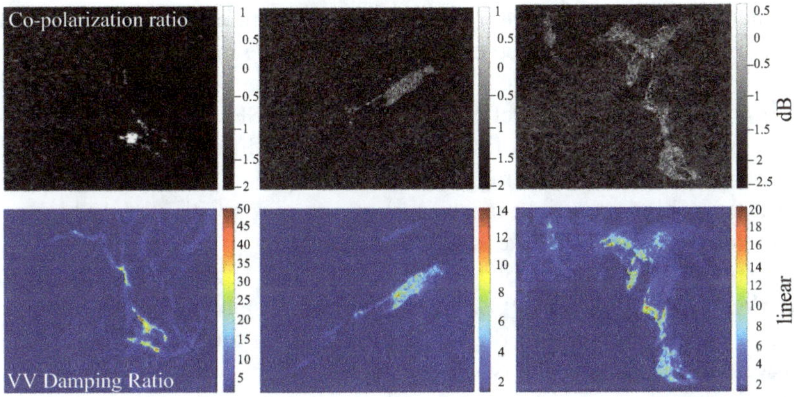

Figure 9.9 (Top) Copolarization ratios in dB corresponding to the panels in Figure 9.4. (Bottom) VV-intensity damping ratio (contrast) in linear units. These are derived from Radarsat-2 C-band quad-polarization data

In the Kudryavtsev model, the ocean wave spectral density term does not cancel in the copolarization ratio:

$$\frac{\sigma_0^{\mathrm{HH}}}{\sigma_0^{\mathrm{VV}}} = \frac{\sigma_{0B}^{\mathrm{HH}} + \sigma_n}{\sigma_{0B}^{\mathrm{VV}} + \sigma_n} \tag{9.8}$$

so the ratio also depends on the nonpolarized scattering component. We note that the Kudryavtsev model explicitly assumes the added scattering mechanism is depolarizing in nature, mixing HH and VV. However, any added scattering component that is non-Bragg, for example, incoherent surface scattering from other wave components, alters the copolarization ratio from the SPM and two-scale models' situation in which ϵ_r can be calculated if only the imaging and surface geometry is known, that is, no information about the wave spectra is required.

Figure 9.9 presents the copolarization ratios corresponding to the three scenes shown in Figure 9.4. The Brage oil production slick, which is produced water (left panel), shows negligible contrast with that of clean water, so according to the SPM, the small fractions of oil contained in the produced water is too low to have an impact on the relative dielectric constant (Oil discharge measured to be 56 kg on day of acquisition; Brage platform, personal communication 2018). Considering the two-scale model, we can also ignore large-scale tilt variation between clean water and the slick. The middle and right panels show the copolarization ratio for the two thicker oil emulsion slicks, which show a clear contrast to the surrounding sea. (Information about the releases in the right panel can be found in [49]). According to the theories described above, this is due to a change in the dielectric properties, damping of the large-scale roughness, or non-Bragg scattering. The small plant oil spill located to the upper left of the large emulsion slick in the right panel exhibits a clear but weaker contrast to the surrounding sea water. Being a monomolecular slick, no change in ε_r is expected. Therefore, the SPM theory does not hold because it predicts no contrast should be visible in the copolarization ratio, which indeed is the case for the plant oil spill in the middle panel (compare with Figure 9.4). The two-scale model predicts the same as SPM for the plant oil slick if the large-scale roughness (facet tilt angles) are not modified between oil and clean sea. Instrument noise or non-Bragg scattering components could account for this. Note that the scene in the right-most panel was acquired with slightly higher incidence angles as compared to the other two scenes.

Figure 9.10 shows the copolarization ratio for two L-band QP scenes acquired with UAVSAR for which the NESZ is well below the slick intensities. In this case, the value for the slick in the North Sea (top), which has thickness 1.3–1.7 μm [50], agrees well with SPM and the two-scale model, but the value for the slick in the Gulf of Mexico shows good contrast with the clean water, even in areas where visual images show sheen. This, too, argues for another factor, either modification of the large-scale roughness and/or a non-Bragg scattering component. It is interesting to note that the North Sea image was acquired at winds of ~10 m/s and with visible wave breaking, but the Gulf image was acquired at wind speed 5 m/s and calm water. This argues for a non-Bragg scattering mechanism not driven solely by wave breaking.

9.3.2.2 Interpreting the damping ratio from real SAR imagery

In line with the SPM scattering theory discussed above, the damping ratio for an oil slick would be expressed as

$$\frac{\sigma_{0,\text{sea}}^{\text{VV}}}{\sigma_{0,\text{oil}}^{\text{VV}}} = \frac{\left|g_{\text{VV}}^{(1)}(\theta, \epsilon_{r,\text{sea}})\right|^2}{\left|g_{\text{VV}}^{(1)}(\theta, \epsilon_{r,\text{oil}})\right|^2} \times \frac{W_{\text{sea}}(0, 2k_r \sin\theta)}{W_{\text{oil}}(0, 2k_r \sin\theta)}$$

Within the two-scale mode, the damping ratio would be expressed as

$$\frac{\sigma_{0,\text{sea}}^{\text{VV}}}{\sigma_{0,\text{oil}}^{\text{VV}}} = \frac{\left|\left(\frac{\sin(\theta+\psi)\cos(\xi)}{\sin(\theta_i)}\right)^2 g_{\text{VV}}^{(1)}(\theta_i, \epsilon_{r,\text{sea}}) + \left(\frac{\sin(\xi)}{\sin(\theta_i)}\right)^2 g_{\text{HH}}^{(1)}(\theta_i, \epsilon_{r,\text{sea}})\right|^2}{\left|\left(\frac{\sin(\theta+\psi)\cos(\xi)}{\sin(\theta_i)}\right)^2 g_{\text{VV}}^{(1)}(\theta_i, \epsilon_{r,\text{oil}}) + \left(\frac{\sin(\xi)}{\sin(\theta_i)}\right)^2 g_{\text{HH}}^{(1)}(\theta_i, \epsilon_{r,\text{oil}})\right|^2}$$

$$\times \frac{W_{\text{sea}}(2k_r \cos(\theta+\psi)\sin(\xi), 2k_r \sin(\theta+\psi))}{W_{\text{oil}}(2k_r \cos(\theta+\psi)\sin(\xi), 2k_r \sin(\theta+\psi))} \tag{9.9}$$

According to both scattering models, for either a monomolecular slick or a thin oil spill with a small thickness compared to the penetration depth of the radar [30], no change in ε_r is expected and the first factor of the damping ratio will hence be unity. Wave damping within the oil slick will be the only contrast driving factor. However, for a thicker slick, ε_r will change between sea and oil and hence both factors will contribute as contrast drivers between the oil slick and the surrounding sea water. Figures 9.9 and 9.10 show the damping ratios along with the

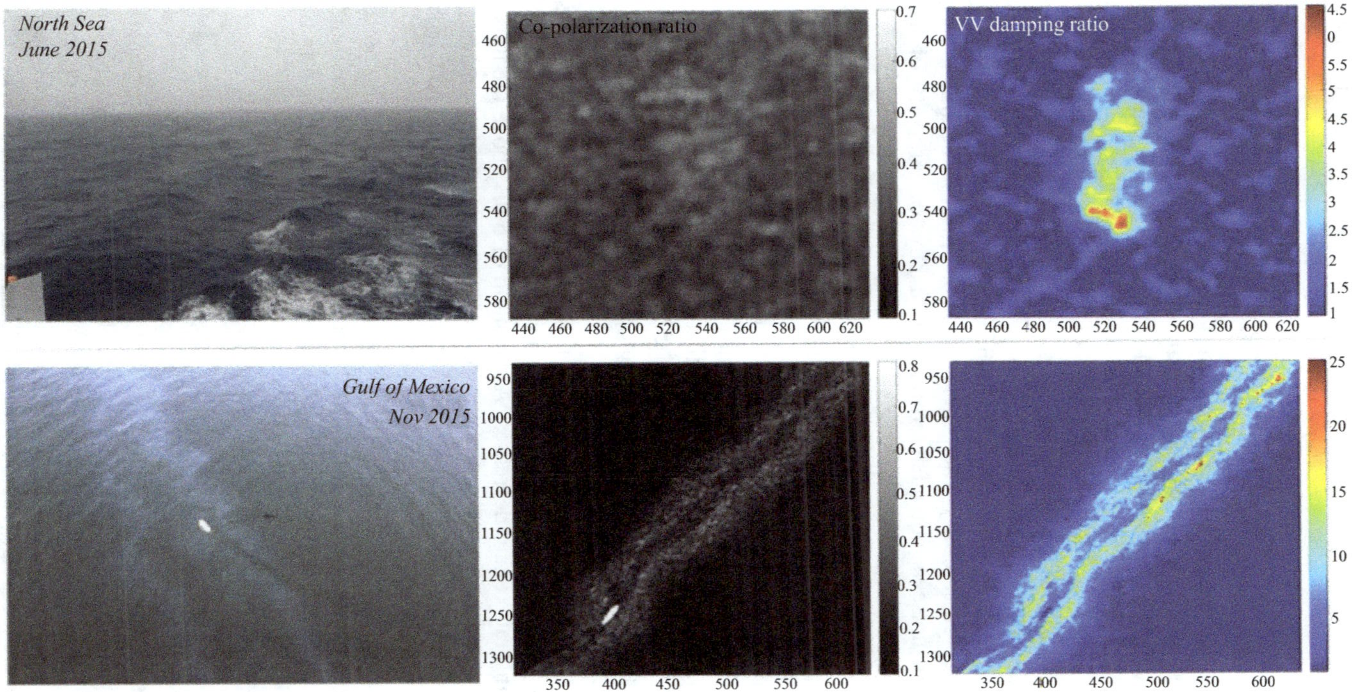

Figure 9.10 *(Left) Photo of sea surface conditions, (middle) copolarization ratios in dB, and (right) VV-intensity damping ratio (contrast) in linear units for acquisitions in the (top row) North Sea and (bottom row) Gulf of Mexico (GOM). The bright object in the GOM copolarization ratio is the ship seen in the photo. Wind speed was 10 m/s (North Sea) and 5 m/s (GOM). In both cases, the oil slick is mainly sheen, but in one case the copolarization ratio shows little contrast with clean water and in the other, it shows 0.5 dB or larger contrast. Products are derived from UAVSAR L-band quad-polarization data*

copolarization ratios for the five oil slick cases discussed above. In all cases, the damping ratio shows contrast within the slick, not just between the slick and clean water, indicative of zones of thicker or more emulsified oil.

9.3.3 Influence of instrument noise

The noise-equivalent sigma zero (NESZ), or *noise floor*, is a measure of the sensitivity of the SAR system to areas with low radar backscatter, and is given by a SNR, that is, the ratio between backscatter level and the sensor NESZ, equal to 1 on a linear scale [51]. The NESZ is a function of slant range [51], and the backscattered signal's proximity to the noise floor has a significant influence on radar-dark measurements, which can lead to misinterpretation of SAR data. This is not an issue for spill detection (as long as the contrast between the oil slick and the surrounding clean sea is sufficient), where location and extent are determined, but the SNR should be considered carefully when slick characterization is the aim so that the contribution from noise is not misinterpreted as a physical property of the scattering medium.

The influence of noise on derived quantities is illustrated using the damping ratio, which is known to increase with increasing Bragg wavenumber [30] and decrease with increasing wind speed [8]. Figure 9.11 shows how the damping ratio is influenced by SNR using data from oil spills measured with TerraSAR-X, Radarsat-2, and UAVSAR. The UAVSAR provides L-band ($\lambda_r = 23.8$cm) quad-polarimetric SAR measurements with an SNR that surpasses many spaceborne missions, including Radarsat-2 and TerraSAR-X [52]. The sensor covers a large range of incidence angles, $19.5^\circ - 67.5^\circ$, in a single scene [52], and the Bragg wavenumbers range from 17.6rad/m – 48.8rad/m, depending upon incidence angle, with higher values in the far range. The measurements show that as the signal from the slick approaches the NESZ, the damping ratio increases, but at very low SNR the damping ratio again decreases. A threshold of 6 dB above the noise floor has previously been suggested as a guard margin for any scattering analysis, where estimated backscatter contributions within or below this range should be ignored [10].

Following [53], for a polarimetric SAR system, and assuming that the noise terms in the different polarization channels are uncorrelated with each other and with the signal, the measurement $\mathcal{M}$ should be related to the scattering matrix of interest as follows:

$$\mathcal{M} = \begin{bmatrix} M_{\mathrm{HH}} & M_{\mathrm{VH}} \\ M_{\mathrm{HV}} & M_{\mathrm{VV}} \end{bmatrix} = \begin{bmatrix} S_{\mathrm{HH}} & S_{\mathrm{VH}} \\ S_{\mathrm{HV}} & S_{\mathrm{VV}} \end{bmatrix} + \begin{bmatrix} n_{\mathrm{HH}} & n_{\mathrm{VH}} \\ n_{\mathrm{HV}} & n_{\mathrm{VV}} \end{bmatrix} \tag{9.10}$$

where n_{tr} represents the additive noise in channel *tr* characterized by a complex zero-mean Gaussian distribution. Many recent oil spill studies are investigating the potential of slick discrimination and characterization in polarimetric SAR systems (see, e.g., [4,10]). A large range of multipolarization SAR features have been explored in the literature [see, e.g., Table III in 55]. Gaining less attention is the fact that the impact of instrument noise will vary depending on how the selected feature

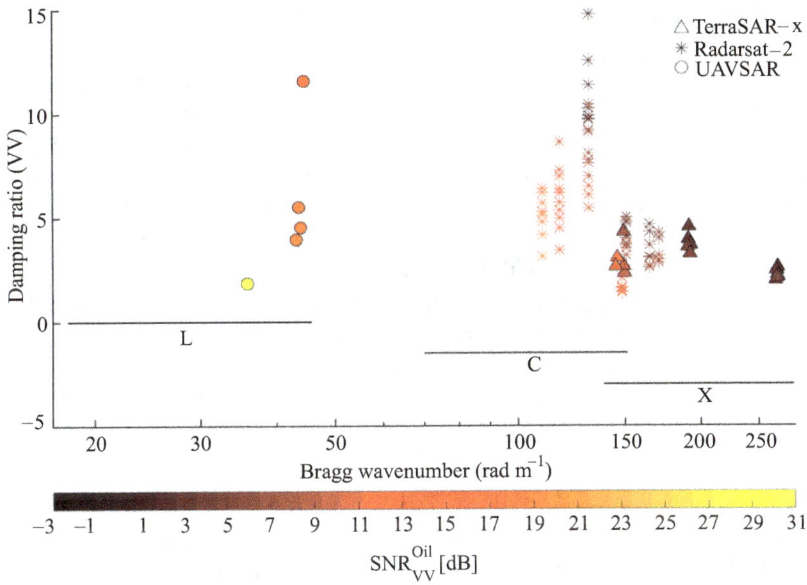

Figure 9.11 *VV polarization clean sea/slick damping ratio on linear scale (y-axis, linear scale) versus Bragg wavenumber (x-axis). Markers are color-coded to indicate the SNR within mineral oil slicks. Each data point represents the mean of 728 pixels, irrespective of image resolution, which varies across the three SAR systems. The data samples are collected from multiple locations within slicks in scenes with different acquisition dates, metocean conditions, incidence angles, and mineral oil properties. Image Courtesy Martine M. Espeseth*

is defined. Features involving more NRCS terms are more sensitive to noise contamination. To demonstrate, we denote the covariance matrix $C_{\mathcal{M}}$ for a linearly polarized system ($d = 3$ when assuming reciprocity), obtained from the outer product of the scattering vector of the measurement:

$$C_{\mathcal{M}} = \begin{bmatrix} \langle M_{\mathrm{HH}} \cdot M^*_{\mathrm{HH}} \rangle & \sqrt{2}\langle M_{\mathrm{HH}} \cdot M^*_{\mathrm{HV}} \rangle & \langle M_{\mathrm{HH}} \cdot M^*_{\mathrm{VV}} \rangle \\ \sqrt{2}\langle M_{\mathrm{HV}} \cdot M^*_{\mathrm{HH}} \rangle & 2\langle |M_{\mathrm{HV}}|^2 \rangle & \sqrt{2}\langle M_{\mathrm{HV}} \cdot M^*_{\mathrm{VV}} \rangle \\ \langle M_{\mathrm{VV}} \cdot M^*_{\mathrm{HH}} \rangle & \sqrt{2}\langle M_{\mathrm{VV}} \cdot M^*_{\mathrm{HV}} \rangle & \langle M_{\mathrm{VV}} \cdot M^*_{\mathrm{VV}} \rangle \end{bmatrix} \tag{9.11}$$

In a QP linearly polarized system, the H-polarization and V-polarization pulses are transmitted sequentially, not simultaneously. Because n_{HH} and n_{VV} are measured at different times, they are uncorrelated. A closer look at some of the matrix elements above shed light upon the influence of the noise term: For the copolarization cross-product $\langle M_{\mathrm{HH}} \cdot M^*_{\mathrm{VV}} \rangle = \langle S_{\mathrm{HH}} \cdot S^*_{\mathrm{VV}} \rangle$ because the additive noise components decorrelate, but $\langle M_{\mathrm{VV}} \cdot M^*_{\mathrm{VV}} \rangle = |S_{\mathrm{VV}}|^2 + \sigma^n_{\mathrm{VV}}$, where σ^n_{VV} is the noise power (NESZ) [53].

9.4 Dark slick detection and characterization techniques

Early quad-polarimetric SAR missions such as the spaceborne imaging radar-C/X-band SAR (SIR-C/X-SAR) system, flown on the space shuttle Endeavor in 1994, paved the way for a new era in studies of sea surface slicks (see, e.g., [8]). However, the research field of oil spill observation by polarimetric SAR has had more rapid growth since the launch of TerraSAR-X in 2007, which first provided coherent dual-polarization (VV and HH) SAR imagery to the scientific community, and the launch of Radarsat-2 with its quad-polarimetic imaging capability later that year. This section mainly covers techniques emerging within the last decade, and most of them are as yet in a preoperational stage. We discuss both slick detection, that is, a binary identification of oil spill versus no oil spill on the sea surface, and slick characterization, that is, quantitative or semiquantitative measurement of a property of the material within the slick.

9.4.1 Slick detection and segmentation

Screening of traditional single-polarization SAR imagery for oil spill detection has commonly been framed into three fundamental steps: dark patch segmentation, feature extraction, and oil spill versus look-alike classification. Traditional methods rely on features extracted from the segmented patches (see, e.g, [54]). The features can be divided into the following categories: geometry and shape of the segmented patch, physical characteristics of the backscatter level of the segmented patch, contextual features, and texture [35]. The features are input into a classifier to separate potential oil spills from other types of look-alike dark patches. Commonly, the segmentation stage involves histogram thresholding or unsupervised segmentation procedures applied to the intensity levels of either the VV or HH polarization SAR channels only [54,55]. More recently, application-generic segmentation approaches taking as input quad-polarimetric SAR data have emerged. Examples are methods involving modeling of the multilooked sample complex covariance matrix [56] or multipolarization SAR feature-based clustering [57]. Figure 9.12 shows a segmentation result of an oil spill derived from an unsupervised K-Wishart mixture model clustering approach incorporating Markov Random Field smoothing [56] to which the user inputs only the number of classes to be found. Before adopting this kind of method operationally, one should consider how sheen regions, which are often diffuse transitions from the oil spill to the ambient sea water, are handled by the various segmentation algorithms available. It is worth noting that any feature describing, for example, the geometry of the segmented patch or the contrast between the average numerical feature values from the interior of the patch and the average numerical feature values from the nearby surrounding clean sea, will be sensitive to the level of constraint placed on the segment perimeter.

9.4.2 Slick type discrimination

In recent years, a large number of studies have been devoted to multipolarization SAR feature selection for sea surface slick detection or characterization. Table 9.6

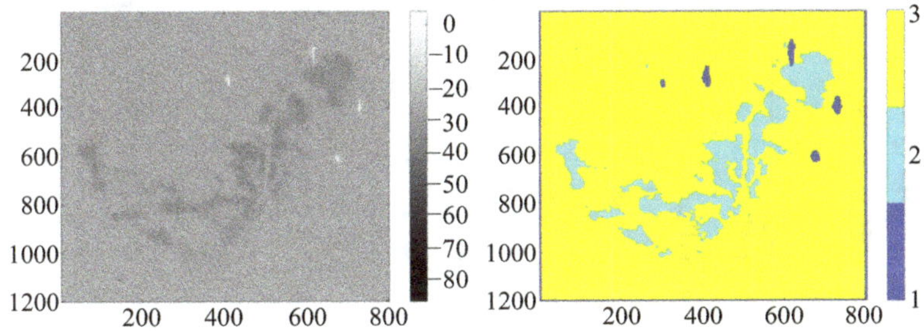

Figure 9.12 (Left) Emulsion oil slick shown in the left panel of Figure 9.2. (Right) Segmentation example from a three class K-Wishart clustering algorithm. Based on auxiliary information, classes are (1) vessels, (2) oil spill, and (3) sea water. Segmentation algorithm is courtesy of Anthony P. Doulgeris. Figure is produced by the authors

gives an overview of a subset of the descriptors discussed in the literature. The features are often computed based on HH and VV polarization channels only, with the cross-polarization channels discarded due to additive noise contamination in low SNR measurements. We follow this approach to emphasize features that are applicable to both airborne (in general high SNR) and spaceborne (in general lower SNR) sensors. A categorization based on each feature's relation to the two-scale Bragg model equations (see Section 9.3.2) has been proposed [58] and is adopted in Table 9.6. Category I contains features that depend on large- and small-scale roughness, incidence angle, and the relative dielectric constant, whereas the features in category II only depend on large-scale roughness, incidence angle, and the relative dielectric constant. These category II features are parameters for which the wave spectrum cancels in a ratio. Note that this categorization is based on the two-scale Bragg model, so the classification of features is not valid outside its validity range.

A feature selection analysis was done in [4], involving features from both categories I and II. Out of eight multipolarization SAR features that were investigated and extracted from Radarsat-2 quad-polarimetric measurements, the study concludes that the real part of the copolarization cross product (F3) and the geometric intensity (F6) are the most powerful features for biogenic versus mineral oil spill discrimination. Both F3 and F6 belong to category I. In theory and from discussions in Section 9.3, category I features should be more useful than category II features for capturing the difference in damping of surface roughness between biogenic and mineral oil slicks. Figure 9.13 shows the result from a clustering algorithm taking as input this particular feature vector $\{F3, F6\}$, and a clear separation between the upper left slick (plant oil) and the two lower slicks (emulsion and crude oil) is observed. For the mineral oil slicks, the boundary areas are assigned to the same class (green) as the monomolecular plant oil slick and are interpreted as thin sheen layers. In addition to this study, the potential of SAR for discriminating oil spills from low wind fields is reported in [43].

Table 9.6 Selected multipolarization SAR features discussed in the oil spill literature. The categorization follows [58]. det$(\cdot)$ *denotes the determinant,* $\boldsymbol{T}_2$ *denote the coherency matrix for the dual-polarization (VV and HH) case (see [4, (6)),* $p_i = \lambda_i/(\lambda_1 + \lambda_2)$ and $\lambda_i(\lambda_1 > \lambda_2)$ *are the eigenvalues of* $\boldsymbol{T}_2$ *, and* $\Re$ *denotes the real part. The superscript ' is used to distinguish the copolarization version of a parameter from the conventional version. The list of references is not exhaustive*

Cate-gory	#	Feature name	Definition	Reference
I	F1	Damping ratio	$\mathrm{DP} = \frac{\langle \lvert S_{VV}\rvert^2\rangle_{sea}}{\langle \lvert S_{VV}\rvert^2\rangle_{oil}}$	[10,58,59]
	F2	Pol. difference	$\mathrm{PD} = \langle \lvert S_{VV}\rvert^2\rangle - \langle \lvert S_{HH}\rvert^2\rangle$	[43,58,60]
	F3	Real part of co-pol. cross-product	$r_{co} = \lvert \Re(\langle S_{HH} \cdot S_{VV}^*\rangle)\rvert$	[4,58,60,61]
	F4	Span	$\mathrm{span}' = \langle \lvert S_{HH}\rvert^2\rangle + \langle \lvert S_{VV}\rvert^2\rangle$	[58,60,62,63]
	F5	Largest eigenvalue	λ_1'	[10,58,60]
	F6	Geometric intensity	$\mu' = (\det(\boldsymbol{T}_2))^{1/2}$	[4,60,61]
II	F7	Co-pol. ratio	$\mathrm{PR} = \frac{\langle \lvert S_{HH}\rvert^2\rangle}{\langle \lvert S_{VV}\rvert^2\rangle}$	[4,10,43,58,60][a]
	F8	Entropy	$H' = -\sum_{i=1}^{2} p_i \log_2 p_i$	[4,10,58,60–62,64,65]
	F9	Anisotropy	$A' = \frac{\lambda_1' - \lambda_2'}{\lambda_1' + \lambda_2'}$	[4,10,58,60–65]
	F10	Std. dev. of co-pol. phase difference	$\sigma_{\phi CO} = \sqrt{\langle(\phi_{HH} - \phi_{VV})^2\rangle - (\langle\phi_{HH} - \phi_{VV}\rangle)^2}$	[4,10,58,60–62,65]

[a]In the literature, PR is either defined as VV/HH or HH/VV.

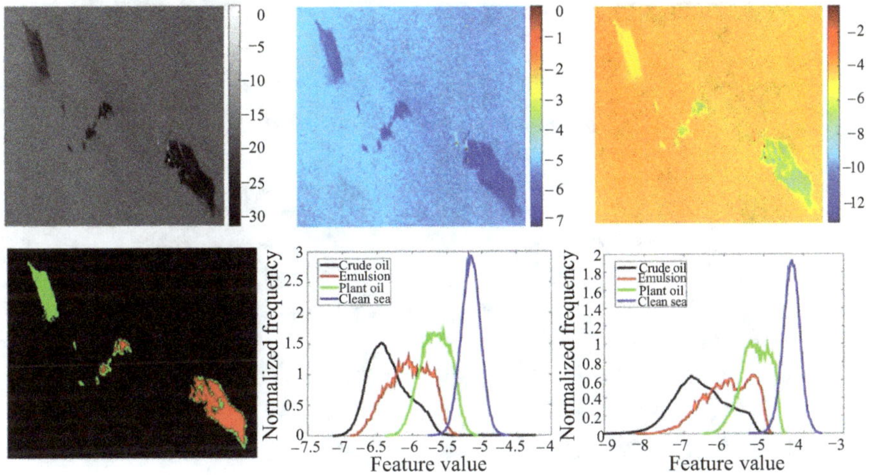

Figure 9.13 Top row, left to right: Intensity of SAR VV polarization channel (dB) containing oil slicks (plant oil (upper left), emulsion (middle) and crude oil (lower right)) from an oil-on-water exercise in 2011; F6; and F3. Bottom row, left to right: Classification of slick areas after ships and sea pixels are masked out; histogram of selected regions from the F6; and histogram of selected regions from F3. RADARSAT-2 Data and Products. Copyright MDA LTD (2011)—All rights reserved. Image Courtesy Stine Skrunes. Modified from [4]

More recent studies using high SNR airborne L-band SAR instruments have identified the most sensitive features for oil slick detection from QP data [58,66]. In both studies, the VV-intensity damping ratio (F1) and the polarization difference (F2) outperformed other investigated features, when considering those only using the copolarized channel data. When including cross-polarization data also, both found the HV-intensity damping ratio to be a good discriminator, and [58], which considered more features, included the QP geometric intensity (F6) in the well-performing set. HP features were synthesized from QP UAVSAR data by [58], and the copolarized and cross-polarized data's damping ratios were the highest performance features. It is interesting to note that most of the best features do not require polarimetric data, but use only the intensities. Oil thickness and concentration are key information needed to effectively target countermeasures during a major oil spill incident. However, thickness has proven to be challenging to obtain by any remote sensor [67] and there are few operational methods available [1]. In infra-red (IR) remote sensing, an oil slick that is solar heated emits IR radiation as oil shows greater emissivity than water; thick oil appears "hot," intermediate thicknesses "cool," and thinner layers like sheen is not detectable [1]. In theory, the contrast drivers of oil spills in SAR measurements discussed in Section 9.3.1 should be affected by thicker regions of oil. In Figure 9.14, airborne

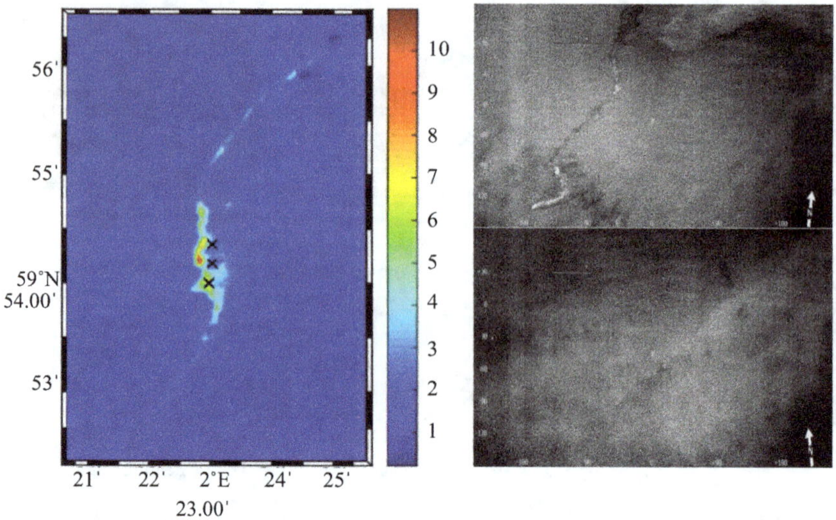

Figure 9.14 (Left) Damping ratio of a RISAT-1 FRS-2 scene (VV polarization channel) containing an oil slick in the North Sea. Positions of thicker oil estimated from aircraft data are indicated by "x." (Right) Aircraft data (infra-red images) provided by the Air Patrol Squadron Finland/SYKE. Image Courtesy Stine Skrunes. Adapted from [68]

IR imagery from a surveillance aircraft is used to spatially identify the thicker regions within the oil slick imaged by space-borne SAR. The positions of the thick oil are estimated from the position of the aircraft and camera angles and plotted as black markers (left panel). The damping ratio of the C-band SAR VV polarization channel exhibits the highest contrast between oil and water at locations where there is spatial overlap with thick oil observed by the IR camera, suggesting there is a potential in detecting thickness variations with SAR [68].

At this point, relative, not quantitative, thickness determination has been accomplished [9,12,13]. Using data from a high SNR instrument, simple thresholds applied to the VV-damping ratio can be used to distinguish relative thickness. Figure 9.15 shows a slick separated into six relative thickness classes, with the evolution of transport and weathering shown in three images acquired at 1-h intervals. Accumulation of oil along fronts is indicated for oil in the thickest two classes. In situ measurements taken in the area and overlapping with the first two acquisitions showed the area to be covered with sheen and patches of true color oil (thickness from a few μm up to ~250 μm). Thickness classification using a neural network and based upon the entropy has also been done [12], but users of this parameter should be cognizant that the entropy is more sensitive to noise than the damping ratio [58].

Research efforts are currently directed toward developing methods for identifying zones of high oil concentration and estimating the volumetric fraction of oil

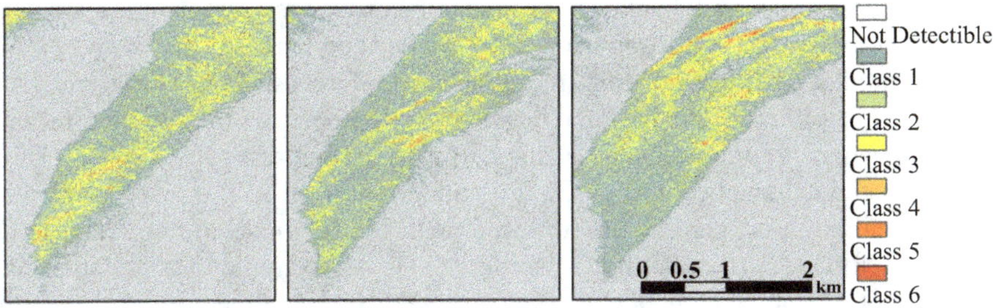

Figure 9.15 Relative thickness classification based upon thresholding the VV-intensity damping ratio of a slick from a seep in the MC20 block of the Gulf of Mexico. The series of images were separated in time by 1 h each, with the earliest image on the left. In situ observations and measurements indicated that this area was mainly covered with sheen from silver to rainbow, with patches of discontinuous true color oil

from quad-polarimetric SAR imagery [10,11]. A method based upon the two-scale model and using the copolarization ratio (see Section 9.3.2.1) was pioneered in 2012 using data from the Deepwater Horizon [10]. The analysis indicated that the material in the slick had between 0.75 and 0.9 oil volumetric fraction. One issue with the method is that the model upon which it is based assumes that there is no contrast between the copolarization ratio for sheen and clean water, which is not always the case (Figure 9.10). A recent refinement to the method using the Universal Weighted Curvature Approximation scattering model [69] obtained 0.49–0.52 oil volumetric fraction for a slick composed of an emulsion of 40:60 oil-to-water ratio upon release in the North Sea [10]. While promising, the method needs validation and refinement to better agree with data for sheen slicks and to estimate the uncertainty in the derived oil volumetric fraction.

9.4.3 Slick transport and evolution

There are demands on both short and long timescales for information about the position of a slick from an oil spill (Table 9.1). In the short-term, the location and extent are most-immediately requested, while on longer timescales an estimate of the probable trajectory is wanted. With respect to nowcasts derived from SAR, a case study is reported in [70] in which a combination of a SAR image pair with 72 h temporal baseline and auxiliary information on sea state and currents were used to determine possible oil spill drift. In general, with this temporal resolution, the spatial shape of the oil spill can be expected to have changed beyond recognition. As pointed out in Section 9.1, drift estimations derived from spatial feature correlation require two or more consecutive SAR images close in time (minutes to few hours). At this timescale, the derived velocity vector can be used as a nowcast product.

In remote areas and harsh environmental conditions, for example, at high latitudes in the Arctic, the response will be slower compared to that at more temperate latitudes, and longer and more accurate forecasts are requested [71]. Numerous oil spill models have been developed over the years to provide forecasts, and it is beyond the scope of this chapter to give a comprehensive review. Three types of methods have been proposed for simulation of oil spill transport in water, all partitioning the oil volume into discrete pieces to be modeled numerically: (1) particle-tracking, (2) tracers, and (3) spillets [71][‡] An example of a pollutant particle ("splot") tracking model is the General NOAA Operational Modeling Environment (GNOME), which is initialized and compared to SAR imagery in [72,73]. In brief, to better understand marine oil spill transport, numerical models are used to simulate both vertical and horizontal transport mechanisms. Recent efforts to integrate numerical models with time-series of SAR measurements have led to several advancements within the field. First, a deeper understanding of the vertical exchange between the submerged oil droplets and the surface slick and how it impacts the slick signature in the SAR imagery has been gained. Second, improved models and prediction capabilities of the oil spill's horizontal transport have been developed [74,75]. Jones *et al.* [74] found that the agreement between theory and measurement (Figure 9.16) is improved significantly by local current measurements and that modeled wind fields are sufficient. The measurements shown were made at wind speeds from 9 to 12 m/s and used to calibrate the OpenOil [https://github.com/knutfrode/opendrift] model's droplet radii and entrainment length.

9.5 Concluding remarks and outlook

In this chapter, we have briefly explained the physical mechanisms behind radar imaging of oil slicks at sea, discussed how scattering models can aid interpretation of SAR imagery, and visited emerging information retrieval techniques for providing some of the information items requested by the community. We discussed the available instruments and the limitations on their use for oil spill observations. C-band SAR has traditionally been used for ocean monitoring from space, for example, ERS-1/-2, ENVISAT, Radarsat-1/-2, and Sentinel-1, but there is increasing utilization of L-band and X-band missions, for example, ALOS-PALSAR, ALOS-2, and TerraSAR-X, soon to be augmented by RCM, SAOCOM-1/2, and NISAR. The most important space-borne missions for operational environmental monitoring of the oceans are currently Radarsat-2, Sentinel-1, and the upcoming RCM, based on coverage, access to the data, operational modes, and noise characteristics.

In Section 9.1, we gave an overview of key information items involving SAR remote sensing. Those items can be split between those requiring *detection* of the spill and those requiring *characterization* of the released material. Information

[‡]See "Notes on further reading" at the end of the chapter for more information about this topic.

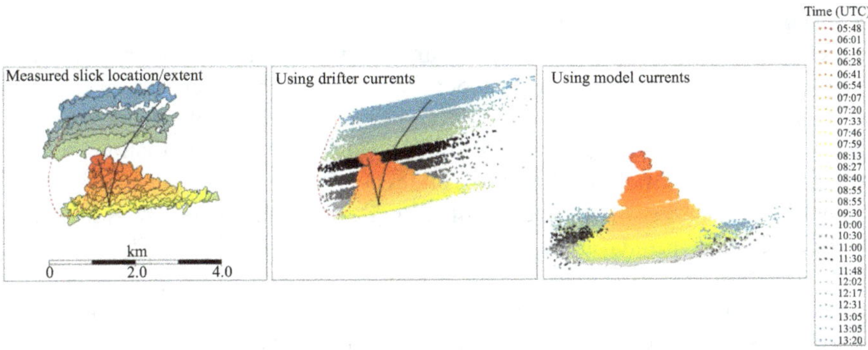

Figure 9.16 Comparison of the OpenOil drift model [https://github.com/knutfrode/opendrift] to measured slick transport using a rapid repeat time series of SAR images. An 0.5 m^3 release of 80:20 oil:water emulsion was measured (left) and modeled (middle, right). Results from modeling based on currents from coreleased drifters (middle) are more accurate than that using currents obtained from an ocean model (right). Modified from [74]

based on slick detection, such as position, extent, and drift forecasting, can be derived utilizing data from any of space-borne SAR missions and do not in general require a particular polarization mode, although the VV mode is preferable because of the higher backscatter levels. Efforts are now being made toward integration of remote sensing information with numerical oil transport models for hind-, now-, and fore-casting, and rapid repeat SAR time series images from airborne SAR have contributed to constraining model parameters [74,75]. The aim is to better predict the transport of the oil to serve coastal monitoring agencies. This work is found valuable and important to continue.

Oil characterization with SAR is not as far advanced insofar as operationalizing the methods, and much of the recent advancement has been made with airborne SAR specifically because the signal's proximity to the noise floor is an issue for space-borne SAR sensors. For oil slick characterization purposes in particular, attention has to be paid to this problem. Three important usages of SAR-based characterization were discussed: look-alike discrimination, thickness measurement, and identification/measurement of emulsified oil. An important advancement in the utilization of SAR for oil spill response will be in characterizing oil within the slick into categories that match recoverable or nonrecoverable oil. At this point, relative thickness, potentially calibrated with in situ or ancillary data, has been demonstrated but not operationalized. Interestingly, contrast features (damping ratio) and the polarization difference are the features most sensitive to variations within slicks, and these require only intensity measurements. Quantifying emulsification via measurement of the oil volumetric fraction (oil:water ratio) requires multi-polarization data to extract information about the dielectric properties of the scattering medium. If we restrict the choice of transmitting and receiving electromagnetic wave polarization to only linear polarization, that is, combinations of horizontal and vertical, information about the electrical characteristics of the slicks can only be obtained if both VV and HH products are available for the scene. A quad-polarimetric SAR system meets this requirement by alternating between transmitting H and V polarized pulses while receiving both. However, this requires twice the pulse-repetition frequency of a single- or dual-polarized system [76], which could cause interference between the received echoes. The upcoming NISAR mission can avoid this problem using for ocean imaging a quasi-quad-polarimetric split-spectrum mode in which V- and H-polarization pulses are transmitted simultaneously at slightly different frequencies. The frequency diversity applied in the quasi-quad-polarimetric mode of NISAR and the multifrequency opportunities from the range of space-borne platforms suggest more spectral investigative and frequency comparison studies in the future.

Progress has clearly been made in the field of oil spill characterization over the last decade or so, but there is still a need for more validation campaigns collecting in situ and auxiliary data, for example, optical remote sensing measurements that could be used to assess the validity range and sensitivity of the algorithms and methods developed. This plus more advanced models with validation across a range of metocean conditions are key advancements needed for operationalization of SAR-based oil spill characterization. Airborne SAR, with its high SNR and rapid repeat capability, could play an important role in response to major spills in the future. However, space-borne SAR will continue to be the main instrument for operational surveillance, particularly as more SAR missions are launched.

Notes on further reading

The two-scale model: See [44] for notes on a small refinement of the model.

Monomolecular slicks: Natural slicks of biogenic origin are common in coastal waters and open ocean, produced by, for example, algae and bacteria. Unlike mineral oil slicks, biogenic surface films consist of surface-active organic compounds that have both a hydrophobic and a hydrophilic component. This combination causes the molecules to spontaneously arrange in a monomolecular layer at the air/water interface, with the hydrophobic part at the air interface and the hydrophilic part at the water interface. This forms a film one molecule thick ($\sim$ 2.4 to 2.7 nm) [77,78]. Crude oil spills consist mainly of alkanes, cycloalkanes, and aromatic compounds, chemicals that are exclusively hydrophobic. Without the combination of hydrophilic and hydrophobic parts, crude oil slicks do not form monomolecular films. Over time crude oil spills spread out into layers of different thickness depending on the amount and viscosity of the oil and on the environmental conditions, with the final thickness on the scale of micrometers to millimeters, and even centimeters for freshly spilled oil [77,78], much thicker than the monomolecular biogenic films.

Numerical drift modeling: In the *particle-tracking* approach, the distribution of particles represents the whole oil spill in a statistical fashion. The oil is parameterized as a finite number of particles that are assigned an initial location and mass and tracked numerically by a trajectory model. For the *tracers*, the tracking area of the oil spill is represented by a fine-resolution grid and the spill occupies the cells that represent its physical extent. The *spillet* method is close to the particle method but adds more degrees of freedom, such as the area or thickness represented by each spillet. This method can be seen as a compromise between the particle and tracer methods. More information about the advantages and drawbacks of the different methods can be found in [71].

Acknowledgments

This work is partly financed by CIRFA (RCN, Grant no. 237906, and partners). The work was carried out in part at the Jet Propulsion Laboratory, California Institute of Technology, under contract with the National Aeronautics and Space Administration. The authors thank Martine Espeseth and Dr. Stine Skrunes for producing some of the figures for this chapter.

References

[1] Fingas, M., and Brown, C. E. "A review of oil spill remote sensing." *Sensors.* 2018;**18**(1):1–18

[2] Carpenter, A. "European maritime safety agency *CleanSeaNet* activities in the North Sea" In A. Carpenter (ed.), *Oil pollution in the North Sea.* Switzerland: Springer; 2016. pp. 33–47

[3] NOAA NESDIS – U.S. National Oceanic and Atmospheric Administration Satellite and Information Service. NESDIS *Marine Pollution Products* [online]. Available from https://www.ospo.noaa.gov/Products/ocean/marinepollution [Accessed 15 October 2018]

[4] Skrunes, S., Brekke, C., and Eltoft, T. "Characterization of marine surface slicks by Radarsat-2 multipolarization features." *IEEE Transactions of Geoscience and Remote Sensing.* 2014; **52**(9):5302–5319

[5] NOAA – National Oceanic and Atmospheric Administration U. S. Department of Commerce. *Characteristics of Response Strategies: A Guide for Spill Response Planning in Marine Environments* [online]. Available from https://response.restoration.noaa.gov/sites/default/files/Characteristics_Response_Strategies.pdf. [Accessed 29 August 2018].

[6] Bonn Agreement. Bonn Agreement Oil Appearance Code, Bonn Agreement Aerial Operations Handbook, Part 3, Annex A. The Bonn Agreement Oil Appearance Code [Online]. Available from https://www.bonnagreement.

org/site/assets/files/1081/special_on_volume_calculation_20160607.docx. [Accessed 13 August 2018].

[7] NOAA Office of Response and Restoration. *Open water oil identification job aid for aerial observation*, Version 3, August 2016 [Online]. Available from https://response.restoration.noaa.gov/sites/default/files/OWJA_2016.pdf [Accessed 29 October 2018]

[8] Gade, M., and Alpers, W. "Imaging of biogenic and anthropogenic ocean surface films by the multifrequency/multipolarization SIR-C/X-SAR." *Journal of Geophysical Research*. 1998;**103**(C9):18851–18866

[9] Jones, C. E., and Holt. B. "Experimental L-band airborne SAR for oil spill response at sea and in coastal waters." *Sensors*. 2018;**18**(2):13 pp.

[10] Minchew, B., Jones, C. E., and Holt, B. "Polarimetric analysis of backscatter from the deepwater horizon oil spill using L-Band synthetic aperture radar." *IEEE Transactions on Geoscience and Remote Sensing*. 2012;**50**(10): 3812–3830

[11] Angelliaume, S., Boisot, O., and Guérin, C.-A. "Dual-polarized L-band SAR imagery for temporal monitoring of marine oil slick concentration." *Remote Sensing*. 2018;**10**(7):17 pp.

[12] Garcia-Pineda, O., MacDonald, I., Hu, C., *et al.* "Detection of floating oil anomalies from the Deepwater Horizon oil spill with synthetic aperture radar." *Oceanography*. 2013;**26**:124–137

[13] Minchew, B. "Determining the mixing of oil and sea water using polarimetric synthetic aperture radar." *Geophysical Research Letters*. 2012;**39**:6 pp.

[14] Yu, F., Li, J., Cui, S., Zhao, Y., Feng, Q., and Chen G. "A hindcast method to simulate oil spill trajectories for Bohai Sea, Norteast China." *Ocean Engineering*. 2016;**124**:363–370

[15] Ly J. M., Alsos G. A., Brekke C., *et al. Norsk oljevernberedskap – rustet for framtiden?* Technical report, Norwegian Ministry of Transportation and Communication, 2015.

[16] Raney, R.K. "Hybrid-polarity SAR architecture." *IEEE Transactions on Geoscience Remote Sensing*. 2007;**45**(11):3397–3404

[17] MacDonald M. D. A. and Dettwiler and Associates Ltd (Canada). *Radarsat-2 product description*. RN-SP-52-1238. Issue 1/13: March 21, 2016.

[18] Espeseth, M., Brekke, C., and Johansson, M. A. "Assessment of RISAT-1 and Radarsat-2 for sea ice observations from a hybrid-polarity perspective." *Remote Sensing*. 2017;**9**(11):1–20

[19] Fritz, T., and Eineder, M. (ed.). *TerraSAR-X ground segment, basic product specification document*. DLR, TX-GS-DD-3302, Issue 1.5, 2008.

[20] Italian Space Agency. *COSMO-SkyMed Mission and Products Description*. ASI, ASI-CSM-PMG-NT-001, Issue 2, 2016

[21] Milillo, P., Fielding, E. J., Schulz, W. H., Delbridge, B. and Bürgmann, R. "COSMO-SkyMed spotlight interferometry over rural areas: The Slumgullion Landslide in Colorado, USA." *IEEE Journal of Selected Topics in Applied Earth Observations and Remote Sensing*. 2014;**7**(7):2919–2926

[22] Misra T., Rana, S.S., Desai, N. M., *et al.* "Synthetic aperture radar payload on-board RISAT-1: Configuration, technology and performance." *Current Science.* 2013;**104**(4):446–461

[23] Dabboor, M., Iris, S., and Singhroy, V. "The RADARSAT constellation mission in support of environmental applications." *Proceedings*. 2018;**2**(7): 5 pp.

[24] Ren, L., Yang, J., Mouche, A., *et al.* "Preliminary analysis of Chinese GF-3 SAR quad-polarization measurements to extract wind in each polarization." *Remote Sensing.* 2017;**9**(1215):1–17

[25] JAXA – Japan Aerospace Exploration Agency. *ALOS-2/PALSAR-2 Calibration and Validation Results Ver. 2018.08.07* [online]. Available from https://www.eorc.jaxa.jp/ALOS-2/en/calval/PALSAR2_CalVal_Results_JAXA_201808.pdf [Accessed 8 November 2018]

[26] Euillades, P. A., Euillades, L. D., Azcueta, M., and Sosa, G. "SAOCOM 1A interferometric error model and analysis." *Proceedings of the Fringe 2015 Workshop*; Frascati, Italy, March 2015. Noordwijk: ESA Communications; 2015. 5 pp.

[27] Holt, B. "SAR imaging of the ocean surface." In C. Jackson and J.Apel (eds.). *Synthetic aperture radar marine user's manual.* Washington, DC: U.S. Department of Commerce, National Oceanic and Atmospheric Administration; 2004. pp. 25–80

[28] Ren, Y., Li, X.-M., and Zhou, G., "Sea surface wind retrievals from SIR-C/X-SAR data: A revisit." *Remote Sensing*. 2015;**7**:3548–3564

[29] Donelan, M. A., and Pierson, W. J. "Radar scattering and equilibrium ranges in wind-generated waves with application to scatterometry." *Journal of Geophysical Research*. 1987;**92**(C5):4971–5029

[30] Gade, M., Alpers, W., Hühnerfuss, H., Wismann, V. R., and Lange, P. A. "On the reduction of the radar backscatter by oceanic surface films: Scatterometer measurements and their theoretical interpretation." *Remote Sensing of Environment.* 1998;**66(**1):52–70

[31] Transportation Research Board and National Research Council. *Oil in the sea III: Inputs, fates, and effects.* Washington, DC: The National Academies Press; 2003

[32] Reed M., Johansen Ø., Brandvik P. J., *et al.* "Oil spill modeling towards the close of the 20th century: Overview of the state of the art." *Spill Science & Technology Bulletin*. 1999;**5**(1):3–16

[33] Hollinger, J. P, and Mennella, R. A. "Oil spills: Measurements of their distributions and volumes by multifrequency microwave radiometry." *Science*, New Series. 1973;**181**(4094):54–56

[34] Visser, A.W. "Using random walk models to simulate the vertical distribution of particles in a turbulent water column." *Marine Ecology Progress Series*. 1997;**158**:275–281

[35] Brekke, C., and Solberg, A. H. S. "Oil spill detection by satellite remote sensing." *Remote Sensing of Environment*. 2005;**95**(1):1–13

[36] Ulaby, F., Moore, R. K., and Fung, A. K. Microwave remote sensing, active and passive. Volume II: Radar remote sensing and emission theory. Norwood, MA: Artech House Inc.; 1986.

[37] Grahn, J., *Multi-frequency radar remote sensing of sea ice*. PhD thesis, UiT The Arctic Univeristy of Norway, 2018.

[38] Hajnsek, I., Pottier, E., and Cloude, S. R. "Inversion of surface parameters from polarimetric SAR." *IEEE Transactions on Geoscience and Remote Sensing*. 2003;**41**(4):727–744

[39] Elachi, C., and Brown, W. E. "Models of radar imaging of the ocean surface waves." *IEEE Transactions on Antennas and Propagation*. 1977; **AP-25** (1):84–95

[40] Hasselmann, K., and Hasselmann, S. "On the nonlinear mapping of an ocean wave spectrum into a synthetic aperture radar image spectrum and its inversion." *JGR-Oceans*, 1991;**96**(C6):10713–10729

[41] Jones, C. E., Minchew, B., Holt, B., and Hensley, S. "Studies of the deep-water horizon oil spill with the UAVSAR radar." In Y. Liu *et al.* (eds). *Monitoring and modeling the deepwater horizon oil spill: A record-breaking enterprise*. Washington, D.C.: American Geophysical Union; 2011. pp. 33–50.

[42] Valenzuela, G. R. "Theories for the interaction of electromagnetic and oceanic waves – A review." *Boundary-Layer Meteorology*. 1978;**13** (1–4):61–85

[43] Kudryavtsev, V. N., Chapron, B., Myasoedev, A. G., Collard, F., and Johannessen, J. "On dual co-polarized SAR measurements of the ocean surface." *IEEE Geoscience and Remote Sensing Letters*. 2013;**10**(4): 761–765.

[44] Salberg, A.-B., Rudjord, Ø., and Solberg, A. H. S. "Oil spill detection in hybrid-polarimetric SAR images." *IEEE Transactions on Geoscience and Remote Sensing*. 2014;**52**(10):6521–6533.

[45] Skrunes, S. Characterization of low backscatter regions in the marine environment by multipolarization C- and X-band synthetic aperture radar data. PhD thesis, UiT The Arctic University of Norway, 2014.

[46] Iodice, A., Natale, A., and Riccio, D. "Retrieval of soil surface parameters via a polarimetric two-scale model." *IEEE Transactions on Geoscience and Remote Sensing*. 2011;**49**(7):2531–2547

[47] Rudjord, Ø., and Salberg, A.-B. "X-Bragg based detection of oil spills using polarimetric SAR." *Proceedings of SeaSAR2012*; Tromsø, Norway, June 2012. Noordwijk, The Netherlands: ESA Communications; 2013. 4 pp.

[48] Yin, J., Yang, J., Zhou, Z.-S., and Song, J. "The extended Bragg scattering model-based method for ship and oil-spill observation using compact polarimetric SAR." *IEEE Journal of Selected Topics in Applied Earth Observation and Remote Sensing*. 2015;**8**(8):3760–3772

[49] Skrunes, S., Brekke, C., and Doulgeris, A. P. "Characterization of low-backscatter ocean features in dual-copolarization SAR using log-cumulants." *IEEE Geoscience and Remote Sensing Letters*. 2015;**12**(4):836–840

[50] Skrunes, S., C. Brekke, C. Jones, and B. Holt. "A multisensor comparison of experimental oil spills in polarimetric SAR for high wind conditions." *Journal of Selected Topics in Applied Earth Observation and Remote Sensing*. 2016;**9**(1):4948–4961.

[51] Younis, M., Huber, S., Patyuchenko, A., Bordoni, F., and Krieger, G. "Performance comparison of reflector – and planar-antenna based digital beam-forming SAR." *International Journal of Antennas and Propagation*. 2009; **2009**: 13 pp.

[52] Skrunes, S., Brekke, C., Jones, C. E., and Holt, B. "A multisensor comparison of experimental oil spills in polarimetric SAR for high wind conditions." *IEEE Journal of Selected Topics in Applied Earth Observations and Remote Sensing*. 2016; **9**(11):4948–4961

[53] Freeman, A. "The effects of noise on polarimetric SAR data." *Proceedings of IGARSS'93 - IEEE International Geoscience and Remote Sensing Symposium*. Tokyo, Japan, August 1993. pp.799–802.

[54] Solberg, A. S., Storvik, G., Solberg, R., and Volden E. "Automatic detection of oil spills in ERS SAR images." *IEEE Transactions on Geoscience and Remote Sensing*. 1999;**37**(4):1916–1924

[55] Derrode, S., and Mercier, G. "Unsupervised multiscale oil slick segmentation from SAR images using a vector HMC mode." *Pattern Recognition*. 2007;**40**:1135–1147

[56] Doulgeris, A. P., Anfinsen, S. N., and Eltoft, T. "Classification with a non-gaussian model for PolSAR data." *IEEE Transactions on Geoscience and Remote Sensing*. 2008;**46**(10):2999–3009.

[57] Doulgeris, A. P. "A simple and extendable segmentation method for mulit-polarization SAR images." *Proceedings of POLinSAR*; Frascati, Rome, Italy, 2013. 8 pp.

[58] Espeseth, M. M., Skrunes, S., Jones, C. E., Brekke, C., Holt, B., and Doulgeris, A. P. "Analysis of evolving oil spills in full-polarimetric and hybrid-polarity SAR." *IEEE Transactions on Geoscience and Remote Sensing*. 2017;**55**(7):4190–4210

[59] Jones, C. E., Espeseth, M. M., Holt, B., Brekke, C., and Skrunes, S. "Characterization and discrimination of evolving mineral and plant oil slicks based on L-band synthetic aperture radar (SAR)." *Proceedings of SPIE*. Edinburg, UK; 2016. 13 pp.

[60] Skrunes, S., Brekke, C., Jones, E. C., Espeseth, M. M., and Holt, B. "Effect on wind direction and incidence angle on polarimetric SAR observations of slicked and unslicked sea surfaces." *Remote Sensing of Environment*. 2018;**213**:73–91

[61] Migliaccio, M., Nunziata, F., and Buono, A. "SAR polarimetry for sea oil slick observation." *International Journal of Remote Sensing*. 2015;**36**(12):3243–3273

[62] Wenguang, W., Fei, L., Peng, W., and Jun, W. "Oil spill detection from polarimetric SAR image." *Proceedings of IEEE 10th International Conference on Signal Processing*; Beijing, China, October 2010. pp. 832–835.

[63] Fortuny-Gausch, J. "Improved oil slick detection and classification with polarimetric SAR." *Proceedings of POLinSAR Workshop on Applications of SAR Polarimetry and Polarimetric Interferometry*; Frascati, Italy, January 2003. 5 pp.

[64] Schuler, D. L., and Lee, J. -S. "Mapping ocean surface features using biogenic slick-fields and SAR polarimetric decomposition techniques." *IEE Proceedings - Radar, Sonar and Navigation*. 2006;**153**(3):260–270

[65] Solberg, A. "Remote sensing of ocean oil-spill pollution." *Proceedings of the IEEE*. 2012;**100**(10):2931–2945

[66] Angelliaume, S., Dubois-Fernandez, P., Jones, C. E., *et al.* "SAR imagery for detecting sea surface slicks: Performance assessment of polarization-dependent parameters." *IEEE Transactions on Geoscience and Remote Sensing*. 2018;**56**(8):4237–4257

[67] Leifer, I., Lehr, W. J., Simecek-Beatty, D., *et al.* "State of the art satellite and airborne marine oil spill remote sensing: Application to the BP Deepwater Horizon oil spill." *Remote Sensing of Environment*. 2012;**124**: 185–209.

[68] Skrunes, S., Brekke, C., Espeseth, M. M. "Assessment of the RISAT-1 FRS-2 mode for oil spill observation." Proceedings of *IGARSS 2017 - IEEE International Geoscience and Remote Sensing Symposium*. Forth Worth Texas, USA, July 2017. pp. 1024–1027.

[69] Guerin, C.-A., Soriano, G., and Chapron, B. "The weighted curvature approximation in scattering from sea surfaces." *Waves Random Complex Media*. 2010;**20**:364–384.

[70] Fan, K., Zhang, Y., and Lin, H. "Satellite SAR analysis and interpretation of oil spill in the offshore water of Hong Kong." *Annals of GIS*. 2010;**16**(4):269–275. https://www.tandfonline.com/loi/tagi20

[71] Drozodowski, A., Nudds, S., Hannah, C. G., Niu, H. Peterson, I. K., and Perrie, W. A. *Review of oil spill trajectory modelling in the presence of ice*. Canadian Technical Report of Hydrography and Ocean Sciences 274, Fisheries and Oceans Canada, 2011.

[72] Xu, Q., Li, X., Wei, Y., Tang, Z., Cheng, Y., and Pichel, W. G. "Satellite observations and modeling of oil spill trajectories in the Bohai Sea." *Marine Pollution Bulletin*. 2013;**71**:107–116

[73] Cheng, Y., Liu, B., Li, X., *et al.* "Monitoring of oil spill trajectories With COSMO-SkyMed X-Band SAR images and model simulation." *IEEE Journal of Selected Topics in Applied Earth Observation and Remote Sensing*. 2014;**7**(7):2895–2901

[74] Jones C. E., Dagestad K.-F., Breivik Ø., *et al.* "Measurement and modelling of oil slick transport." *Journal of Geophysical Research: Oceans*. 2016; **121**(10):7759–7775

[75] Röhrs J., Dagestad K.-F., Asbjørnsen H., *et al.* "The effect of vertical mixing on the horizontal drift of oil spills." *Ocean Science*. 2018; **14**:1581–1601

[76] Raney, R. K. "A perspective on compact polarimetry." *IEEE Geoscience and Remote Sensing Newsletter*. 2011;**160**:12–18

[77] Alpers, W., and Espedal, H. A. "Oils and surfactants." In C. Jackson and J. Apel (eds.). *Synthetic aperture radar marine user's manual.* U.S. Department of Commerce, Washington, DC: National Oceanic and Atmospheric Administration; 2004. pp. 263–275

[78] Hühnerfuss, H. "Basic physicochemical principles of monomolecular sea slicks and crude oil spills." In M. Gade, H. Hühnerfuss, and G. M. Korenowski (eds.). *Marine surface films*. Berlin, Germany: Springer-Verlag; 2006. pp. 21–35.

Chapter 10

Joint use of SAR and collaborative signals

Raffaella Guida[1], Pasquale Iervolino[1] and Maximilian Rodger[1]

10.1 Interoperability opportunities in the maritime scenario

Maritime surveillance is a topic of growing interest and can be regarded as key for its strategic importance and potential benefits for the future of all the countries. Indeed, according to [1], the global seaborne trade was expanding at a rate of 4% in 2017, the fastest growth in 5 years with total volumes of 10.7 billion tonnes. Prospects for the sector are also positive and United Nations (UN) is forecasting a 3.8% compound annual growth rate between 2018 and 2023. Regarding the European Union (EU), it has been estimated that almost 90% of the external freight trade is seaborne, and short sea shipping represents 40% of the intra-EU exchanges and more than 400 million passengers embark and disembark in European ports each year. The gross weight of goods handled in the main EU ports exceeded 3.9 billion tonnes of goods in 2017 [2]. A total of 90% of oil is transported by sea, while there is an increasing trend toward transport of natural gas in a liquefied form by tankers [3]. In this scenario, all the nations have to guarantee the safety of navigation and the security of all marine activities and they need more reliable operational tools to operate. Several ways to monitor and track ships (thus controlling their activities) exist, but no single mean can effectively operate in all the practical scenarios assuring ship detection and identification along with a very accurate geo-location operating in all weather and light conditions. The potential of Earth Observation data to support national authorities in the maritime domain is only just being realized. Unlike land monitoring, activities in the maritime domain have not been subject to the same level of oversight, until nowadays. The increasing availability of satellite data delivered by services such as the Copernicus Maritime Surveillance Service means that even the most remote areas at sea will never be completely out of sight.

In this framework, monitoring techniques aim at fusing multiple sources from both satellite and terrestrial monitoring data for a better understanding of the marine picture for near real time applications. SAR sensors are a valid monitoring

[1]Electronic Engineering Department, University of Surrey, Guildford, UK

technique, often considered a complementary means to the traditional coastal-based ones. SARs are able to acquire images independently from daylight, meteorological conditions, and national borders; in addition, modern sensors also offer a wide spatial coverage and usually operate in constellation thus reducing the revisit time and allowing, consequently, the control in open sea areas. The main issue concerned with SAR is the impossibility to identify ships, although high-resolution SAR (HR-SAR) with its submeter spatial resolution can recognize at least the ship type. In addition, these sensors are usually characterized by high mission costs compared with optical sensors.

Regarding the terrestrial techniques, the most widely used monitoring system is certainly the automatic identification system (AIS). As part of the coastal-based surveillance systems, the AIS was originally designed for safety reasons and mainly for collision avoidance and is basically a shipborne system by which ships inform each other and the coastal receivers about their position, course, speed and name. However, the system is able to cover up to only about 40 km off the coast (due to the Earth's curvature) and requires, obviously, ships with AIS on-board working correctly [4]. The main drawbacks concerning the AIS system are limited spatial coverage, impossibility of berth-to-berth tracking, and need for land-based means.

Traditionally, AIS is employed as a ground truth reference in order to validate SAR detection maps of cooperative ships. In [5–8], for example, the SAR/AIS position match is obtained based on a minimum distance algorithm which is always problematic in high traffic areas. It has to be underlined that AIS data are not always reliable due to the presence of human-based errors in the reports which cause several uncertainties in the retrieval of ship features making more difficult the SAR/AIS match [9,10]. For example, according to [10], 47% of the ships displayed incorrect length and 18% incorrect beam (width) in AIS reports. Considering only the length feature, 6.4% shows a length of 0 m, 36.3% presents an error between 1 m and 5 m, and 4.3% an error more than 5 m. For these reasons, more recent studies aim at cooperative integration [11] between SAR and AIS in order to extract more reliable information as the final output. For example, in [12], relevant ship features (length, width and position) are extracted from both SAR and AIS data sources before being fused together based on a weights method to produce the outcomes. In [13], instead, SAR and AIS are fused from a different perspective: tracking noncooperative vessels in open sea scenarios. It is here assumed that the latest AIS position before the uncooperative ship switches off the transceiver is known and that the heading angle is unchanged between the SAR and AIS acquisitions; in these hypotheses, it is possible to identify the area in which the noncooperative ship is located allowing the tasking of the SAR sensor for future acquisitions.

Fusing heterogeneous data coming from different sensors is never a trivial task. First of all, the source data have to be aligned and coregistered to ensure that measurements are properly connected to each other and to the respective object of interest. In the SAR/AIS fusion context scenario, this means taking in account the different acquisition geometries and the "intrinsic artifacts" of each technology, in particular, SAR images are acquired in azimuth and slant range projection, while AIS positions are provided in latitude/longitude coordinates; in addition, SAR moving targets are Doppler shifted (in azimuth direction) due to their radial velocity and, as a consequence, they appear in a different azimuth position from their wake when visible. Furthermore, the temporal gap between the different acquisitions cannot be neglected (especially when dealing with moving targets) and the datasets have to be also temporally aligned by, for example, projecting the SAR targets in latitude/longitude coordinates at the AIS acquisition time by assuming a constant heading angle and target velocity between SAR and AIS acquisitions. When all these preparatory steps are performed and the input datasets are geometrically and temporally aligned, the proper fusion strategy can be implemented according to the desired fusion level: (1) observation-level fusion (or pixel level); (2) feature-level fusion, and (3) decision-level fusion [14]. While observation-level implies the direct fusion of raw data, feature-level fusion involves a preliminary extraction of representative features from the original sensor data. Finally, decision-level fusion is used only after a first determination of the target's attributes of interest by each sensor alone has already been achieved [11].

Obviously, SAR and AIS are not the only techniques which can be fused for the monitoring of the maritime scenario. Others can be employed to overcome the drawbacks of both techniques. For example, the berth-to-berth tracking can be performed by SatAIS (Satellite AIS) and long-range identification and tracking (LRIT) as they work from spaceborne platforms. Recently, it has been demonstrated that AIS signals can be positively received using spaceborne systems [4]. The ship AIS signal detection is possible in the entire radio visibility range of a satellite equipped with an AIS payload; for example, for a satellite at an altitude of 650 km, the average field of view is above 20 million square kilometers [4]. The main advantage of this technique is the better global coverage compared to the traditional AIS. In 2011, a SatAIS system was launched by the German Aerospace Centre (DLR) in order to enable observations of worldwide ship movements by means of AIS. In the next couple of years, space agencies are planning to launch further SatAIS payloads to enhance the reception of AIS messages. At the moment, there are tracking systems exclusively based on the received AIS signals. For example, Spire Global Inc. (a startup born in San Francisco in 2012) has designed and developed a system (Spire Sense) able to detect and monitor ships based on more than 60 satellites [15]. The system provides the most frequently refreshed global ship tracking data collecting 80 million messages per day from vessels and, in future, it aims to fuse the AIS data with additional data streams.

A different spaceborne monitoring system is provided by LRIT. The International Maritime Organization (IMO) adopted the LRIT system as amendments to Chapter V of the 1974 SOLAS Convention and the LRIT system entered in force on 1st January 2009 [16]. The LRIT was set-up to provide a global system for identification and tracking of ships that extends the monitoring beyond those areas covered by existing AIS coastal network (short range). LRIT system is mandatory for all passenger ships, high-speed craft, mobile or shore drilling units, and cargo ships of over 300 gross tonnes. These ships must report their position between four and six times a day. The LRIT shipborne equipment transmits position information to the Communication Service Provider via satellite link (Inmarsat and Iridium constellations); the LRIT data are then transmitted to the LRIT Data Centre and stored. Finally, the data distribution plan, managed by IMO, defines rules and access rights to the LRIT data [16]. The main disadvantages of SatAIS and LRIT are that the transmitter system may break down and there could be some noncooperative ships (traffickers, pirates, and smugglers) which, intentionally, do not transmit any signal to be tracked.

Other kinds of data that can be used conjunctly or fused with SAR images are the vessel monitoring system (VMS) data or the most recent video from space (i.e., VividX2 constellation). VMS is a satellite tool for fisheries monitoring and control. The EU legislation, for example, requires that all coastal EU countries should set up systems that are compatible with each other, so that countries can share data and the Commission can check that rules are respected. EU funding is available for Member States to acquire state-of-the-art equipment and to train their people to use it [17]. VMS can be used for the identification and the tracking of non-cooperative fishing vessels in areas already known for overfishing or poaching. Regarding the video from space, the UK young company Earth-i has already launched in January 2018 the first batch (five satellites) of a constellation of 15 satellites. Each satellite operates in the optical band with 1 m spatial resolution and is able to acquire video from space; the full constellation of 15 satellites will be able to acquire twice a day every point on the Earth [18]. At the moment, no application based on the exploitation of VivdX2 video has been developed yet; however, the very high-resolution RGB video from space will certainly enhance the knowledge of the maritime sector helping to tackle some crucial issues: piracy, fisheries monitoring and protection, environmental protection, and oil-spill detection with the identification of polluters.

It is clear that a single monitoring means cannot address all the needed requirements for the monitoring and the preservation of the maritime environment. For example, on one hand, coastal AIS can identify a ship but provides a limited spatial coverage; on the other hand, the SAR sensors present better spatial coverage but they can only recognize the ship-type in the best case. In this scenario, some novel monitoring techniques have already been developed merging the data coming from different sources and combining their advantages. For example, the European Maritime Safety Agency (EMSA), instead, has developed tools for both oil spills identification (CleanSeaNet) [19] and ship detection (SafeSeaNet) [20], which are both based on SAR images. CleanSeaNet and SafeSeaNet services aim to strengthen operational responses to accidental and deliberate discharges from ships and cargoes assisting the EU States to locate and identify polluters in areas under their jurisdiction [19,20]. In particular, the services employ three polar-orbiting SAR systems: the Canadian RADARSAT-2, the Italian CosmoSkyMed, and the European Union's ENVISAT (used until May 2012), which was replaced by the novel Sentinel-1 from late 2015 [19].

The remaining of the chapter presents a more in-depth overview of the signals and the technologies cooperating with the SAR and a discussion of current progress in application development.

10.2 Collaborative signals

In this section, we explore some of the most widely used self-reporting monitoring and tracking systems in the area of maritime surveillance, namely the AIS, VMS, and LRIT system.

These systems are used internationally and share the same purpose of assisting in the identification and location of ships to ensure, amongst other things, safe navigation, protection of the marine environment, and maritime security. This information can also be exploited to complement current navigational systems such as terrestrial radar and spaceborne SAR.

The fusion of the above systems with SAR means that ships are more likely to be detected. For example, cooperative ships can be identified, thanks to the self-reporting data of these system(s) and non-cooperative ships, which have either lost their navigational signal or are intentionally avoiding detection to perform illicit activities, can be detected (but not identified) thanks to the independent, observation-based data of the SAR sensor. This provides a comprehensive picture of human activities at sea, which is of great significance to the maritime community, especially those involved in maritime safety and security.

Next, a brief description of each system is given followed by a comparison of their similarities and differences.

10.2.1 Automatic identification system (AIS)

The AIS is an important tool in contributing to maritime safety and security. Its purpose is to help identify and locate vessels, assist in traffic monitoring and target tracking, simplify information exchange, and provide additional information to assist in situation awareness and collision avoidance decision making [21]. The radio-based communication system is designed to exchange information from ship to ship and also to competent authorities such as the coastguard and search and rescue services, similar to the relationship seen between aircraft and air traffic control services.

In detail, the radio-based system is expected to meet the standards set by the International Maritime Organization (IMO) in accordance with the International Convention for the Safety of Life at Sea (SOLAS), adopted in 1974 [22]. As well as other safety standards, the SOLAS specifies that AIS should automatically provide information (e.g., ship identity, type, position, course, speed, etc.) to shore-based stations, other ships and aircraft; similarly, receive such information from other AIS-fitted ships; and finally, exchange information with shore-based stations such as vessel traffic service (VTS) centers.

Since 2004, AIS transponders are installed on all ships of 300 gross tonnages (GT) and upwards engaged on international voyages, cargo ships of 500 GT and upwards not engaged on international voyages, and passenger ships irrespective of size [22]. In a typical day, over 160,000 ships can be monitored and tracked globally using AIS making it one of the most ubiquitous and important maritime technologies to date [23].

However, not all ships are required to operate AIS. Specific vessels such as military and government-owned ships are not required to be fitted with AIS and small vessels such as pleasure craft and fishing boats are also exempt from carrying AIS [21]. In fact, ships fitted with AIS may be switched off entirely, either at the discretion of the captain, in a situation where the safety and/or security of the ship could be compromised or, in the unlawful case, to perform illicit activities such as illegal fishing. For this reason, AIS is regarded as a cooperative system, that is, requiring cooperation from the ship. Therefore, AIS may not always provide a complete maritime picture, thus demonstrating one of its inherent limitations. This has led to the use of SAR and AIS in a collaborative way which is discussed in Section 10.3.

In total, there are three different types of information* that can be transmitted (and received) by AIS transponders on a ship, including static, dynamic, and voyage-related information and, depending on the ship's speed and activity, the transponder will transmit at different rates [21]. For example, if a ship is moving relatively fast or changing course, its information will be broadcast at a higher rate compared to a ship on a straight-line course and moving relatively slowly. Table 10.1 provides details of the three different types of information broadcast by ships equipped with AIS.

Moreover, the system operates primarily on two dedicated VHF channels and is commonly referred to as coastal or terrestrial-based AIS (Terr-AIS). The system has a reported typical range of 20–30 nautical miles but can reach up to 50 nautical miles or more [24,25]. The main limitation is due to the fact that AIS is a line-of-sight communication method, meaning that its horizontal range is limited by the curvature of the Earth. However, the variation in terrestrial range will also depend, amongst other things, on the antenna properties, such as its quality and sensitivity, and its height above sea level. In addition, the range may also be affected by the surrounding topography and, to some degree, by the atmospheric conditions [26].

Motivated to overcome the limitations of the original system design, AIS receivers have since been launched into space aiming to provide a global picture of human activities at sea. The possibility of receiving AIS signals in space was first presented in the open literature in 2003 [27]. Subsequently, in 2004, ORBCOMM in partnership with the U.S. Coast Guard demonstrated in practice that shipborne AIS information was able to be received from low-Earth orbit (LEO). Since then, interest in satellite-based AIS (Sat-AIS) has grown significantly, owing to its superior ship tracking as well as its long-range monitoring capabilities. This makes Sat-AIS the perfect candidate to fill the gap of acquiring shipping information over open oceans and provides a full space-based maritime surveillance system (see Figure 10.1).

Today, satellite-based AIS receivers are both commercially and institutionally available with proven space-flight heritage. Spaceborne radar missions, including the UK's first spaceborne SAR mission, NovaSAR-1 [28] (launched September

*There are also short safety-related messages which can be transmitted to either an individual MMSI or to all ships in a certain area.

Table 10.1 AIS broadcast information. Adapted from [21]

Static information	Update rate	Notes
MMSI (Maritime Mobile Service Identity) number	Six minutes (or on request)	A unique nine-digit identification number manually entered upon installation which may need amending if ownership of the ship changes
Radio call sign and name		Manually entered upon installation, which may need amending if ownership of the ship changes
IMO vessel number		A unique seven-digit identification number manually entered upon installation.
Length and beam		Manually entered upon installation (or if changed)
Type of ship		Selected from a preinstalled list (e.g., tanker, passenger, etc.)
Location of position-fixing antenna		Manually entered upon installation, which may be changed depending on type and number of antennae
Dynamic information		
Position	2–12 s (moving[b]) or 3 min (anchored)	Latitude and longitude coordinates automatically updated from the ship's position sensor, which is connected to AIS[a]
Position time stamp		Automatically updated from the ship's position sensor (UTC)
Course over ground (COG)		Automatically updated from the ship's position sensor (if available)
Speed over ground (SOG)		Automatically updated from the ship's position sensor (if available)
Heading		Automatically updated from the ship's heading sensor
Navigational status		Manually entered and updated as necessary. Examples include "underway using engine," "at anchor," and "not under command," etc.
Rate of turn (RoT)		Automatically updated from the ship's RoT sensor or derived from the ship's gyroscope (if available)
Voyage-related information		
Draught	Six minutes (or on request)	Manually entered at the start of the voyage
Hazardous cargo (type)		Manually entered at the start of the voyage to confirm if hazardous cargo is being carried such as dangerous goods (DG), harmful substances (HS), and marine pollutants (MP)
Destination and ETA		Manually entered at the start of the voyage and updated as necessary
Route plan (waypoints)		Manually entered at the start of the voyage and updated as necessary

[a]The AIS positions are derived from the ship's Global Navigation Satellite System (GNSS), which is usually GPS.
[b]The dynamic information update rate depends on both the speed and course of the ship while under way.

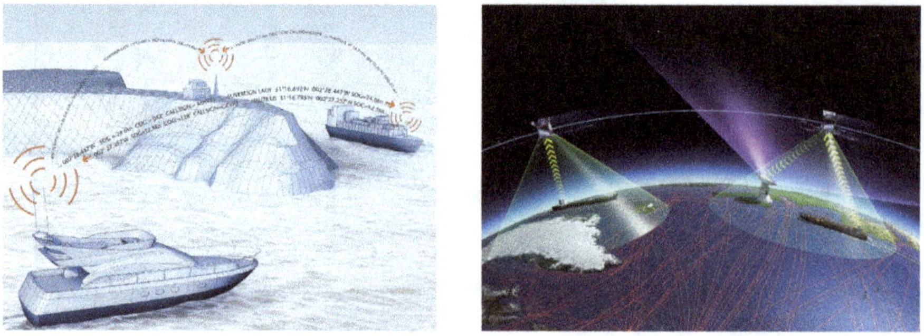

Figure 10.1 Computer-generated images of the two forms of AIS in action: terrestrial-based AIS (Terr-AIS) (Image Courtesy: IMO) and satellite-based AIS (Sat-AIS) (Image Courtesy: ESA)

2018), Spain's PAZ mission [29] (launched February 2018), and Canada's upcoming RADARSAT Constellation mission (RCM) [30] all share the innovative concept and capability of having both the SAR and AIS payloads colocated on the same platform. Therefore, the temporal difference (or so-called "temporal gap") between respective SAR and AIS sensor acquisitions is mitigated. In other words, the near-contemporaneous acquisition of SAR imagery and AIS data makes the challenge of data fusion with temporally diverse data far less difficult.

10.2.2 Vessel monitoring system (VMS)

The VMS is a satellite-based fishing vessel monitoring system that provides data at regular intervals to the fisheries authorities on the location, course, and speed of vessels [31]. The VMS allows for the detection and identification of fishing vessels at all times, allowing environmental and fisheries authorities to monitor and track fishing fleets to ensure compliance with regulations and procedures for the protection of the marine environment.

The system was originally developed by the European Commission, the executive of the European Union, and it is now an EU requirement that all fishing vessels of length 12 m and over must transmit their position and related data automatically to the fisheries monitoring center (FMC) of the Member State every 2 h when at sea (see Figure 10.2). The positions are derived from the fishing vessel's GPS.

The FMC of some Member States, however, may require this information at shorter time intervals and also have the capacity of polling the actual position of each of its fishing vessels. In Norwegian waters, for example, the VMS device must transmit at least once an hour, even in port [32]. In addition, some Member States may impose additional national requirements, for example, in the UK access to certain areas may be subject to VMS monitoring, such as marine protected areas [33].

Additionally, as well as the most recent geographical position of the fishing vessel, the transmitted data or information must include the fishing vessel's identification, date and time (expressed in coordinated universal time (UTC)) of the

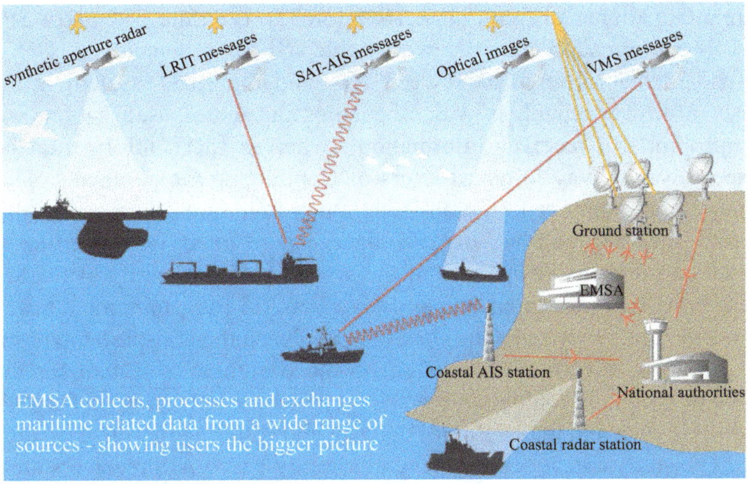

Figure 10.2 An illustration showing the various maritime systems and services available, including the vessel monitoring system (VMS) and long-range identification (LRIT) system (Image Courtesy: EMSA)

said position, and the speed and course of the vessel [31]. Importantly, the position and related data must be transmitted via satellite which has an associated cost. Vessel owners are billed for all VMS transmissions sent via satellite, which in the UK are billed as pay as you go (PAYG) with a typical rate of $0.11 (or approximately £0.08) per report [33].

10.2.3 Long-range identification tracking (LRIT)

The LRIT system was originally designed for maritime security and, like AIS, is expected to meet the standards set by the IMO in accordance with Chapter V of the SOLAS Convention. As described in the regulation [34], ships engaged on international voyages, including passenger ships, cargo ships of 300 GT and upwards, and mobile offshore drilling units are required to carry LRIT equipment to allow the competent authorities to undertake the LRIT of ships. Most ships already fitted with AIS, however, are not required to comply with the provisions of this regulation.

Moreover, the above-mentioned types of ships are required to automatically transmit the ship's identity, position (derived from the ship's GNSS), and the date and time of the said position (UTC) at 6-h intervals to an LRIT Data Centre [34,35]. This time interval can be changed to a maximum of once every 15 min upon user request [36]. Specifically, the LRIT information is first transmitted to a telecommunication satellite, which is operated by a communication service provider (CSP). Next, the LRIT information is forwarded to an application service provider (ASP) that adds additional information to the LRIT message, before passing the expanded message onto its associated LRIT Data Centre (DC) that finally distributes the information to

Contracting Governments† (see Figure 10.2) [36,37]. Both the CSP and ASP ensure that the LRIT information is routed in a reliable and secure manner.

A fundamental difference between AIS and the LRIT system is that AIS broadcasts information publicly with other ships and shore-based stations within its range (and similarly receives information), whereas LRIT information is transmitted securely one way to data centers of Contracting Governments. Therefore, AIS is more commonly reported on in the open literature compared to the LRIT system and this accessibility of AIS data has resulted in numerous studies, including a comprehensive analysis of the system [38], a security evaluation [39], and a feasibility and performance analysis of Sat-AIS [40], to name a few.

A notable study by Greidanus *et al.* [41] analyzed the real-time monitoring and tracking of ships, specifically focusing on the design of a system for maritime authorities in Africa. A multisource system was set up consisting of Terr-AIS, Sat-AIS, and LRIT data and the quality of the data and resulting maritime pictures were assessed. The study concluded that several satellite-based AIS receivers are needed to provide the desired spatial coverage and data refresh rates for continuous maritime situation awareness.

In conclusion, the AIS, VMS, and LRIT systems each have their own advantages and disadvantages and no single best solution exists. For example, when comparing the vessels required to carry the AIS and LRIT systems (see Table 10.2), a vessel transmitting LRIT information is likely to also be transmitting AIS, but the opposite is not true. In other words, ships required to use an LRIT system mostly overlap with those required to use AIS, but there are numerous cases where ships required to use AIS are not required to use the LRIT system.

This implies that AIS is a more widely used system; however, compared to LRIT and VMS systems, AIS signals, specifically Sat-AIS signals, are not guaranteed to be received due to the overwhelming number of users in the satellite's field of view causing message collisions at the Sat-AIS receiver [40]. This makes AIS less reliable. In addition, although AIS provides more information, the majority is manually entered (see Table 10.1) which is prone to human error [9], whereas LRIT systems have dedicated service providers (i.e., ASPs) who add additional information such as the identity of the ship (IMO number) and ensure secure subsequent transfer to the data center.

It is important to note that each of these systems is a means to an end, not an end in itself. The real opportunity lies in using all of these systems together in a complementary way to provide the most complete maritime picture and maximize maritime situation awareness.

10.2.4 VHF data exchange system (VDES)

Maritime communications have not historically followed the same path of the terrestrial communication. The main reason is the simplicity in designing traditional

†A Contracting Government (or flag state) is the name given to any government of a nation that is a signatory to the SOLAS Convention. There are currently over 100 Contracting Governments.

Table 10.2 Comparison of cooperative ship self-reporting systems

	AIS	VMS	LRIT
System type	Terrestrial-based or Satellite-based	Satellite-based	Satellite-based
Provider	Open (nonproprietary) Publicly available (nonsecure transfer of info.)	Closed (proprietary) Authorized users only (secure transfer of info.)	Closed (proprietary) Authorized users only (secure transfer of info.)
Communication method	Terr-AIS (two way): Ship-to-ship Ship-to-shore (vice versa) Sat-AIS (one way): Ship-to-satellite	Ship-to-satellite (one way)	Ship-to-satellite (one way)
Range	Terr-AIS:20–30 nm (approx.)Sat-AIS: Line-of-sight with satellite	Line-of-sight with satellite	Line-of-sight with satelliteLRIT information is available to authorized users whenever a vessel is within 1,000 nm of the coast of the flag state [45]
Applicability	Mandatory for: • All ships of 300 GT and upwards engaged on international voyages • Cargo ships of 500 GT and upwards not engaged on international voyages • Passenger ships irrespective of size	Mandatory for: • Fishing vessels of length 12 m and over	Mandatory for ships engaged on international voyages: • Passenger ships • Cargo ships of 300 GT and upwards • Mobile offshore drilling units
	Regulations are implemented under government legislation of participating member or flag states, but it is important to note that some ships are exempt		
User base	High 160,000+ users per day (globally)	Low 327 users (England) [46]	Medium Some of the largest data centers in the LRIT system tracking over 8,000 ships per day [45]
Transmitting information	Static, dynamic (e.g., position[a], course, speed, etc.) and voyage-related (automatically) Timestamped (UTC)[b]	Position[a], ID, speed, and course (automatically) Timestamped (UTC)	Position[a] and ID (automatically) Timestamped (UTC)[b]

(Continues)

Table 10.2 (Continued)

	AIS	VMS	LRIT
Typical transmission rate	2 s to 6 min (or on request) Depends on both the speed and course of the ship while under way	1–2 h (or on request)	6 h (or a max. of 15 min on user request and also following receipt of a poll request)
Transmission cost	Free—free of charge to ships following the provisions of AIS [47]	Paid—vessel owners billed for all VMS transmissions sent via satellite (shared)	Free—flag states bear all costs associated with any LRIT information they request and receive [34]

[a]All three systems' GNSS position is based on the WGS84 geodetic datum.
[b]The date is to be added by receiving equipment.

maritime communication systems (i.e., AIS systems), while the terrestrial communication is heading to the opposite direction of increasingly sophisticated and costly solutions [42]. Recently, the AIS has experienced a critical overload of its very high frequency (VHF) channels, originally designed for the exclusive exchange of vessels movement information to prevent collisions [43]. While the increased traffic load certainly played a role, the current congestion is mainly caused by the use of the VHF band for noncritical messaging applications, especially in regions densely populated by vessels. The situation called for the development of a new maritime communication system and, for this reason, the VHF Data Exchange System (VDES) has been recently proposed and developed by IMO, International Telecommunication Union (ITU) and other organizations [44]. VDE overcomes the limited spatial coverage and the spectral inefficiency of coastal AIS communication systems by allowing an additional 150 kHz for both up- and downlink communication from space-borne platforms [43]. In addition, VDE enriches the communications channels by offering new ship-to-satellite and satellite-to-ship communication links. In future, VDE will enhance vessels' situational awareness, allowing mariners to downlink updated charts of ice and other hazards and to transmit relevant ship's information to shore. VDES protocol (operational frequency, bandwidth, and channels) is currently under discussion and the system is expected to start operating from the end of 2021.

10.3 Applications

In the following, examples of maritime applications based on cooperation of SAR and other signals are discussed. Of the collaborative signals listed above, AIS is the most used for applications in the maritime domain. Some of the most interesting applications are described in this section with particular emphasis on surveillance and safety.

10.3.1 Ship detection and tracking

Disregarding environmental applications and concentrating on the surveillance emphasis of this book, it is possible to agree that all applications in the sector finally bring back to a problem of ship detection and/or tracking. The only distinction is then made on whether the ship is cooperative or not, meaning for this its ability to transmit signals to inform other vessels or monitoring systems about its position, course, etc. In this sense, a vessel with a faulty system and a vessel in which the identification system has been intentionally turned off are both considered noncooperative. The following discussion is then organized grouping the most interesting applications in this way.

Being ship detection at the core of all the applications, the cooperation of systems/signals is here predominantly meant as SAR-AIS cooperation. When both employed, SAR and AIS are used in two possible ways. The most diffused is validation, with the SAR ship detections verified against AIS messages. In a second case, less common, the two datasets are integrated into a fusion framework to improve the information that could separately be extracted from each dataset [48]. An application for ship detection and tracking, based on SAR and AIS technologies, may be unsuccessful if simply one of the following conditions verifies:

- High density of ships, especially for spaceborne AIS
- Large temporal gap between the acquisition times of SAR and AIS datasets
- AIS data incomplete or unreliable

As Section 10.2.1 highlighted, technological developments and recent launches of new satellite missions with both SAR and AIS on board are going to address the problem of temporal gap between the two datasets drastically reducing it down to the order of a few minutes.

Over a high traffic density area, collision of AIS messages at the spaceborne receiver is highly likely. This is because, differently from a terrestrial AIS, a spaceborne AIS receives signals from a large area on Earth. While within a limited area, called cell, the allocation of AIS transmission slots is self-coordinated, it happens that a satellite covers many of these cells simultaneously with the consequent possibility that messages transmitted at the same time and from the same distance, but from different cells, collide [49]. The same may happen also if the AIS messages were originally transmitted in adjacent time slots but the different path distances between each transmitting ship and the spaceborne AIS receiver result in the AIS messages reaching the receiver at the same time.

It is not difficult to understand, then, that the AIS data relevant to a single ship might be incomplete (if, despite transmitting, a time slot could not be allocated to a ship) or unreliable (depending on how the collision is treated). It is worth remembering, as previously mentioned, that unreliable AIS data do not always depend on the system but on human operators [9].

10.3.1.1 Cooperative ships

Iceberg avoidance and navigation safety

In 1914, a few years after the Titanic collided with an iceberg and sank, the International Ice Patrol (IIP) was established with the aim of contributing to safety of life at sea, safety and efficiency of navigation and protection of the marine environment in the North Atlantic [50] by locating icebergs in the region, predicting their drifts and informing passing ships. The Patrol still operates today and is especially active during the ice season (from middle of February till beginning of July each year) in which reports are transmitted to ships twice a day. While in the past the iceberg position was detected (and retransmitted to ships) by other ships or aircrafts, today icebergs reports may originate from different sources including also satellite radars. Key information is normally represented by the limit of the iceberg region and the iceberg density distribution within the limit.

Navigation in the Southern oceans is not equally safeguarded despite the increase in routes at high southern latitudes.

SAR capability in iceberg detection has been already demonstrated [51–53] even in situations where it becomes particularly challenging as the case of small icebergs in ice-infested waters [51,52]. However, as the IIP itself reports [54], the iceberg size and shape can limit the radar detection in significant sea state (see also Chapter 8).

In addition, the discrimination from other similar radar targets, like ships, can be hard and it can be reasonably argued if a SAR alone can detect and discriminate both ships and icebergs at sea.

The challenge of the task is due to the high backscattering coefficient that both targets, for different reasons, produce in SAR imagery. Advanced discrimination algorithms are required in this case and ships/icebergs can be discriminated on the base of additional features such as their morphology, radar cross section, and polarization state.

In [53], polarization information is considered fundamental for a potential discrimination between ships and icebergs and airborne data are the only additional data to SAR images but mainly for validation of purposes.

It appears more common, though, that the simultaneous consideration of different sources is mandatory for navigation safety. Radar on ships are used for the detection of "stationary radar targets" which the ships are recommended to report despite whether they could be or not icebergs; sonar is also used to record underwater sounds that mechanical stresses and cracking in icebergs can produce up to 2 km away; satellite radar, like SAR, is employed for a big picture of the oceans and potentially the mapping of large icebergs or sea ice; AIS remains the main source of information to localize ships in such regions.

Also, in [55], the authors acknowledge that the SAR discrimination might not always be accurate, especially when high-resolution data are not provided. This led them to discuss the combination of SAR and AIS to monitor navigation in iceberg-prone waters, adding in the problem of detecting/monitoring icebergs and discriminating them from ships.

Dead reckoning is the basic approach they use to predict (and successively associate) the ship locations dead reckoned from AIS report with the SAR target locations.

As for other applications, also in this case the main challenge may derive by the use in general of AIS in high-density areas and the fusion of AIS with SAR when a large temporal gap between the two exist. The former does not actually constitute a real problem for the majority of regions of interest. As Figure 10.3 shows, the iceberg region limit in the North Atlantic, in the charts updated daily by the IIP during the dangerous ice season (an example in the left side) delineate areas of substantially low traffic and normally destined to fishing, as for the Marine Traffic website [23] (a corresponding snapshot in the right side).

As regards the temporal gap between SAR and spaceborne AIS data, instead, it is quickly approaching the order of a few minutes in new missions thus guaranteeing proper association between datasets.

Current results then show that, for navigation safety purposes, a monitoring service based on SAR and AIS can be sufficiently accurate and recently launched satellites with both SAR and AIS payloads should be able to address the problem of temporal gap between datasets.

10.3.1.2 Noncooperative ships

Polluters identification

Discharge of oil in sea waters has sadly become a common practice especially in some areas where prosecution is not performed or means to find responsible parties are lacking or not accessible by some countries. While the situation and the circumstances of a large oil spill accident are such that liable parties declare the accident themselves or can be easily found, more problematic is the scenario in which small boats (not carrying an AIS on board) "clean" their boats from oil residuals at sea. In these cases, the synergy might more easily come from multisource SAR. Studies in literature have demonstrated the potential of SAR in detecting vessels (of dimensions in the order or bigger than the sensor spatial resolution) as well as oil spills. Looking at the two applications together has quite immediately highlighted the possibility that, once vessels and spills are detected in a SAR image, associating spills to vessels (hence the responsible parties) can be performed as well to some extent. Many of the image requirements for a similar application can probably be found in the same acquisition system. Detection of vessels is more successful in HH polarization while VV is the recommended channel for oil spill detection. A quad-pol system should be able to address this but, as well known, the share of the synthetic aperture may halve (dual-pol) or reduce of a quarter (quad-pol) the final spatial resolution. The trade-off between spatial resolution and coverage is still problematic in SAR until new technological solutions are implemented. In the scenario mentioned, the need for both high resolution (to detect small boats) and wide coverage (to scan a large geographical area or to monitor the evolution of larger spills) is better addressed with multiple SAR sources. There is however a lack in literature about how to specifically address it for the scenario at issue. An operational solution or, more simply, an algorithm has

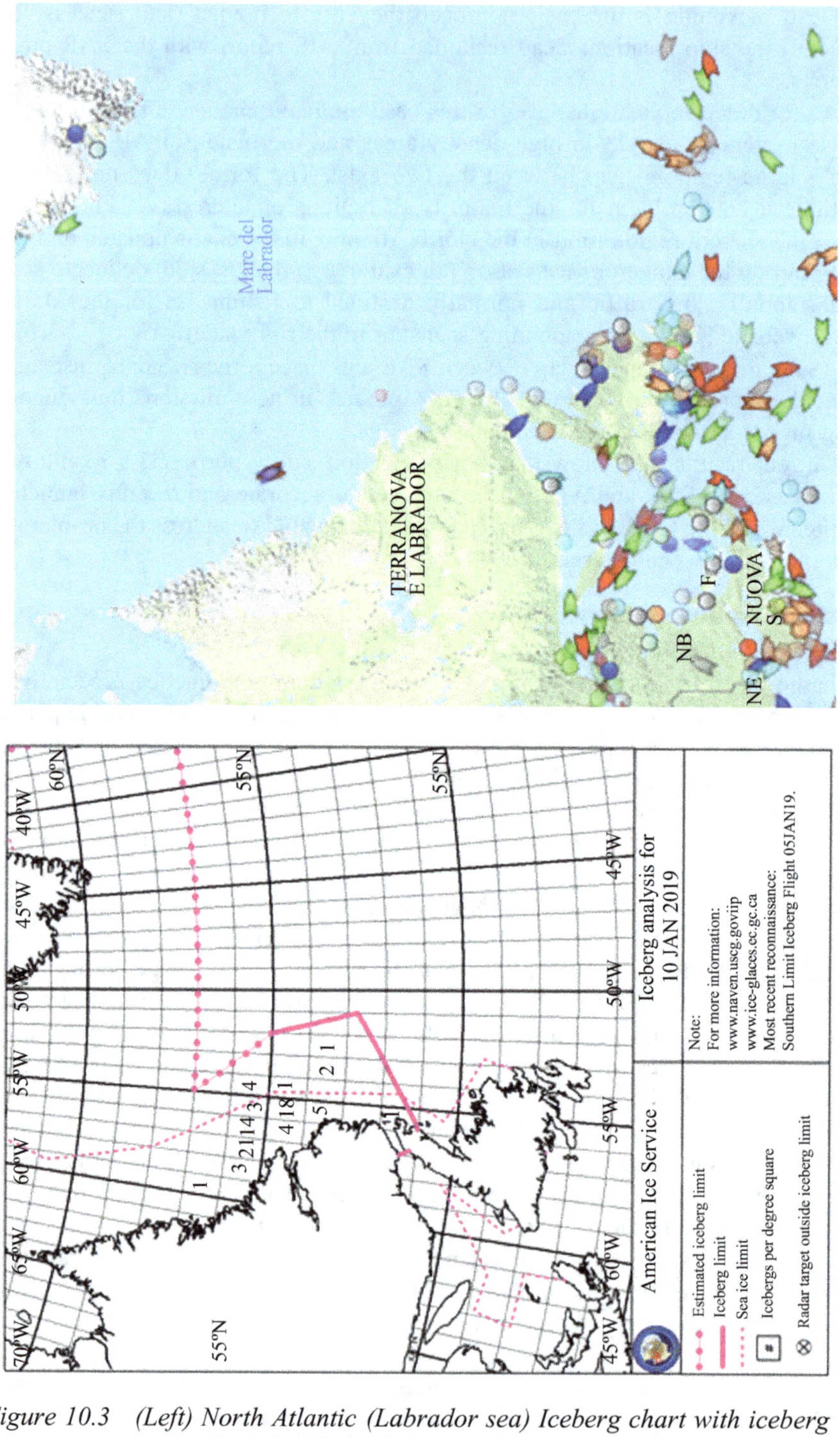

Figure 10.3 *(Left) North Atlantic (Labrador sea) Iceberg chart with iceberg region limit highlighted (Image Courtesy: NAIS) and (right) corresponding daily ship traffic in the Labrador sea (Image Courtesy: Marine Traffic)*

not yet been conceived. In [56], the authors discuss the weather conditions under which a similar application can be successful when based on SAR images, which are the same as for oil spill detection (wind speed between 3 m/s and 10 m/s) and, instead of using the SAR again, recommends the use of AIS for the identification of the vessels. It also highlights that in case of high-density traffic, the association slick-ship cannot be performed. In case of low-density traffic, the association can be performed by visual inspection after combining hindcast with AIS data and observing some features in the image like spill's and wake's trajectory coinciding (the latter, however, may be a wrong clue to use if the trajectories are part of high traffic routes). No algorithm or technical solution is provided for the association so far.

Search and rescue

Maritime search and rescue[‡] operations are devolved to locating persons in distress at sea and rescuing them. They are regulated at national levels with programs of cooperation among countries. Search and rescue services based on satellite imagery and collaborative signals have just started being conceived and, due to technological constraints, do not regard the detection of single human being at sea but, for example, passenger ships. It is again a problem of ship detection and positioning with high likelihood that the collaborative signals one would normally rely on are no more available and satellite imagery, SAR in particular, is the only option left. This is the case of the several ships that in the past years were hijacked by pirates, stopped any communication and, in the lucky cases, reappeared after captain and crew took control back. Such situations require immediate intervention and, even if signals like AIS are not available after the hijacking started, new studies are exploring how the cooperation between SAR and AIS till the moment when the AIS disappeared could help in predicting the area where to concentrate searching operations in the first hours after the AIS signal disappeared. This piece of information is extremely valuable for any scanning system, especially if airborne. But it would be equally important for spaceborne sensors, like SAR, to optimize the satellite tasking operations. This is the idea suggested in [13] where the authors elaborate this concept for the recently launched first UK SAR mission NovaSAR-1 (with AIS system as additional payload) and discuss the tracking opportunities it may offer once a full constellation would be operative. As per their statement, the procedure is however general and can be applied to any combination of SAR/AIS systems. In an emergency scenario, it would be life-saving to adopt the very first SAR satellite orbiting over the identified search area despite which SAR/AIS combination was used till that point for standard tracking operations. The SAR tasking, as suggested in [13], would use the information of max velocity, last position and heading from the latest AIS available for the missing ship in order to center the image acquisition around its new expected position at the time of the SAR acquisition.

[‡]Note that very often the acronym SAR, in this book used for synthetic aperture radar, is widely used in literature also for search and rescue operations. To avoid confusion, in this book the acronym SAR will only be used for synthetic aperture radar.

Detection of illegal fishing

Fishing activities are regulated differently in every country. For the European countries, the EU regulation against illegal, unregular, and unreported (IUU) fishing entered into force on 1st January 2010. It defines as illegal any fishing activity that is not authorized, against conservation and management measures or against national laws or international obligations.

There is a close cooperation between EU and non-EU countries to promote compliance with international rules. When a third country not fulfilling international rules is identified, the EU can list the country and ban all the fisheries products caught by fishing vessels from that country, unless the country makes an effort in improving the situation and addressing the problem.

It results then that the identification of a ship nationality and its activities (information generally made available by the AIS of a cooperative ship) is important to update listing by the EU. Then, detection of illegal fishing can be considered another particular case of ship detection and tracking application with SAR. Contextual information such as fishing routes and authorized waters for every nationality should be used jointly with ship detection maps from SAR to trigger alerts. Collaborative signals like AIS, when present, can increase the confidence in the detection maps as previously seen. Alerts should also be triggered when an AIS, normally transmitted by a ship along its route, disappears at the transition between waters of a different nationality. The illegal activities can actually be hidden/masked often by legal ones.

In remote sensing literature, the studies focusing specifically on illegal fishing monitoring based on SAR solutions are only a few and not actually ready for operational environments. In [57], for example, the authors consider potential synergies of multifrequency SAR (C- and X-band) for maritime target detection. Terrestrial AIS is used but only to validate findings. Together with monitoring of harbor area and analysis of ships and ship wake signatures, the surveillance of small boats without AIS is tackled as potential application. By analyzing the signature of a tanker ship, the paper shows partial agreement with findings of previous papers relative to the signature of fishing vessels but also identify deviations from standard patterns. It seems to suggest then that, despite some obvious discrepancies, identifying the main activity of a ship by its radar signature is not yet a mature study.

A few other applications of particular complexity are treated with multiple sources, including SAR, like oil spill detection and tracking but, for their relevance, they are discussed in a separate chapter of this book (see Chapter 9).

10.4 Main challenges

It can be concluded that, in maritime surveillance, many are the challenges that would require an advanced technological solution to be addressed. Some of the most relevant are as follows:

Detection and tracking of small objects. The limited spatial resolution of sensors affects heavily the detection rate when, for instance, small boats are part of the investigation. In specific situations, the detection is further challenged by the materials the boats are made of (as the case of go-fast-boats, transparent to radar, and used for the discharge in sea of drugs or other illegal activities). High-resolution SARs are imperative technological solutions which can detect small boats (if material is detectable by SAR) or can detect their wakes (when material is transparent). In the case of non-cooperative ships, it is unlikely that AIS or similar information is available and a fusion of SAR information with contextual data is probably the only way to support the ability to distinguish between threats and legitimate activities.

Lack of wide-area, persistent, and continuous surveillance. Constellations of satellites are necessary for a monitoring and tracking service thought as an unblinking eye in the sky. While current SAR systems are more often conceived in constellations, especially those with also AIS payload on board, the revisit time is still far from guaranteeing a persistent service.

Limited interoperability between sector stakeholders and systems. Some modern satellite constellations, like the Copernicus one, represent an extraordinary technological effort to create the conditions for a greater interoperability between sensors in several applicative areas though the maritime ones may have been penalized a bit, from the integration point of view, due to the lack of a spaceborne AIS system in the constellation. Other examples, however, like the Radarsat-AIS constellation, promise to at least fill the gap of interoperability between these two systems.

Lack of early warning systems. Due to the very different rates at which upstream and downstream activities developed in the space and earth observation sector (with the latter barely and slowly following the former), the majority of applications did not reach an operational environment or were not conceived as services, maritime surveillance included, and are still at low technology readiness levels. Any early warning solution based on cooperation between SAR and other signals looks the way forward but much research in the understanding of how to integrate signals and systems of different nature is still required.

References

[1] United Nations (UN). *Review of maritime transport*. New York and Geneva: UN; 2018.

[2] Eurostat. *Country level - gross weight of goods handled in all ports* [online]. 2018. Available from https://ec.europa.eu/eurostat/ [Accessed May 2019].

[3] European Commission (EC). *Energy policy and maritime policy: Ensuring a better fit* [online]. 2007. Available from https://ec.europa.eu/tranparency/regdoc/ [Accessed May 2019]

[4] Brusch S., Lehner S., Fritz T., Soccorsi M., Soloviev A., and van Schie B. “Ship surveillance with TerraSAR-X.” *IEEE Transactions on Geoscience and Remote Sensing*. 2011; 49(3): 1092–1103

[5] Pelich R., Longépé N., Mercier G., Hajduch G., and Garello R. “AIS-based valuation of target detectors and SAR sensors characteristics for maritime

surveillance." *IEEE Journal of Selected Topics in Applied Earth Observations and Remote Sensing*. 2015; 8(8): 3892–3901

[6] Iervolino P., Guida R., and Whittaker P. "A model for the backscattering from a canonical ship in SAR imagery." *IEEE Journal of Selected Topics in Applied Earth Observations and Remote Sensing*. 2016; 9(3): 1163–1175

[7] Iervolino P. Guida R., Lumsdon P., *et al*. "Ship detection in SAR imagery: A comparison study." *Proceedings of the 2017 IEEE International Geoscience and Remote Sensing Symposium (IGARSS)*; IEEE, New York, NY, USA, July 2017. pp. 2050–2053.

[8] Iervolino P., and Guida R. "A novel ship detector based on the generalized-likelihood ratio test for SAR imagery." *IEEE Journal of Selected Topics in Applied Earth Observations and Remote Sensing*. 2017; 10(8): 3616–3630

[9] Harati-Mokhtari A., Wall A., Brooks P., and Wang J. "Automatic identification system (AIS): Data reliability and human error implications." *Journal of Navigation*. 2007; 60(3):373–389.

[10] Bošnjak R., Šimunović L., and Kavran Z. "Automatic identification system in maritime traffic and error analysis." *Transactions on Maritime Science*. 2012; 1(2): 77–84

[11] Schmitt M., and Zhu X. X. "Data fusion and remote sensing: An ever-growing relationship." *IEEE Geoscience and Remote Sensing Magazine*. 2016; 4(4): 6–23

[12] Achiri L., Guida R., and Iervolino P. "SAR and AIS fusion for maritime surveillance." *Proceedings of the IEEE 4th International Forum on Research and Technology for Society and Industry (RTSI);* IEEE, New York, NY, USA, September 2018. pp. 1–4.

[13] Achiri L., Guida R., and Iervolino P. "Collaborative use of SAR and AIS data from NovaSAR-S for maritime surveillance." *Proceedings of 12th European Conference on Synthetic Aperture Radar (EUSAR)*; IEEE, New York, NY, USA, June 2018. pp. 1–6.

[14] Hall D. L., and Llinas J. "An introduction to multisensory data fusion." *Proceedings of the IEEE*. 1997; 85(1): 6–23

[15] Spire Global. *Spire Sense* [online] 2015. Available from https://spire.com/products/sense/. [Accessed May 2019]

[16] International Maritime Organization (IMO). *Long-range identification and tracking (LRIT)*. 2014. Available from http://www.imo.org/ourwork/safety/navigation/pages/lrit.aspx [Accessed May 2019]

[17] European Commission (EC). *Vessel monitoring system (VMS)* [online]. Available from https://ec.europa.eu/fisheries/cfp/control/technologies/vms_en [Accessed May 2019]

[18] Earth-i. *Constellations* [online]. Available from https://earthi.space/constellations/ [Accessed May 2019]

[19] European Maritime Safety Agency (EMSA). *Cleanseanet* [online]. 2013. Available from http://www.emsa.europa.eu/csn-menu.html [Accessed May 2019]

[20] European Maritime Safety Agency (EMSA). *Vessel traffic monitoring in EU waters (Safe-SeaNet)* [online]. 2013. Available from http://www.emsa.europa.eu/ssn-main.html [Accessed May 2019]

[21] International Maritime Organization (IMO). *International Convention for the Safety of Life at Sea (SOLAS), 1974 (Chapter V, Regulation 19, Annex 17)* [online]. Available from http://solasv.mcga.gov.uk/Annexes/Annex17.htm [Accessed January 2019]

[22] International Maritime Organization (IMO). *International Convention for the Safety of Life at Sea (SOLAS), 1974 (Chapter V, Regulation 19)* [online]. Available from http://solasv.mcga.gov.uk/ [Accessed January 2019]

[23] MarineTraffic. *Live map* [online]. Available from https://www.marinetraffic.com/ [Accessed January 2019]

[24] exactEarth. *Should I track global fishing activity with VMS or AIS?* [online]. 2015. Available from http://blog.exactearth.com/blog/should-i-track-global-fishing-activity-with-vms-or-ais [Accessed January 2019]

[25] MarineTraffic. *Seven things you should know about AIS* [online]. 2018. Available from https://www.marinetraffic.com/blog/seven-things-know-ais/ [Accessed January 2019]

[26] Maritime and Coastguard Agency (MCA). *Marine guidance note (MGN) 324 (M+F): Radio: Operational guidance on the use of VHF radio and automatic identification systems (AIS) at sea*. 2006

[27] Wahl T., Høye G. K., Lyngyi A., and Narheim B. T. "New possible roles of small satellites in maritime surveillance." *Acta Astronautica.* 2005; 56 (1):273–277

[28] Surrey Satellite Technology Limited (SSTL). *NovaSAR-1* [online]. 2018. Available from https://www.sstl.co.uk/space-portfolio/launched-missions/2010-2018/novasar-1 [Accessed January 2019]

[29] eoPortal. *eoPortal Directory: PAZ SAR satellite mission of Spain* [online]. 2018. Available from https://directory.eoportal.org/web/eoportal/satellite-missions/p/paz [Accessed January 2019]

[30] Canadian Space Agency. *RADARSAT Constellation Mission* [online]. 2018. Available from http://www.asc-csa.gc.ca/eng/satellites/radarsat/default.asp [Accessed January 2019]

[31] European Commission. *Commission Implementing Regulation (EU) No 404/2011*. 2011

[32] Marine Management Organisation. *Device requirements* [online]. 2016. Available from https://www.gov.uk/government/publications/vessel-monitoring-system-devices/device-requirements [Accessed January 2019]

[33] Marine Management Organisation. *Vessel monitoring system (VMS+) guidance* [online]. 2013. Available from https://www.gov.uk/government/publications/vessel-monitoring-system-devices [Accessed January 2019]

[34] International Maritime Organization (IMO). *International convention for the safety of life at sea (SOLAS), 1974 (Chapter V, Regulation 19-1)* [online]. Available from http://solasv.mcga.gov.uk/ [Accessed January 2019]

[35] Maritime and Coastguard Agency (MCA). *Marine guidance note (MGN) 441 (M+F)*. 2011.
[36] European Maritime Safety Agency (EMSA). *LRIT CDC - How it works* [online]. Available from http://www.emsa.europa.eu/lrit-main/lrit-home/how-it-works.html [Accessed January 2019].
[37] International Maritime Organization (IMO). *LRIT technical documentation (Part I) (MSC.1/Circ.1259/Rev.7)*. 2017.
[38] Last P., Bahlke C., Hering-Bertram M., and Linsen L. "Comprehensive analysis of automatic identification system (AIS) data in regard to vessel movement prediction." *Journal of Navigation*. 2014; 67(5):791–809
[39] Balduzzi M., Pasta A., and Wilhoit K. "A security evaluation of AIS automated identification system." *Proceedings of the 30th Annual Computer Security Applications Conference*; Association for computing Machinery, New York, NY, USA, December 2014. pp.436–445.
[40] Cervera M. A., Ginesi A., and Eckstein K. "Satellite-based vessel automatic identification system: A feasibility and performance analysis." *International Journal of Satellite Communications and Networking*. 2011; 29(2): 117–142
[41] Greidanus H., Alvarez Alvarez M., Eriksen T., *et al.* "Basin-wide maritime awareness from multi-source ship reporting data." *TransNav, the International Journal on Marine Navigation and Safety of Sea Transportation*. 2013; 7(2):185–192.
[42] Xia T., Wang M. M., and You X. "Satellite machine-type communication for maritime Internet of things: An interference perspective." *IEEE Access*. 2019. (Early Access paper).
[43] Clazzer F., Munari A., and Giorgi F. "Asynchronous random access schemes for the VDES satellite uplink." *OCEANS 2017 Conference*; IEEE, New York, NY, USA, 2017, pp. 1–7.
[44] Wong D. T. C., Chen Q., Peng X., and Chin F. "Analysis of a future VDES uplink slot carrier-sense TDMA MAC protocol with decollision algorithm." *23rd Asia-Pacific Conference on Communications (APCC)*, IEEE, New York, NY, USA, 2017, pp. 1–6.
[45] European Maritime Safety Agency (EMSA). *LRIT cooperative data centre* [online]. Available from http://www.emsa.europa.eu/lrit-main/lrit-home.html [Accessed January 2019]
[46] Department for Environment, Food and Rural Affairs (Defra). *Consultation on the introduction of inshore vessel monitoring systems for all licensed British fishing boats under 12 metres in length operating in English waters*. 2018.
[47] International Maritime Organization (IMO). *International convention for the safety of life at sea (SOLAS), 1974 (Chapter V, Regulation 11)* [online]. Available from http://solasv.mcga.gov.uk/ [Accessed January 2019]
[48] Guerriero M., Willett P., Coraluppi S., and Carthel C. "Radar/AIS data fusion and SAR tasking for Maritime Surveillance." *Proceedings of the 11th International Conference on Information Fusion*; IEEE, New York, NY, USA, June 2008. pp. 1–5.

[49] Swetha G. M., Hemavathy K, and Natarajan S. "Overcome Message Collisions in Satellite Automatic ID Systems" [online]. Available from https://www.mwrf.com/systems/ [Accessed January 2019]

[50] International Maritime Organization (IMO). *International Convention for the Safety of Life at Sea (SOLAS), 1974 (Chapter V, Regulation 6)* [online]. Available from http://solasv.mcga.gov.uk/ [Accessed January 2019]

[51] Marino A., Dierking W., and Wesche C. "A depolarization ratio anomaly detector to identify icebergs in sea ice using dual-polarization SAR images." *IEEE Transactions on Geoscience and Remote Sensing*. 2016; 54 (9): 5602–5615.

[52] Akbari V., and Brekke C. "Iceberg detection in open and ice-infested waters using C-band polarimetric synthetic aperture radar." *IEEE Transactions on Geoscience and Remote Sensing*. 2018; 56(1): 407–421

[53] Howell C., Power D., Lynch M., *et al.* "Dual polarization detection of ships and icebergs - Recent results with ENVISAT ASAR and data simulations of RADARSAT-2." *Proceedings of the IEEE International Geoscience and Remote Sensing Symposium (IGARSS)*; IEEE, New York, NY, USA, July 2008. pp. III - 206-III – 209.

[54] U.S. Coast Guard Navigation Center (USCG NAVCEN). *Announcement of 2018 International Ice Patrol Services* [online] Available from: https://www.navcen.uscg.gov/pdf/iip/Announcement_of_Services_2018.pdf [Accessed January 2019]

[55] English J., Hewitt R., Power D., and Tunaley J. "ICE-SAIS – Space-based AIS and SAR for improved ship and iceberg monitoring." *Proceedings of the IEEE Radar Conference (RadarCon13)*; IEEE, New York, NY, USA, April 2013. pp. 1–6.

[56] Uiboupin R., Raudsepp U., and Sipelgas L. "Detection of oil spills on SAR images, identification of polluters and forecast of the slicks trajectory." *Proceedings of the IEEE/OES US/EU-Baltic International Symposium*; IEEE, New York, NY, USA, EE, May 2008. pp. 1–5.

[57] Velotto D., Bentes C., Tings B., and Lehner S. "First comparison of Sentinel-1 and TerraSAR-X data in the framework of maritime targets detection: South Italy case." *IEEE Journal of Oceanic Engineering*.2016; 41 (4): 993–1006

Chapter 11

Sea state and wind speed

Gerardo Di Martino[1] *and Antonio Iodice*[1]

11.1 Introduction

Monitoring of sea state and sea surface wind is certainly of fundamental importance for studies on climatology, meteorology, weather forecast, and ocean circulation. However, it has a great relevance also in the framework of maritime surveillance, particularly for navigation safety. In fact, of course navigation safety strongly depends on weather forecasts; in addition, knowledge of sea surface winds and sea currents allows predicting paths followed by floating sea ice and icebergs, which constitute a serious danger for navigation. Also, knowledge of wind speed and surface currents allows predicting the spreading and movements of oil spills over the sea surface. Microwave remote sensing systems, such as synthetic aperture radar (SAR) systems, have been recognized since their first operational uses as the ideal sensors for the monitoring of sea state and sea surface winds.

First experiments of microwave measurement of sea state and wind speed from space date back to 1974, when a radiometer and a scatterometer flew on Skylab. Then, in 1978, a scatterometer and a SAR were borne by Seasat. More systematic monitoring of sea state and wind speed from space started in the 1990s, with the launch of C-band ERS1/2 and Radarsat satellites, then continued with ENVISAT and currently is active with Sentinel-1, in addition to L-band (ALOS PALSAR) and X-Band (TerraSAR and Cosmo SkyMed) satellite sensors. At the beginning, wind speed measurements relied almost exclusively on scatterometer data, while SAR systems were mainly considered as tools to retrieve sea long-wave spectra. However, it was soon realized that SAR images bring useful information on wind speed, too, and at a higher resolution than scatterometers. Therefore, today, SAR data are operatively employed to provide both sea spectra and wind speed measurements. Accordingly, in this last chapter of the book, we illustrate the use of SAR sensors for the monitoring of such important phenomena.

The chapter is organized as follows. First of all, a statistical description of the dynamics of sea surface is provided that allows linking some statistical parameters

[1]Dipartimento di Ingegneria Elettrica e delle Tecnologie dell'Informazione, Università di Napoli Federico II, Naples, Italy

(e.g., significant wave height, sea spectrum) to sea state and wind speed and direction. Then, the mechanism of formation of SAR images of the sea surface is presented. Scattering models, see Chapter 2, are recalled, relating the scattered electromagnetic field to sea statistical parameters. In addition, the effect of sea dynamics on SAR image intensity is taken into account. Available methods to derive the sea surface spectrum from the SAR image intensity are recalled, and finally, a survey of techniques for wind speed retrieval from SAR images is provided.

11.2 Sea surface statistical description

11.2.1 Sea surface waves

The sea surface height deviation z with respect to the mean sea level is spatially and temporally varying so that it can be represented as

$$z = z(x, y, t) \tag{11.1}$$

where x and y are the Cartesian coordinates in the plane of the mean sea level (for the sake of simplicity, we here are locally approximating the geoid with its tangent plane) and t is the time.

Let us consider a water body of uniform and constant depth h_0 only subject to gravity and surface tension, and assumed uncompressible (which is a good assumption as far as sound velocity in water is much larger than sea surface wave velocity). It can be demonstrated that the height deviation of its free surface propagates as a wave with the following dispersion relation:

$$\omega(k_x, k_y) = \omega(k) = \sqrt{\left(gk + \frac{\tau}{\rho}k^3\right)\tanh(kh_0)} \tag{11.2}$$

where ω is the angular frequency (i.e., $2\pi/T$, with T being the period), k_x and k_y are the x and y components of the wavenumber vector and $k = \sqrt{k_x^2 + k_y^2} = 2\pi/\lambda$, with λ being the wavelength, is its modulus, $g \cong 9.8$ m/s^2 is the gravity acceleration, $\tau \cong 0.072$ N/m is the water surface tension, and $\rho \cong 1000$ kg/m^3 is the water density.

Accordingly, under the above-cited hypotheses and in the absence of other external forces (i.e., in the absence of wind), the sea surface height deviation can be expressed as a superposition of sinusoidal plane waves:

$$z(x, y, t) = \frac{1}{4\pi^2}\iint Z(k_x, k_y)\, e^{j[k_x x + k_y y - \omega(k_x, k_y)t]} dk_x dk_y \tag{11.3}$$

where

$$Z(k_x, k_y) = \iint z(x, y, 0)\, e^{-j[k_x x + k_y y]} dx dy \tag{11.4}$$

is the two-dimensional (2D) Fourier transform (FT) of $z(x, y, 0)$. Therefore, in principle from (11.3)–(11.4), given the sea surface at an initial time $t = 0$, one can compute the sea surface at any time. However, in practice, this ideal picture is not realistic because of the presence of wind and, above all, because it is impossible to precisely know the surface deviations at a given time. Therefore, a statistical description is needed, and $z(x,y,t)$ must be modeled as a stochastic process. However, before moving to this statistical modeling, some considerations on the dispersion relation (11.2) can be conveniently made.

First of all, for water depth much larger than wavelength, $kh_0 \gg 1$ and $\tanh(kh_0) \cong 1$, so that (11.2) can be rewritten as

$$\omega(k_x, k_y) = \omega(k) \cong \sqrt{gk\left(1 + \frac{k^2}{k_m^2}\right)} \tag{11.5}$$

where

$$k_m = \sqrt{\frac{g\rho}{\tau}} \cong 370 \ \text{m}^{-1} \tag{11.6}$$

so that for $k << k_m$, that is, wavelength larger than about 20 cm, $\omega(k) \cong \sqrt{gk}$ and sea wave propagation is dominated by gravity. Therefore, waves longer than about 20 cm are said "gravity waves." Conversely, for $k >> k_m$, that is, wavelength smaller than about 2 mm, $\omega(k) \cong \sqrt{\tau k^3/\rho}$ and sea wave propagation is dominated by surface tension. Therefore, waves shorter than about 2 mm are said "capillary waves." Finally, for k of the order of k_m, that is, wavelength between about 2 mm and about 20 cm, both gravity and surface tension govern sea wave propagation. Therefore, waves longer than about 2 mm and smaller than about 20 cm are called "gravity-capillary waves."

Finally, the phase (i.e., propagation) velocity of a sinusoidal plane wave at wavenumber k can be computed from the dispersion relation (11.2) as

$$v_p(k) = \frac{\omega(k)}{k} = \sqrt{\frac{g}{k}\left(1 + \frac{k^2}{k_m^2}\right)\tanh(kh_0)} \tag{11.7}$$

11.2.2 Sea surface modeled as a stochastic process

As mentioned in Section 11.2.1, the sea surface height deviation $z(x,y,t)$ must be modeled as a stochastic process. Its first-order characterization is provided by the probability density function (pdf) $p_z(\zeta; x, y, t)$, whose meaning is that $p_z(\zeta; x, y, t)d\zeta$ is the probability that $\zeta < z(x, y, t) < \zeta + d\zeta$.

An assumption that is usually made is that, at least on areas of linear size of the order of the kilometer and over time intervals of the order of some minutes, the process is statistically stationary, that is, $p_z(\zeta; x, y, t) = p_z(\zeta)$. Another usual assumption is that the pdf is a zero-mean, σ^2-variance Gaussian:

$$p_z(\zeta) = \frac{1}{\sqrt{2\pi}\sigma} e^{-\frac{\zeta^2}{2\sigma^2}} \tag{11.8}$$

By moving to higher order characterization, it is worth assuming that the stochastic process is Gaussian and exploiting the property that a zero-mean Gaussian process is completely characterized by its second order description, provided by its autocorrelation function. More specifically, we can define a spatial autocorrelation function

$$c_{xy}(\Delta x, \Delta y; x, y, t) = \langle z(x, y, t) z(x + \Delta x, y + \Delta y, t) \rangle \tag{11.9}$$

with $\langle \cdot \rangle$ indicating the statistical mean operation, and a temporal autocorrelation function

$$c_t(\Delta t; x, y, t) = \langle z(x, y, t) z(x, y, t + \Delta t) \rangle \tag{11.10}$$

The statistical stationarity assumption implies that

$$c_{xy}(\Delta x, \Delta y; x, y, t) = c_{xy}(\Delta x, \Delta y) \tag{11.11}$$

and

$$c_t(\Delta t; x, y, t) = c_t(\Delta t) \tag{11.12}$$

The second-order statistical characterization of the sea surface is more often provided in the frequency domain, in terms of the power spectral density (PSD), which is the FT of the autocorrelation function, or, equivalently (Wiener–Khinchin theorem), the mean square modulus of the FT of the sea surface deviation z, divided by the considered (large) sea surface area A_S, or time interval T_I. Specifically, we can define the 2D PSD, or *directional spectrum*, as

$$S(k_x, k_y) = \text{2D-FT}\{c_{xy}(\Delta x, \Delta y)\} = \frac{1}{A_S} \langle |Z(k_x, k_y)|^2 \rangle \tag{11.13}$$

The directional spectrum is more often expressed in polar coordinates:

$$\widetilde{S}(k, \varphi) = S(k \cos \varphi, k \sin \varphi) k \tag{11.14}$$

The 1D PSD, or *scalar spectrum*, can be defined as

$$E(\omega) = \text{1D-FT}\{c_t(\Delta t)\} = \frac{1}{T_I} \langle |Z(\omega)|^2 \rangle \text{ or } \widetilde{E}(f) = E(2\pi f) \tag{11.15}$$

f being the frequency. The relation between directional and scalar spectra is given by

$$E(\omega) = \frac{1}{2\pi} \int_{-\pi}^{\pi} \widetilde{S}[k(\omega), \varphi] \frac{dk}{d\omega} d\varphi \tag{11.16}$$

where $dk/d\omega$ can be obtained from the dispersion relation, (11.2) or (11.5):

$$\frac{dk}{d\omega} = \left(\frac{d\omega(k)}{dk}\right)^{-1} \cong 2\sqrt{\frac{k}{g}} \tag{11.17}$$

In the last approximate equality of (11.17), we have used the approximate dispersion relation of gravity waves.

The variance of sea surface deviation σ^2 can be computed by integrating the sea spectrum:

$$\begin{aligned}\sigma^2 = \langle |z(x,y,t)|^2 \rangle &= \frac{1}{4\pi^2}\iint S(k_x,k_y)dk_x dk_y = \frac{1}{4\pi^2}\int_{-\infty}^{+\infty}\int_{-\pi}^{+\pi}\widetilde{S}(k,\varphi)d\varphi dk \\ &= \frac{1}{2\pi}\int_{-\infty}^{+\infty} E(\omega)d\omega = \int_{-\infty}^{+\infty}\widetilde{E}(f)df\end{aligned} \tag{11.18}$$

Its square root, that is, the root mean square (rms) height σ, is a measure of the wave height. In fact, although other substantially equivalent definitions exist [1], the *significant wave height* H_s is defined as [1]

$$H_s = 4\sigma \tag{11.19}$$

This means that for a significant wave height equal to H_s about one-tenth of the observed waves have a trough to crest height larger than H_s.

At this point, it is important to distinguish between *wind sea* and *swell.* The wind sea is the motion of the waves generated by the wind blowing directly on the observed sea area or in its immediate vicinity; conversely, swell waves are generated by winds blowing over a distant sea area, which travel rapidly with a regular period and flat crests. The latter have usually a very narrow spectrum, concentrated around a well-defined wavenumber vector, that is, well-defined direction of propagation and wavelength. Therefore, they can be simply characterized by their length, height, and direction of propagation. Conversely, wind sea has a wider spectrum, whose shape depends on wind speed and direction, and on the size of the area over which the wind blows (*fetch*).

Several both theoretical and experimental studies have been conducted to determine the wind sea surface spectrum and to relate it to wind speed, direction, and fetch [2–17]. First of all, it turns out that the directional spectrum $\widetilde{S}(k,\varphi)$ can be factorized in an omnidirectional spectrum $S_k(k)$ and a direction dispersion function $f(k,\varphi)$:

$$\widetilde{S}(k,\varphi) = S_k(k)f(k,\varphi) \tag{11.20}$$

The direction dispersion function has the following general expression:

$$f(k,\varphi) = 1 + b(k,u_{10})\cos\,[2(\varphi-\varphi_w)] \tag{11.21}$$

where u_{10} is the wind speed measured at a height of 10 m over the sea level and φ_w is the angle between wind direction and x-axis. A possible expression for $b(k, u_{10})$ is provided, for example, in [17] and in (18.8) of [18]. Examples of direction dispersion functions for different values of wind speed are depicted in Figure 11.1.

By using (11.21), the sea surface deviation variance (11.18) can be expressed as

$$\sigma^2 = \frac{1}{2\pi}\int_{-\infty}^{+\infty} S_k(k)dk \tag{11.22}$$

and the omnidirectional spectrum can be related to the scalar spectrum as

$$S_k(k) = E(\omega)\frac{d\omega}{dk} \tag{11.23}$$

For the omnidirectional spectrum, two expressions are commonly used in literature: the Pierson–Moskowitz (PM) spectrum [3] and the Joint North Sea Wave Project (JONSWAP) spectrum [6,8]. The former can be used for a fully-developed sea, that is, in situations where the wind has constantly blown for several hours over a large area of linear size of the order of the hundreds of kilometers, and is given by

$$S_k^{\mathrm{PM}}(k) = \frac{2\pi a}{k^3} e^{\frac{-0.74g^2}{k^2 u_{19.5}^4}} \tag{11.24}$$

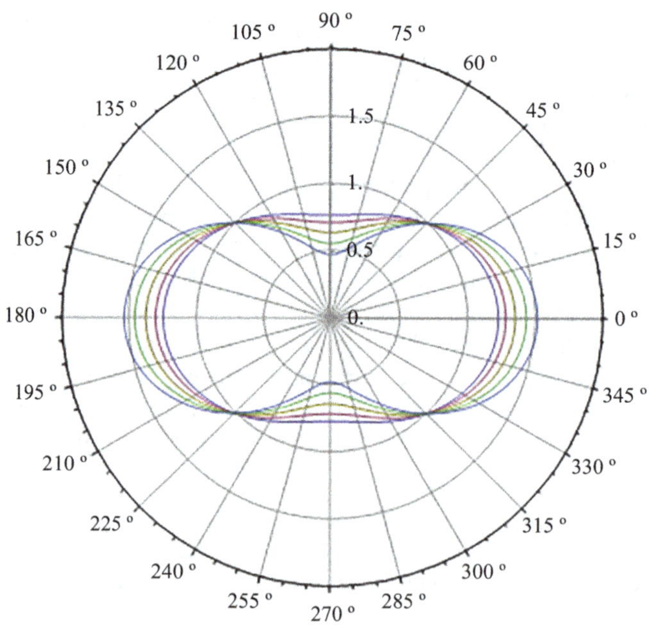

Figure 11.1 Direction dispersion function versus angle from wind direction for $k = 150\ m^{-1}$ and $u_{10} = 5, 10, 15, 20,$ and 25 m/s

where $u_{19.5}$ is the wind speed measured at 19.5 m height ($u_{19.5} \cong 1.026\ u_{10}$) and $a = 4.0510^{-3}$, if SI units are used, as we will assume in this chapter. By integrating this spectrum, it turns out that the significant wave height is related to wind speed as follows:

$$H_s = 4\sigma \cong 0.22 \frac{u_{19.5}^2}{g} \cong 0.23 \frac{u_{10}^2}{g} \tag{11.25}$$

In addition, the peak wavenumber (i.e., the wavenumber for which the spectrum reaches its maximum value) is

$$k_p \cong 0.7 \frac{g}{u_{10}^2} \tag{11.26}$$

and phase and group velocities at the peak wavenumber are

$$v_p(k_p) = \frac{\omega}{k_p} \cong \sqrt{\frac{g}{k_p}} \cong \frac{u_{10}}{\sqrt{0.7}} \tag{11.27}$$

and

$$v_g(k_p) = \frac{d\omega}{dk}\bigg|_{k=k_p} \cong \frac{1}{2}\sqrt{\frac{g}{k_p}} \cong \frac{u_{10}}{2\sqrt{0.7}} \tag{11.28}$$

If the sea is not fully developed, the sea spectrum also depends on the *fetch*, which is the linear size of the area over which the wind has blown with constant intensity and direction. In this case, the JONSWAP spectrum is often used in literature [6,8,17]:

$$S_k^J(k) = \frac{\pi\alpha_p}{k^3} \gamma^r e^{-\frac{5}{4}\left(\frac{k_p}{k}\right)^2} \tag{11.29}$$

where

$$r = e^{-\frac{\left(\sqrt{\frac{k}{k_p}}-1\right)^2}{2s^2}} \tag{11.30}$$

and

$$\begin{aligned} k_p &= \frac{g}{u_{10}^2}\Omega_c^2, s = 0.08\left(1 + 4\Omega_c^{-3}\right), \alpha_p = 0.006\sqrt{\Omega_c}, \\ \gamma &= 1.7 + 6 \max\{0,\ \log(\Omega_c)\} \end{aligned} \tag{11.31}$$

The so-called inverse wave age Ω_c can be related to the fetch F by the following relation, proposed in [17]:

$$\Omega_c = 0.84 \tanh\left[\left(\frac{g\,F}{22\,u_{10}^2}10^{-3}\right)^{0.4}\right]^{-0.75} \tag{11.32}$$

By using (11.32), it turns out that, for very large fetch, peak wavelength and phase and group velocities are still given by (11.26)–(11.28). In addition, the significant wave height turns out to be [17]

$$H_s \cong 0.26\frac{u_{10}^2}{g}\tanh\left[\left(\frac{g\,F}{22\,u_{10}^2}10^{-3}\right)^{0.4}\right]^{1.5} \tag{11.33}$$

which, for very large fetch, is in reasonable agreement with (11.25).

PM and JONSWAP spectra accurately describe the sea surface gravity waves. However, as shown in Chapter 2 and as recalled in Section 11.3, microwave scattering from the sea surface strongly depends on the gravity-capillary part of the sea spectrum. Therefore, proper modifications to these spectra have been proposed in literature to better describe this short-wave part of the spectrum. A modified PM spectrum is reported for instance in [11]. However, at present, the most widespread formulation for this purpose is the unified Elfouhaily spectrum [17], which substantially coincides with the JONSWAP spectrum for long waves, that is, for $k < 10\ k_p$, and conversely differs from it for short waves, that is, for $k > 10\ k_p$:

$$S_k^E(k) = \frac{v_p(k_p)}{v_p(k)}S_k^J(k)T(k) + \frac{v_p(k_m)}{v_p(k)}S_k^{\mathrm{SW}}(k)[1 - T(k)] \tag{11.34}$$

where $v_p(k_m) \cong 0.23$ m/s and

$$S_k^{\mathrm{SW}}(k) = \frac{\pi\alpha_m}{k^3}e^{-\frac{1}{4}\left(\frac{k}{k_m}-1\right)^2} \tag{11.35}$$

with

$$\alpha_m = \begin{cases} 0.01\left[1 + \ln\left(u* / v_p(k_m)\right)\right] & \text{for} \quad u* \le v_p(k_m) \\ 0.01\left[1 + 3\ln\left(u* / v_p(k_m)\right)\right] & \text{for} \quad u* > v_p(k_m) \end{cases} \tag{11.36}$$

The friction velocity $u*$ can be expressed in terms of wind speed as follows [17–19]:

$$u* = \sqrt{C_d}\ u_{10} \tag{11.37}$$

with

$$C_d = \begin{cases} 1.205 \cdot 10^{-3} & \text{for } 4 \le u_{10} < 11\text{m/s} \\ (0.49 + 0.065u_{10})10^{-3} & \text{for } 11 \le u_{10} \le 25\text{m/s} \end{cases} \tag{11.38}$$

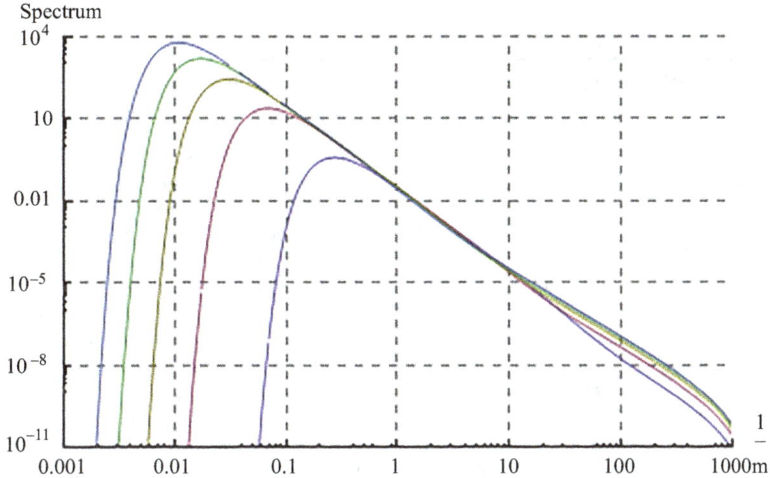

Figure 11.2 Unified Elfouhaily omnidirectional spectra (m^3) versus k (m^{-1}), for very large fetch and u_{10} = 5, 10, 15, 20, and 25 m/s

Finally, the function $T(k)$ is a transition, or cut-off, function that is equal to about unity for $k < 10k_p$ and to about zero for $k > 10k_p$. By slightly modifying the function employed in [17], we can set

$$T(k) = \min\left\{1,\ \exp\left[-\frac{\Omega_c}{\sqrt{10}}\left(\sqrt{\frac{k}{k_p}} - 1\right)\right]\right\} \tag{11.39}$$

It is worth noticing that, as already noted in Chapter 2, in the wavelength interval between 10 k_p and k_m the spectrum in (11.34) exhibits a power-law decay $\sim k^{-2.5}$ typical of fractal surfaces with fractal dimension equal to 2.25 [18].

Examples of unified Elfouhaily omnidirectional spectra are depicted in Figure 11.2 for different values of wind speed.

For the sake of completeness, we conclude this section by reporting the Beaufort scale of wind force (Table 11.1) [20] and the Douglas scales of sea state for both wind sea (Table 11.2) and swell (Table 11.3) [21].

11.3 SAR images of the sea surface

The SAR sensors' resolution can vary from 1 m to several tens of meters so that they can actually *image* only relatively long gravity waves. However, the electromagnetic signal backscattered by the sea (and therefore the SAR image pixel intensity) is strongly dependent on the shortest gravity-capillary waves, that have a wavelength comparable to the one of the incident electromagnetic wave. In addition, it must be noted that the sea moves during the SAR image acquisition time,

Table 11.1 Beaufort Scale

Wind force	Description	km/h	m/s	knots	Specifications
0	Calm	<1	< 0.3	<1	Smoke rises vertically
1	Light Air	1–5	0.3–1.5	1–3	Direction shown by smoke drift but not by wind vanes
2	Light Breeze	6–11	1.6–3.3	4–6	Wind felt on face; leaves rustle; wind vane moved by wind
3	Gentle Breeze	12–19	3.4–5.4	7–10	Leaves and small twigs in constant motion; light flags extended
4	Moderate Breeze	20–28	5.5–7.9	11–16	Raises dust and loose paper; small branches moved
5	Fresh Breeze	29–38	8.0–10.7	17–21	Small trees in leaf begin to sway; crested wavelets form on inland waters
6	Strong Breeze	38–49	10.8–13.8	22–27	Large branches in motion; whistling heard in telegraph wires; umbrellas used with difficulty
7	Near Gale	50–61	13.9–17.1	28–33	Whole trees in motion; inconvenience felt when walking against the wind
8	Gale	62–74	17.2–20.7	34–40	Twigs break off trees; generally impedes progress
9	Strong Gale	75–88	20.8–24.4	41–47	Slight structural damage (chimney pots and slates removed)
10	Storm	89–102	24.5–28.4	48–55	Seldom experienced inland; trees uprooted; considerable structural damage
11	Violent Storm	103–117	28.5–32.6	56–63	Very rarely experienced; accompanied by widespread damage
12	Hurricane	> 117	> 32.7	> 63	Devastation

Table 11.2 Douglas Scale for wind sea

Description term		Waves average height
State of the sea		
0	Calm (glassy)	—
1	Calm (rippled)	0–0.10 m
2	Smooth	0.10–0.50 m
3	Slight	0.50–1.25 m
4	Moderate	1.25–2.50 m
5	Rough	2.50–4 m
6	Very rough	4–6 m
7	High	6–9 m
8	Very high	9–14 m
9	Phenomenal	Over 14 m

Table 11.3 Douglas Scale for swell

	Wave length and height	Specification	Meters
0	No swell	Short wave	<100
		Average wave	100–200
		Long wave	>200
1	Very low (short or low wave)	Low wave	<2
2	Low (long and low wave)	Moderate wave	2–4
		Heavy wave	>4
3	Light (short and moderate wave)		
4	Moderate (average and moderate wave)	Note: The swell reports also comprise the wave direction according to the eight main directions of the wind rose expressed in the English notation (N, NE, E, SE, S, SW, W, and NW). For instance: Swell 2 from SW or Low swell from NW	
5	Moderate rough (long and moderate wave)		
6	Rough (short and heavy wave)		
7	High (average and heavy wave)		
8	Very high (long and heavy wave)		
9	Confused (wave length and height indefinable)		

and this renders the SAR image formation process of sea surfaces very involved [22,23].

We assume that, due to the long waves, the sea mean plane in an SAR resolution cell is tilted and has slopes s_a and s_r along azimuth and range directions, respectively. Here we will assume that the x-axis is the azimuth direction, whereas the y-axis is along the ground range direction so that s_a and s_r are the derivatives with respect to x and y of the long-wave component of the sea surface, respectively. The electromagnetic field backscattered by the resolution cell can be then computed as the field backscattered by a rough tilted facet using the small perturbation method (SPM) [18,24–28]:

$$E^s_{pq} = \frac{E^i_p\, k^2_{em} \cos^2\vartheta_l}{\pi r} \chi_{\mathrm{pq}}(\vartheta_l, \beta)\, Z(0, 2k_{em} \sin\, \vartheta_l) \exp(-jkr) \tag{11.40}$$

where E^i_p is the incident field, $k_{em} = 2\pi/\lambda_{em}$ is the electromagnetic wavenumber, r is the radar-to-target distance, and p and q are the polarizations of the incident and scattered field, respectively, and can each stand for H (horizontal) or V (vertical), ϑ_l is the local incidence angle (i.e., the angle between the SAR look direction and the normal to the tilted facet, as opposed to the global incidence angle ϑ, that is, the angle between the SAR look direction and the vertical direction), β is the incidence plane rotation angle (i.e., the angle between the incidence plane over the tilted facet and the incidence plane over the horizontal plane), and $\chi_{pq}(\vartheta_l,\beta)$ are the elements of the matrix:

$$\underline{\underline{\chi}}(\vartheta_l,\beta) = \underline{\underline{R_2}}(\beta)\cdot\begin{pmatrix} F_H(\vartheta_l) & 0 \\ 0 & F_V(\vartheta_l)\end{pmatrix}\cdot\underline{\underline{R_2}}^{-1}(\beta) \tag{11.41}$$

$$\underline{\underline{R_2}}(\beta) = \begin{pmatrix} \cos\beta & \sin\beta \\ -\sin\beta & \cos\beta\end{pmatrix} \tag{11.42}$$

is the 2 × 2 unitary rotation matrix, and F_H and F_V are the SPM (or Bragg) coefficients for H and V polarization, respectively:

$$\begin{cases} F_H = \dfrac{\cos\vartheta_l - \sqrt{\varepsilon - \sin^2\vartheta_l}}{\cos\vartheta_l + \sqrt{\varepsilon - \sin^2\vartheta_l}} \\[2ex] F_V = (\varepsilon - 1)\dfrac{\sin^2\vartheta_l - \varepsilon\left(1 + \sin^2\vartheta_l\right)}{\left(\varepsilon\cos\vartheta_l + \sqrt{\varepsilon - \sin^2\vartheta_l}\right)^2}\end{cases} \tag{11.43}$$

with ε being the sea water *relative complex dielectric constant* (or *relative complex permittivity*). The rotation angle and the local incidence angle are related to the global incidence angle and to the facet's slopes as [18,29]

$$\begin{cases} \tan\beta = \dfrac{s_a}{\sin\vartheta - s_r\cos\vartheta} \\[2ex] \cos\vartheta_l = \dfrac{\cos\vartheta + s_r\sin\vartheta}{\sqrt{1 + s_a{}^2 + s_r{}^2}}\end{cases} \tag{11.44}$$

The mean square value of the backscattered field is proportional to the normalized radar cross section (NRCS), which turns out to be

$$\sigma^0_{\text{pq}} = \frac{4}{\pi}k^4_{\text{em}}\cos^4\vartheta_l\left|\chi_{\text{pq}}\right|^2 S(0, 2k_{\text{em}}\sin\vartheta_l) \tag{11.45}$$

Equations (11.40) and (11.45) clearly show that the return from the resolution cell depends on the intensity of the sea spectrum at the so-called Bragg resonance wavenumber $2k_{em}\sin\vartheta_l$, which in turn depends on the electromagnetic frequency and on the local incidence angle, and it is in the range of gravity-capillary waves. Variations of the local incidence angle due to local slopes variations produce variations of SAR image intensity, the so-called "tilt modulation," that actually allow seeing long waves in SAR images. Tilt modulation is also enhanced by the foreshortening effect, that in the sea SAR imaging literature is sometimes called "range bunching" [30]. Actually, a further image intensity modulation is due to the fact that the sea spectrum intensity of gravity-capillary waves at the Bragg resonance wavenumber may depend on the part of the long wave where they are superimposed (front, rear, trough, and crest). This nonlinear effect, called hydrodynamic modulation, is usually smaller than tilt modulation and it is currently still not fully understood, in spite of the fact that it was first observed about 40 years ago.

Tilt and hydrodynamic modulation are "instantaneous" phenomena that affect each single backscattered SAR pulse. In addition, during the SAR integration time the sea surface moves, and this of course further affects the final image. It is known that the sea surface particle velocity range component causes an azimuth displacement in the final SAR image that is directly proportional to the particle range velocity itself. This causes zones where returns from sea particles are "bunched" and zones where they are "dispersed": the former zones appear brighter on the final image, whereas the latter appear darker, and this causes a further image intensity modulation called "velocity bunching" effect. Sea movement also causes defocusing in the final image.

Finally, it must be recalled that the image intensity resulting from the modulation effects mentioned above is multiplied by the speckle noise, see Chapter 1.

An example of SAR image of the ocean surface is shown in Figure 11.3.

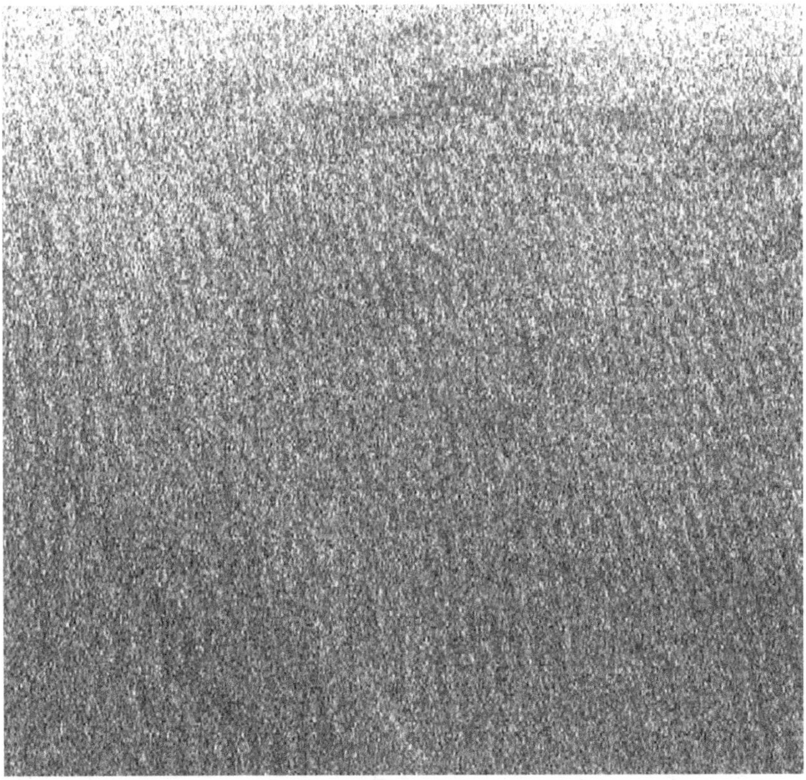

Figure 11.3 SAR image of the ocean surface: AIRSAR P-band image of the Pacific Ocean, off the coast of California, with a swell (Image Courtesy: NASA)

11.4 Sea surface spectra retrieval using SAR images

Sea SAR image intensity fluctuations can be related to the long wave sea spectrum so that the latter can be retrieved from SAR images. The rationale of sea spectra retrieval from SAR images can be understood by considering the SAR image formation mechanism described in Section 11.3. In fact, all mentioned modulation effects are functions of first derivatives of the long-wave components $z^{LW}(x,y,t)$ of the surface height: sea surface local slopes or sea particles velocity. If a linear approximation is used, then the FT of the image fluctuations is proportional to the 2D FT of $z^{LW}(x,y,t)$ and its mean square modulus is proportional to the sea long-wave spectrum. Let us see in detail how this works in the case of the tilt modulation. Substituting (11.44) in (11.45) and expanding the obtained expression around zero slopes up to the first order, we get, for the copolarized NRCSs

$$\sigma^0_{\mathrm{pp}} \cong \frac{4}{\pi} k^4_{\mathrm{em}} \cos^4\vartheta \, |F_p|^2 S(0, 2k_{\mathrm{em}} \sin\vartheta) + s_r \cdot \frac{\partial\sigma^0_{\mathrm{pp}}}{\partial s_r}\Big|_{s_r=s_a=0} \tag{11.46}$$

where we have exploited the fact that the first derivative of copolarized NRCSs with respect to the azimuth slope s_a is zero at zero slopes [18]. It must be noted that the cross-polarized NRCS is second order infinitesimal with respect to slopes [18]:

$$\sigma^0_{\mathrm{hv}} \cong \frac{4}{\pi} k^4_{\mathrm{em}} \cos^4\vartheta \left|\frac{F_V - F_H}{\sin\vartheta}\right|^2 S(0, 2k_{\mathrm{em}} \sin\vartheta) s_a^2 \tag{11.47}$$

so that it is not used for sea spectrum retrieval. In order to obtain more manageable explicit expressions, we can let ε tend to infinity, which is a reasonable approximation for sea water, so obtaining

$$\begin{cases} F_H = -1 \\ F_V = -\dfrac{1 + \sin^2\vartheta_l}{\cos^2\vartheta_l} \end{cases} \tag{11.48}$$

and approximate the short-wave spectrum as $S(k) \cong S_0 k^{-\alpha}$, with $\alpha = 3.5$ for the Elfouhaily spectrum and S_0 being a parameter depending on wind speed. Under these hypotheses, (11.46) and (11.47) can be rewritten as

$$\sigma^0_{\mathrm{vv,hh}} \cong \frac{4S_0 k^4_{\mathrm{em}} (1 \pm \sin^2\vartheta)^2}{\pi (2k_{\mathrm{em}} \sin\vartheta)^{\alpha}} \left\{ 1 + \frac{\cot\vartheta \left[4 - (4 - \alpha)\left(1 \pm \sin^2\vartheta\right)\right]}{1 \pm \sin^2\vartheta} s_r \right\} \tag{11.49}$$

and

$$\sigma^0_{\mathrm{hv}} \cong \frac{4S_0 k^2_{\mathrm{em}} (2k_{\mathrm{em}} \sin\vartheta)^{2-\alpha}}{\pi} s_a^2 \tag{11.50}$$

The statistical mean of the SAR intensity image $i(x,y)$ (in practice, the SAR intensity image after speckle removal) of a sea area with constant wind, after radiometric calibration, is equal to the NRCS, so that, for copolarized channels, by using (11.49), we can write

$$
\begin{aligned}
i(x,y) &\cong \frac{4S_0k_{\mathrm{em}}^4(1\pm\sin^2\vartheta)^2}{\pi(2k_{\mathrm{em}}\sin\vartheta)^\alpha}\left\{1+\frac{\cot\vartheta\left[4-(4-\alpha)\left(1\pm\sin^2\vartheta\right)\right]}{1\pm\sin^2\vartheta}s_r(x,y)\right\}\\
&= i_{av}\left\{1+A\cdot\frac{\partial z^{\mathrm{LW}}(x,y,0)}{\partial y}\right\}
\end{aligned}
\tag{11.51}
$$

where i_{av} is the average of the intensity $i(x,y)$ over the whole image, which depends on wind speed via S_0, and A is a constant whose expression is implicitly defined by (11.51).

After dividing $i(x,y)$ by its average i_{av} and subtracting 1, we get the normalized, zero-mean intensity image $\tilde{i}(x,y)$:

$$\tilde{i}(x,y)=\frac{i(x,y)}{i_{\mathrm{av}}}-1\cong A\cdot\frac{\partial z^{\mathrm{LW}}(x,y,0)}{\partial y} \tag{11.52}$$

By performing the 2D FT of $\tilde{i}(x,y)$, we get

$$\tilde{I}(k_x,k_y)\cong -jk_yA\cdot Z^{\mathrm{LW}}(k_x,k_y)=G_{\mathrm{TM}}(k_x,k_y)\cdot Z^{\mathrm{LW}}(k_x,k_y) \tag{11.53}$$

where $G_{TM}(k_x,k_y)$ is termed the tilt modulation transfer function [30], given by

$$G_{\mathrm{TM}}(k_x,k_y)=-jk_y\frac{\cot\vartheta\left[4-(4-\alpha)\left(1\pm\sin^2\vartheta\right)\right]}{1\pm\sin^2\vartheta} \tag{11.54}$$

Similar procedures can be followed for the other modulation effects mentioned in Section 11.3, and for each of them an expression of the kind

$$i(x,y)=i_{\mathrm{av}}\left(1+\mathscr{F}\left\{\frac{\partial z^{\mathrm{LW}}(x,y,0)}{\partial x},\frac{\partial z^{\mathrm{LW}}(x,y,0)}{\partial y},\left.\frac{\partial z^{\mathrm{LW}}(x,y,t)}{\partial t}\right|_{t=0}\right\}\right) \tag{11.55}$$

can be written, with $\mathscr{F}$ being a linearized function. Accordingly, in the linear approximation, a transfer function for each effect can be obtained. In addition, defocusing due to the sea surface movement must be accounted for. The 2D FT of $\tilde{i}(x,y)$ can be then expressed as [30,31]

$$\tilde{I}(k_x,k_y)\cong\exp\left[-\left(\frac{k_x^2}{2d_x^2}+\frac{k_y^2}{2d_y^2}\right)\right]G(k_x,k_y)\cdot Z^{LW}(k_x,k_y) \tag{11.56}$$

where d_x and d_y are wind-depending parameters that govern azimuth and range defocusing caused by sea surface movement, and

$$G(k_x,k_y)=G_{\mathrm{TM}}(k_x,k_y)+G_{RB}(k_x,k_y)+G_{\mathrm{HM}}(k_x,k_y)+G_{VB}(k_x,k_y) \tag{11.57}$$

is the overall transfer function, with $G_{\mathrm{RB}}(k_x, k_y)$, $G_{\mathrm{HM}}(k_x, k_y)$, and $G_{\mathrm{VB}}(k_x, k_y)$ being the transfer functions of range bunching, hydrodynamic modulation, and velocity bunching, respectively. Possible forms of their expressions are available in literature [30]:

$$\begin{cases} G_{\mathrm{RB}}(k_x, k_y) = -jk_y \cot \vartheta \\ G_{\mathrm{HM}}(k_x, k_y) = 4\dfrac{k_y^2}{k} \\ G_{\mathrm{VB}}(k_x, k_y) = -jk_x \dfrac{r}{V}\sqrt{gk}\left(\dfrac{k_y}{k} \sin \vartheta - j \cos \vartheta\right) \end{cases} \tag{11.58}$$

where V is the SAR sensor velocity.

Finally, the mean square value of $\widetilde{I}(k_x, k_y)$, that is, the PSD of the normalized SAR intensity image, turns out to be directly proportional to the sea surface long-wave 2D directional spectrum:

$$\langle |\widetilde{I}(k_x, k_y)|^2 \rangle \cong A_S \exp\left[-\left(\frac{k_x^2}{d_x^2} + \frac{k_y^2}{d_y^2}\right)\right] |G(k_x, k_y)|^2 \cdot S^{\mathrm{LW}}(k_x, k_y) \tag{11.59}$$

Therefore, dividing the PSD of the normalized SAR intensity by the square modulus of the transfer function, a first estimate of the sea surface long-wave 2D directional spectrum can be obtained. This is an approximate estimate, due to the employed linear approximation. However, iterative methods have been devised to refine this first estimate. The interested reader is referred to the available literature on these methods [30–35].

It must be noted that the image spectrum (similarly to the sea 2D directional spectrum) is a real-valued symmetric function so that it has inherently a 180° ambiguity in identifying the dominant wave propagation direction. To resolve this ambiguity, some methods have been devised. A popular one is the "cross-spectrum" method [36,37]: two different images of the same area are formed by using two different parts of the SAR integration time (i.e., two different looks are used) and their cross-spectrum is computed, that is, the FT of one image is multiplied by the complex conjugate of the FT of the other image and then the mean value is computed. The cross spectrum is a complex-valued function, whose real part is symmetric and whose imaginary part is antisymmetric. The positive peak of the imaginary part indicates the direction of propagation of the dominant wave so that the 180° direction ambiguity is resolved.

Finally, some methods have been proposed to retrieve sea long-wave spectra from polarimetric SAR data [28,38]. The main advantage of the use of polarimetric data is that some proper combinations of polarimetric channels are very simply related to sea surface slopes. In particular, a proper combination of the elements of the polarimetric coherency matrix **T**, defined in (4.12) of Chapter 4, is directly related to the rotation angle β, as it can be proven by using (11.40)–(11.42), and hence to the azimuth slope via (11.44):

$$R = \frac{-2\,\mathrm{Re}\{T_{23}\}}{-T_{22} + 4T_{33}} = \tan(4\beta) \cong \frac{4s_a}{\sin\vartheta} \tag{11.60}$$

Conversely, the element T_{11} of the coherency matrix is completely independent of the rotation angle, so that it is mainly related to the range slope. Accordingly, a combination of the images of R and T_{11} may allow reconstruction of the whole long-wave spectrum.

Experiments of ocean long-wave spectra retrieval were conducted since the first airborne and spaceborne SAR missions, and results are reported in the relevant literature [22,23,30–38]. By now, long-wave spectra retrieval methods are at an operative level. For instance, among the level-2 Sentinel-1 products routinely provided by European Space Agency (ESA), there is the "ocean swell spectra" (OSW), which is a two-dimensional ocean surface swell spectrum [39,40]. In particular, spectra and cross-spectra are computed over areas of 20 km × 20 km both from stripmap acquisitions (see Chapter 3), in which case thirty-two 20 km × 20 km spectra are obtained from each stripmap 170 km × 80 km image, and from wave-mode acquisitions (see Chapter 3), in which case one 20 km × 20 km spectrum is obtained from each wave-mode imagette.

An example of ocean wave spectrum obtained from Sentinel-1 data is displayed in Figure 11.4.

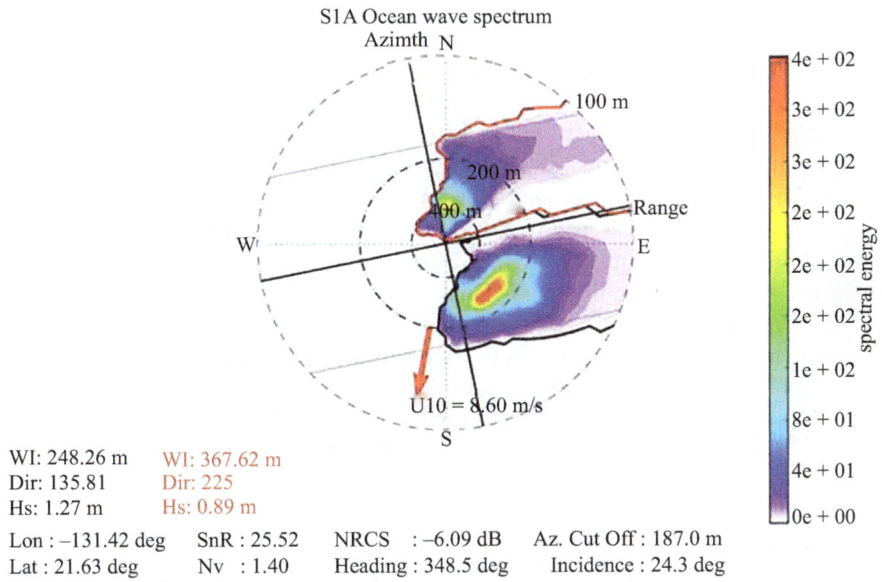

Figure 11.4 Ocean wave spectrum obtained from Sentinel-1 data (Image Courtesy: ESA)

11.5 Wind speed retrieval using SAR images

In order to obtain a more direct link between SAR image intensity and wind speed, averages on areas of linear size larger than the dominant wavelength can be performed. This is the so-called "scatterometry approach" to wind retrieval [41]. In this way, both speckle and intensity modulation effects described in Section 11.4 are averaged out, and obtained SAR image intensity for co- and cross-polarized channels can be expressed by the average, over surface slopes, of (11.46)–(11.47):

$$\sigma^0_{\mathrm{pp}} \cong \frac{4}{\pi} k^4_{\mathrm{em}} \cos^4\vartheta \, |F_p|^2 S(0, 2k_{\mathrm{em}} \sin\vartheta) \tag{11.61}$$

$$\sigma^0_{\mathrm{hv}} \cong \frac{4}{\pi} k^4_{\mathrm{em}} \cos^4\vartheta \left|\frac{F_V - F_H}{\sin\vartheta}\right|^2 S(0, 2k_{\mathrm{em}} \sin\vartheta)\sigma^2_a \tag{11.62}$$

where σ^2_a is the variance of the azimuth slope of long waves, which is related to both wind speed and wind direction. Since the cross-polarized return is very small, low signal-to-noise ratio (SNR) is usually obtained and until recently this polarimetric channel has not been used for wind retrieval. Therefore, for the moment being we will focus on the copolarized returns and will come back to the cross-polarized one later in this section.

By writing more explicitly the 2D directional spectrum in (11.61) in terms of omnidirectional Elfouhaily short-wave spectrum and direction dispersion function, we get

$$\sigma^0_{\mathrm{pp}}(u_{10}, \vartheta, \varphi_w) \cong \frac{4}{\pi} k^4_{\mathrm{em}} \cos^4\vartheta \, |F_p(\vartheta)|^2 \frac{S_0(u_{10})}{(2k_{\mathrm{em}} \sin\vartheta)^{3.5}} [1 + b(u_{10}, \vartheta)\cos(2\varphi_w)] \tag{11.63}$$

where now φ_w is the angle between wind and range directions. The relation linking S_0 to u_{10} could be obtained by using the theoretical relationships reported in Section 11.2. However, better wind speed retrieval results are obtained by using an empirical, nonlinear relationship. In addition, (11.63) predicts equal returns for radar pointing upwind and downwind, whereas, in practice, the return is stronger for winds blowing toward the radar. This is probably due to the combined effect of breaking waves and hydrodynamic modulation, and it is heuristically accounted for by adding a term proportional to cos φ_w in the square brackets of (11.63), so that we obtain the following relationship:

$$\sigma^0_{\mathrm{pp}}(u_{10}, \vartheta, \varphi_w) \cong A(\vartheta) u_{10}^{\gamma(\vartheta)} [1 + c(u_{10}, \vartheta)\cos\varphi_w + b(u_{10}, \vartheta)\cos 2\varphi_w] \tag{11.64}$$

The functions $A(\vartheta)$, $\gamma(\vartheta)$, $c(u_{10}, \vartheta)$, and $b(u_{10}, \vartheta)$ define a "geophysical model function" (GMF), which is different for sensors at different frequencies, and sometimes for different investigators. Thanks to ESA satellites ERS1/2, ENVISAT, and now Sentinel-1, a large amount of both scatterometer and SAR data has been available since 1991 at C band. Therefore, a GMF for C band, known as CMOD (i.e., C-band MODel) [41,42] was developed by researchers, that has been progressively refined, moving from the original one to CMOD2 [43], CMOD4 [44], and CMOD5 [45]. However, subsequently also L- and X-band GMFs have been developed to retrieve ocean surface wind speeds under moderate sea states [46–49].

It is clear that the two unknowns wind speed u_{10} and wind direction φ_w cannot be simultaneously estimated by a single radar measurement. This problem is solved by scatterometers by using multiple antennas pointing in different directions, or a single antenna able to produce beams pointing in different directions. In this way, the same sea area can be sensed from different azimuth directions at (almost) the same time and wind speed and direction can be both retrieved. This is not possible for SAR systems, for which the viewing angle for a single acquisition is fixed. The usual strategy in this case is to obtain an *a priori* estimate of the wind direction from an independent source and then retrieve wind speed from SAR data by using (11.64) with the proper GMF. The more obvious external sources to use are numerical weather prediction (NWP) models for atmospheric circulation. Such an approach has been used for operational wind retrieval for a long time, since ERS1/2 data up to, currently, Sentinel-1 data [50,51]. Another possibility is to use wind direction data obtained from a scatterometer flying on the same satellite of the SAR sensor [41].

These solutions have the disadvantage that both NWP and scatterometer data have a resolution significantly worse than the SAR one, so that high spatial frequency fronts can be missed or displaced in these data. Therefore, high-resolution capabilities of SAR systems are partially lost with this approach.

A different approach can be used, that exploits information on wind direction contained in the SAR image itself. In fact, it was soon realized during the first SAR missions that SAR images of the ocean sometimes show linear features, termed wind rows by some investigators [51], which are aligned with wind direction [41,42,51]. These patterns may be produced by boundary layer rolls (BLRs), which are determined by thermal and dynamic air-sea instability, occurring with typical wind speed values of about 15 m/s [51–54]. Another possibility is constituted by the so-called wind streaks, that are due to the presence of either elongated convective cells, wind-driven Langmuir cells, orography inhomogeneity, or wind distributed surfactants [51,54–56]. The disadvantage is that such linear features are not always visible on SAR images of the sea, their reported occurrence varying from 35% to 48% [51]. In spite of this, some successful wind direction retrieval results based on this approach have been reported [57–61].

More recently, it has been experimentally observed that the cross-polarized return is more sensitive to wind speed than copolarized ones and that it is much less dependent on wind direction so that a priori information on wind direction is not strictly needed [42,62,63]. Therefore, apart from very low wind speed, for which the cross-polarized channel SNR is too low, the latter can be used to retrieve wind speed when knowledge of wind direction is not required. This experimental observation can be theoretically explained by using (11.62). By writing more explicitly the 2D directional spectrum in (11.62) in terms of omnidirectional

Elfouhaily short-wave spectrum and direction dispersion function, and relating the azimuth slope variance to upwind and crosswind slope variances as in [18], we get

$$\sigma^0_{\mathrm{hv}}(u_{10},\vartheta,\varphi_w) \cong \frac{4}{\pi}k^4_{\mathrm{em}}\cos^4\vartheta\left|\frac{F_V(\vartheta)-F_H(\vartheta)}{\sin\vartheta}\right|^2\frac{S_0(u_{10})}{(2k_{\mathrm{em}}\sin\vartheta)^{3.5}}\cdot$$
$$[1+b(u_{10},\vartheta)\cos(2\varphi_w)]\frac{\sigma^2_{\mathrm{up}}(u_{10})+\sigma^2_{\mathrm{cross}}(u_{10})}{2}\left[1-\frac{\sigma^2_{\mathrm{up}}(u_{10})-\sigma^2_{\mathrm{cross}}(u_{10})}{\sigma^2_{\mathrm{up}}(u_{10})+\sigma^2_{\mathrm{cross}}(u_{10})}\cos 2\varphi_w\right] \tag{11.65}$$

It can be noted that dependence on wind speed in (11.65) is not only in the factor $S_0(u_{10})$, as it is the case for the copolarized channels but also in the arithmetic average of upwind and crosswind variances so that the higher sensitivity of the cross-polarized channel to wind speed is explained. As for wind direction, the oscillations of kind $\cos(2\varphi_w)$ due to the spectrum direction dispersion function are damped by the presence of the analogous oscillations due to the azimuth slope variance, which, as shown in (11.65), introduce another dependence of the kind cos $(2\varphi_w)$, but with opposite sign. This explains the reduced sensitivity of the cross-polarized return on wind direction.

Approaches to wind retrieval from SAR images alternative to the scatterometry one are also available. The first one is the "kinematic-based approach" [41], based on the defocusing effect caused by sea motion. In fact, as mentioned in Section 11.4, sea-movement-induced defocusing can be modeled as a low-pass filter applied to the image spectrum, whose transfer function can be represented by the exponential function in (11.56) and (11.59). The azimuth and range cutoff spatial frequencies d_x and d_y are dependent on wind speed so that the latter can be estimated by evaluating these cutoff frequencies. For instance, for ERS1/2 data, an empirical linear relationship was found by [41,64] between the inverse of d_x and wind speed.

Another possibility is to retrieve wind speed from Doppler centroid estimates. Doppler shift is related to the scattering surface movements along the range direction. The main contribution to Doppler shift is provided by earth rotation, but this contribution is known, and it can be removed. The remaining term, called Doppler centroid anomaly, is related to sea surface range movement and hence to wind speed and direction. Actually, exact measurement of the Doppler centroid is not an easy task, and, in order to obtain sufficient frequency accuracy, a large number of image pixels are needed. However, Doppler centroid maps are currently routinely generated with Sentinel-1 data and they are available as level-2 products. In particular, a Doppler standard deviation between 2.49 Hz and 2.89 Hz is achieved at a spatial resolution of 2×2 km^2 for homogenous ocean areas acquired in swath S2 and S3. Over mixed coastal scene acquired in swath S4, the achieved standard deviation is 3.81 Hz. Accordingly, an accuracy of less than 0.2 m/s in the corresponding radial velocity projected on the ocean surface is obtained [65].

Wind velocity retrieval has been also obtained by using the cross-spectrum method. In fact, the correlation between two looks depends on the movement of short wind waves over the long ones, so that a relation between the cross-spectrum phase-plane tilt and wind speed has been established [41,66].

Finally, recently the use of polarimetric data for strong winds monitoring has been proposed. In particular, empirical relationships between the copolarized ratio, that is, the ratio of VV and HH channels intensities, and wind speed and direction have been proposed [67].

In conclusion, we can state that, by now, wind retrieval methods are at an operative level. For instance, among the level-2 Sentinel-1 products routinely provided by European Space Agency (ESA), there is the "ocean wind field" (OWI), which is composed by two-dimensional wind-speed and wind-direction maps with a resolution of 1 km × 1 km [50]. The accuracy of wind speed is about 2 m/s for wind velocity smaller than 25 m/s, while the accuracy on wind direction is about 30°. These maps are produced from stripmap acquisitions (see Chapter 3), covering areas of 170 km × 80 km, and from TOPSAR acquisitions (see Chapter 3), covering areas up to 400 km × 400 km.

An example of SAR wind measurement over Gibraltar, obtained with Sentinel-1 data, is shown in Figure 11.5.

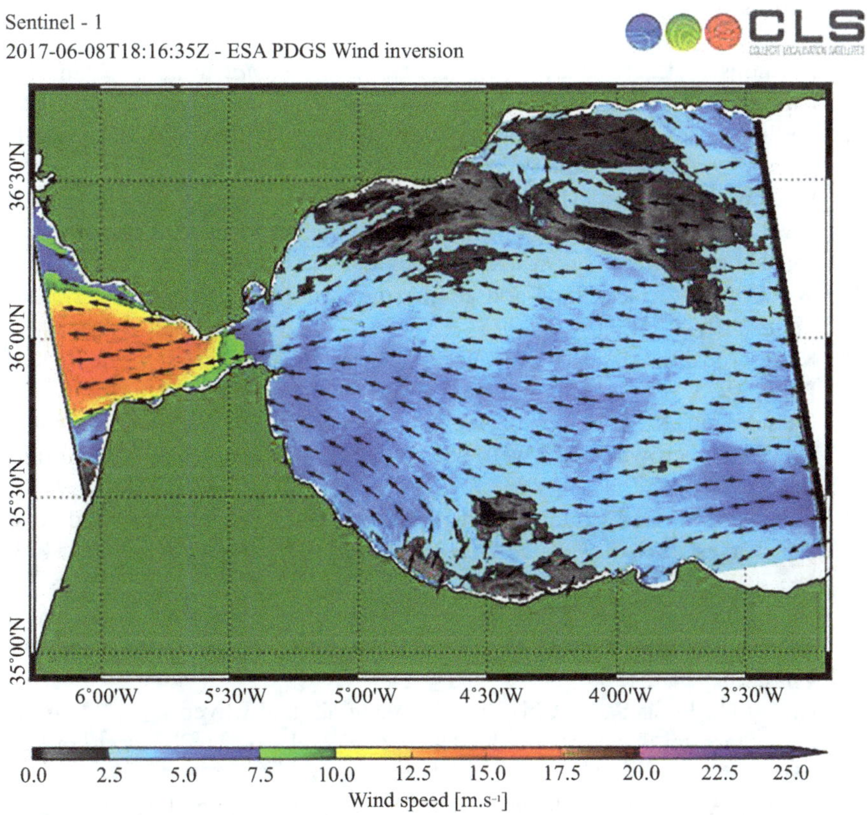

Figure 11.5 Sentinel-1 wind measurement over Gibraltar (Image Courtesy: ESA)

11.6 Concluding remarks and ocean monitoring further applications

In this chapter, an overview of the main methods for the SAR monitoring of sea state and wind speed has been provided. It can be stated that the main fundamental algorithms for sea spectra measurements and wind evaluation were devised in the last 20 years of the last century, and are still employed, by now at a fully operational level. In the last 20 years, these methods have been refined, and new techniques, exploiting the availability of polarimetric SAR data, have been developed. Currently, many of this SAR-derived information on sea state and wind speed are routinely employed by the institutions that are responsible for producing weather forecasts and, at a larger time scale, for modeling climate change.

It is important to note that SAR remote sensing of the ocean surface is not limited to wind monitoring and sea spectra evaluation: other important applications, such as monitoring of ocean currents [68] and internal waves [69], must be mentioned. However, they are out of the scope of the present chapter.

References

[1] Holthuijsen L. H. *Waves in oceanic and coastal waters.* Cambridge University Press, Cambridge, Uk, 2007.

[2] Phillips O. M. "The equilibrium range in the spectrum of wind generated waves." *Journal of Fluid Mechanics.* 1958; 4: 426–434.

[3] Pierson W. J., and Moskowitz L. "A proposed spectral form for fully developed wind sea based on the similarity theory of S. A. Kitaigorodskii." *Journal of Geophysical Research.* 1964;69: 5181–5190.

[4] Wu J. "Sea-surfaces lope and equilibrium wind-wave spectra." *Physics of Fluids.* 1972; 15: 741–747.

[5] Kitaigorodskii S. A., The physics of air-sea interactions. Jerusalem: Keter Press; 1973. pp. 9–73.

[6] Hasselmann, K., Hasselmann, K., and Olbers, D., *et al.* "Measurements of wind-wave growth and swell during the Joint North Sea Wave Project (JONSWAP) ." *Dtsch. Hydrogr. Z.* 1973; 12: 95.

[7] Mitsuyasu H., and Honda T. "The high-frequency spectrum of wind generated waves." *Journal of the Oceanographical Society of Japan.* 1974; 30: 185–198.

[8] Hasselmann D. E., Kunckel M., and Ewing J. A. "Directional wave spectra observed during JONSWAP 1973." *Journal of Physical Oceanography.* 1980; 10, 1264–1280.

[9] Huang N., Long S. R., Chi-Chao T., Yuen Y., and Bliven F. L. "A unified two-parameter wave spectral model for a general sea state." *Journal of Fluid Mechanics.* 1981; 112, 203–224.

[10] Kahma K. "A study of the growth of the wave spectrum with fetch." *Journal of Physical Oceanography.* 1981; 11: 1503–1515.

[11] Fung A. K., and Lee K. K. "A semi-empirical sea-spectrum model for scattering coefficient estimation." *IEEE Journal of Oceanic Engineering*. 1982; OE-7(4): 166–176.

[12] Donelan M. A., Hamilton J., and Hui W. H. "Directional spectra of wind generated waves." *Philosophical Transactions of the Royal Society A*. 1985; 315: 509–562.

[13] Shemdin H., and Hwang P. A. "Comparison of measured and predicted sea surface spectra of short waves." *Journal of Geophysical Research*. 1988; 93 (13): 883–913, 890.

[14] Banner M. L. "Equilibrium spectra of wind waves." *Journal of Physical Oceanography*. 1990; 20: 966–984.

[15] Jähne B., and Riemer K. S. "Two-dimensional wave number spectra of small-scale water surface waves." *Journal of Geophysical Research*. 1990; 95(11): 531–11,546.

[16] Apel J. R. "An improved model of the ocean surface wave vector spectrum and its effects on radar backscatter." *Journal of Geophysical Research*. 1994; 99(16): 269–216, 291.

[17] Elfouhaily T., Chapron B., Katsaros K., and Vandemark D. "A unified directional spectrum for long and short wind-driven waves." *Journal of Geophysical Research*. 1997; 102(C7): 15781–15796.

[18] Di Martino G., Iodice A., and Riccio D. "Closed-form anisotropic polarimetric two-scale model for fast evaluation of sea surface backscattering." *IEEE Transactions on Geoscience and Remote Sensing*. 2019; 57(8): 6182–6194.

[19] Dhanak M. R. and Xiros N. I. (eds). *Handbook of ocean engineering*. Cham, Switzerland: Springer; 2016.

[20] Royal Meteorological Society. Available from https://www.rmets.org/resource/beaufort-scale. [Accessed February 25, 2020].

[21] Eurometeo. Available from http://www.eurometeo.com/english/read/doc_douglas. [Accessed February 25, 2020].

[22] Brown Jr., W. E., Elachi C., and Thompson T. W. "Radar imaging of ocean surface patterns." *Journal of Geophysical Research*. 1976; 81: 2656–2667.

[23] Alpers W. and Rufenach C. L. "The effect of orbital motions on synthetic aperture radar imagery of ocean waves." *IEEE Transactions on Antennas and Propagation*. 1979; 27: 685– 690.

[24] Wright J. W. "A new model for sea clutter." *I IEEE Transactions on Antennas and Propagation*. 1968; AP-16(2): 217–223.

[25] Valenzuela G. R. "Scattering of electromagnetic waves from a tilted slightly rough surface." *Radio Science*. 1968; 3(11): 1057–1066.

[26] Schuler D. L. and Lee J. S. "A microwave technique to improve the measurement of directional ocean wave spectra." *International Journal of Remote Sensing*. 1995; 16(2): 199–215.

[27] Yueh S. H. "Modeling of wind direction signals in polarimetric sea surface brightness temperatures." *IEEE Transactions on Geoscience and Remote Sensing*. 35(6): 1400–1418, 1997.

[28] He Y., Perrie W., Xie T., and Zou Q. "Ocean wave spectra from a linear polarimetric SAR." *IEEE Transactions on Geoscience and Remote Sensing.* 42(11): 2623–2631, 2004.

[29] Lee J. S., Schuler D. L., and Ainsworth T. L. "Polarimetric SAR data compensation for terrain azimuth slope variation." *IEEE Transactions on Geoscience and Remote Sensing.* 38(5): 2153–2163, 2000.

[30] Engen G., Johnsen H., Krogstad H. E., and Barstow S . F. "Directional wave spectra by inversion of ERS-1 synthetic aperture radar ocean imagery." *IEEE Transactions on Geoscience and Remote Sensing.* 1994; 32(2): 340–352.

[31] Monaldo F. M. and Lyzenga D. R. "On the estimation of wave slope- and height-variance spectra from SAR imagery." *IEEE Transactions on Geoscience and Remote Sensing.* 1986; 24(4): 543–551.

[32] Goldfinger A. D. "Estimation of spectra from speckled images." *IEEE Transactions on Aerospace and Electronic Systems.* 1982; AES-18(5): 675–681.

[33] Hasselmann K. and Hasselmann S. "On the nonlinear mapping of an ocean wave spectrum into a synthetic aperture radar image spectrum and its inversion." *Journal of Geophysical Research.* 1991; 96(C6): 10713–10729.

[34] Krogstad H. E. "A simple derivation of Hasselmann's nonlinear ocean-synthetic aperture radar transform." *Journal of Geophysical Research.* 1992; 77(C2): 2421–2425.

[35] Hasselmann S., Bruning C., Hasselmann K., and Heimbach P. "An improved algorithm for the retrieval of ocean wave spectra from synthetic aperture radar image spectra." *Journal of Geophysical Research.* 1996; 101: 16615–16629.

[36] Engen G. and Johnsen H. "SAR-ocean wave inversion using image cross-spectra" *IEEE Transactions on Geoscience and Remote Sensing.* 1995; 33: 1047–1056.

[37] Lehner S., Schulz-Stellenfleth J., Schattler B., Breit H., and Horstmann J. "Wind and wave measurements using complex ERS-2 SAR wavemode data." *IEEE Transactions on Geoscience and Remote Sensing.* 2000; 38(5): 2246–2257.

[38] Schuler D. L., Lee J. S., Kasilingam D., and Pottier E. "Measurement of ocean surface slopes and wave spectra using polarimetric SAR image data." *Remote Sensing of Environment.* 2004; 91: 198– 211.

[39] European Space Agency. Available from https://sentinel.esa.int/web/sentinel/user-guides/sentinel-1-sar/product-types-processing-levels/level-2. [Accessed March 03, 2020].

[40] European Space Agency. Available from https://sentinel.esa.int/web/sentinel/technical-guides/sentinel-1-sar/products-algorithms/level-2/products/ocean-swell-spectra-component. [Accessed March 03, 2020].

[41] Monaldo F., Kerbaol V., Clemente-Colon P., *et al.* "The SAR measurement of ocean surface wind: An overview." *Proceedings of the 2ndWorkshop on Coastal and Marine Applications of SAR,* Longyearbyen, Spitsbergen, Norway, September 8–12, 2003.

[42] Dagestad K. -F., Horstmann J., Mouche A., *et al.* "Wind retrieval from synthetic aperture radar: An overview." *Proceedings of SEASAR 2012* European Space Agency, TromsØ, Norway, 18-22 June 2012. ESA - SP, No. 709, 2013.

[43] Attema E.P.W. "An experimental campaign for the determination of the radar signature of the ocean at C-band." *Proceedings of the Third International Colloquium on Spectral Signatures of Objects in Remote Sensing*, 16-20 December 1985, Les Arcs, France, ESA, SP-247, 791–799, 1986.

[44] Stoffelen A. C. M. and Anderson D. L. T. "Scatterometer data interpretation: Derivation of the transfer function CMOD4." *Journal of Geophysical Research.* 1997; 102(C3): 5767– 5780.

[45] Hersbach H., Stoffelen A., and de Haan S. "An improved C-band scatterometer ocean geophysical model function: CMOD5." *Journal of Geophysical Research.* 2007; 112: C03006.

[46] Isoguchi O. and Shimada M.,"An L-band ocean geophysical model function derived from PALSAR." *IEEE Transactions on Geoscience and Remote Sensing.* 47: 1925–1936, 2009.

[47] Ren Y., Lehner S., Brusch S., Li X., and He M. "An algorithm for the retrieval of sea surface wind fields using X-band TerraSAR-X data." *International Journal of Remote Sensing.* 2012; 33: 7310–7336.

[48] Thompson D. R., Horstmann J., Mouche A., Winstead N. S., Sterner R., and Monaldo F. M. "Comparison of high-resolution wind fields extracted from Terrasar-X SAR imagery with predictions from the WRF mesoscale model." *Journal of Geophysical Research.* 2012; 117: C02035.

[49] Li X. and Lehner S. "Algorithm for sea surface wind retrieval from Terrasar-X and Tandem-X data." *IEEE Transactions on Geoscience and Remote Sensing.* 2014; 52: 2928–2939.

[50] European Space Agency. Available from https://sentinel.esa.int/web/sentinel/ocean-wind-field-component. [Accessed March 03, 2020].

[51] Rana F. M., Adamo M., Lucas R., and Blonda P. "Sea surface wind retrieval in coastal areas by means of Sentinel-1 and numerical weather prediction model data." *Remote Sensing of Environment.* 2019; 225: 379–391.

[52] Levy G. "Boundary layer roll statistics from SAR." *Geophysical Research Letters.* 2001; 28(10), 1993–1995.

[53] Drobinski P., and Foster R. C. "On the origin of near-surface streaks in the neutrally stratified planetary boundary layer." *Boundary-Layer Meteorology.* 2003; 108(2): 247–256.

[54] Svensson N., Sahlée E., Bergström H., Nilsson E., Badger M., and Rutgersson A. "A case study of offshore advection of boundary layer rolls over a stably stratified sea surface." *Advances in Meteorology.* 2017, 9015891.

[55] Dankert H., Horstmann J., and Rosenthal W. "Ocean wind fields retrieved from radarimage sequences." *Journal of Geophysical Research: Oceans.* 2003; (C11): 108.

[56] Koch W., and Feser F. "Relationship between SAR-derived wind vectors and wind at 10-m height represented by a mesoscale model." *Monthly Weather Review.* 2006; 134(5): 1505–1517.

[57] Wackerman C. C., Pichel W. G., and Clemente-Colon P. "Automated estimation of wind vectors from SAR." *Proceedings of the 12th Conference on

Interactions of the Sea and Atmosphere, Long Beach, California, USA, 9-13 February 2003.
[58] Du Y., Vachon P. W., and Wolfe J. "Wind direction estimation from SAR images of the ocean using wavelet analysis." *Canadian Journal of Remote Sensing.* 2002; 28(3), 498–509.
[59] Fichaux N. and Ranchin T. "Combined extraction of high spatial resolution wind speed and wind direction from SAR images: A new approach using wavelet transform." *Canadian Journal of Remote Sensing.* 2002; 28(3), 510–516.
[60] Zecchetto S. and De Biasio F. "A wavelet-based technique for sea wind extraction from SAR images." *IEEE Transactions on Geoscience and Remote Sensing.* 2008; 46(10): 2983–2989.
[61] Leite G. C., Ushizima D. M., Medeiros F. N., and De Lima G. G. "Wavelet analysis for wind fields estimation." *Sensors*. 2010; 10(6), 5994–6016.
[62] Vachon P. W., and Wolfe J. "C-band cross-polarization wind speed retrieval." *IEEE Geoscience and Remote Sensing Letter.* 2011; 8: 456–459.
[63] Hwang P. A., Zhang B., and Perrie W. "Depolarized radar return for breaking wave measurement and hurricane wind retrieval." *Geophysical Research Letters*. 2010; 37: L01604.
[64] Kerbaol V., Chapron B., and Queffeulou P. "Analysis of the wind field during the Vend′ee Globe race: A kinematic SAR wind speed algorithm." *Earth Observatory Quarterly.* 1998; 59: 16.
[65] European Space Agency. Available from https://sentinel.esa.int/web/sentinel/technical-guides/sentinel-1-sar/products-algorithms/level-2/products/surface-radial-velocity-component. [Accessed March 03, 2020].
[66] Engen G., Hoegda K. A., and Johnsen H. "A new method for wind field retrieval from SAR data." *Proceedings of the CEOS SAR Workshop*, Noordwijk, The Netherlands, 3-6 February. 1998; WPP-138: 43–51.
[67] Zhang B. and Perrie W. "Recent progress on high wind-speed retrieval from multi-polarization SAR imagery: A review." *International Journal of Remote Sensing.* 2014; 35(11–12): 4031–4045.
[68] Goldstein R. M. and Zebker H. A. "Interferometric radar map of ocean currents." *Nature*. 1987; 328: 707–709.
[69] Klemas V. "Remote sensing of ocean internal waves: An overview." *Journal of Coastal Research*. 2012; 28(3): 540–546.

Index

www.ingramcontent.com/pod-product-compliance
Lightning Source LLC
LaVergne TN
LVHW011845200926
848484LV00002B/401
* 9 7 8 1 7 8 5 6 1 6 0 1 3 *